National 5 PHYSICS

Student Book

Steven Devine, Paul Ferguson,
David McLean, Stephen Smith

© 2018 Leckie

001/20062018

10 9 8 7 6

All rights reserved. No part of this publication may be reproduced, stored in a retrieval system, or transmitted in any form or by any means, electronic, mechanical, photocopying, recording or otherwise, without the prior written permission of the Publisher or a licence permitting restricted copying in the United Kingdom issued by the Copyright Licensing Agency Ltd, 5th Floor, Shackleton House, 4 Battle Bridge Lane, London SE1 2HX

ISBN 9780008282097

Published by
Leckie
An imprint of HarperCollinsPublishers
Westerhill Road, Bishopbriggs, Glasgow, G64 2QT
T: 0844 576 8126 F: 0844 576 8131

leckiescotland@harpercollins.co.uk www.leckiescotland.co.uk

HarperCollins Publishers
Macken House, 39/40 Mayor Street Upper, Dublin 1 D01 C9W8 Ireland

Special thanks to
Jennifer Richards (project management); Jouve (layout); Jill Laidlaw (copy edit); Roda Morrison (proofread); Lucy Hadfield (proofread)

A CIP Catalogue record for this book is available from the British Library.

Leckie would like to thank the following copyright holders for permission to reproduce their material:
Book Images: © Shutterstock /Thinkstock Fig 5.19.49 Change (www.change.org.bd); Fig 1.4.51: NASA; Figs 1.1.2 and 1.1.6 uses map data licensed from Ordnance Survey © Crown copyright and database right (2013) Ordnance Survey (100018598)

Acknowledgements
Whilst every effort has been made to trace the copyright holders, in cases where this has been unsuccessful, or if any have inadvertently been overlooked, the Publishers would gladly receive any information enabling them to rectify any error or omission at the first opportunity.

Printed in Great Britain by Ashford Colour Press Ltd.

This book contains FSC™ certified paper and other controlled sources to ensure responsible forest management.

For more information visit: www.harpercollins.co.uk/green

INTRODUCTION

AREA 1 – DYNAMICS

1 Scalars and vectors 14
- *Definition and identification of scalar and vector quantities* 14
- *Distance and displacement* 15
- *Experiment to measure average speed* 19
- *Experiment to measure instantaneous speed* 23
- *Speed and velocity* 26
- *The resultant of two vector quantities* 29

2 Velocity-time graphs 39
- *Drawing velocity-time graphs* 39
- *Displacement of a velocity-time graph* 42

3 Acceleration 47
- *Definition of acceleration* 47
- *Experiment to measure acceleration* 48
- *Acceleration from a velocity-time graph* 52

4 Newton's laws 64
- *Balanced forces and Newton's first law* 64
- *Free-fall and terminal velocity* 66
- *Unbalanced forces and Newton's second law* 72
- *Newton's third law* 85

5 Energy 95
- *Energy conservation, conversion and transfer* 95
- *Gravitational potential energy* 110
- *Kinetic energy* 110

6 Projectile motion 121
- *Horizontally launched projectiles* 124
- *Explaining satellite orbits* 132
- *Benefits of satellite technology* 133
- *Geostationary satellite* 134

AREA 2 – SPACE EXPLORATION AND COSMOLOGY

7 Space exploration 144
- *Current understanding of the universe* 144
- *Space terminology* 144
- *The challenges of space travel* 149
- *The risks associated with manned space exploration* 150
- *Optical telescopes* 152

8 Cosmology 155
- *Big Bang theory* 158
- *Age of the universe* 158
- *Using the e-m spectrum to analyse astronomical objects* 158
- *Line spectra* 159
- *Identifying star composition* 160

AREA 3 – ELECTRICITY

9 Electrical charge carriers 164
- *Definition of electrical current* 168
- *Alternating and direct current* 171
- *Instruments used to identify a.c. or d.c. sources* 172

10 Potential difference (voltage) 176
- *Definition of potential difference (voltage)* 176
- *Force on charged particles in an electric field* 181

11 Ohms law 183
- *Resistance* 183
- *Experiment to verify Ohm's law* 185
- *V-I graphs* 186
- *Temperature and resistance* 189

12 Practical electricity and electronic circuits 191
- *Electrical symbols* 191
- *Measurement of current, potential difference (voltage) and resistance* 193
- *Resistance in series circuits* 197
- *Resistance in parallel circuits* 205
- *The potential divider circuit* 213
- *Transistor switching circuits* 220
- *The capacitor* 228

13 Electrical power 232
- *Definition of electrical power* 232
- *The effect of potential difference (voltage) and resistance on current in and power developed across components in a circuit* 235
- *Fuse ratings for electrical appliances* 239

AREA 4 – PROPERTIES OF MATTER

14 Specific heat capacity 242
- *The temperature of a substance and the mean kinetic energy of its particles* 242
- *The relationship between temperature, mass, heat energy and specific heat capacity* 243
- *Using the principle of conservation of energy to determine heat transfer* 247

15 Specific latent heat 252
- *Quantity of heat required to change the state of unit mass* 252
- *Latent heat of fusion and latent heat of vaporisation* 252

16 Gas laws and the kinetic model 259
- *Defining pressure* 259
- *Definition of pressure in terms of force and area* 259
- *Kinetic model and pressure of a gas* 264
- *Relationship between Kelvin and degrees Celsius* 265
- *Boyle's law experiment* 267
- *Charles' law experiment* 271
- *Guy-Lussac's law experiment* 274
- *Explanation of the gas laws in terms of the kinetic model* 277

AREA 5 – WAVES

17 Wave parameters and behaviours 286
- *Transverse and longitudinal waves* 286
- *Frequency, period, wavelength, amplitude and wave speed of longitudinal and transverse waves* 289
- *Diffraction* 300

18 Electromagnetic spectrum 303
- *Relative wavelength and frequency of bands of the electromagnetic spectrum* 304
- *The nature of electromagnetic waves* 304
- *Sources of electromagnetic waves* 307

19 Refraction of light 309
- *Definition of refraction* 309
- *How refraction means a change in speed, wavelength and sometimes diffraction of light waves* 310
- *How light refracts when it enters a different medium* 312
- *Terminology: incident ray, refracted ray, angle of incidence, angle of refraction and the normal* 313
- *Applications of refraction* 317
- *Some applications of refraction – lenses and optical fibres* 317
- *Total internal reflection* 324
- *Dispersion of light waves* 325

AREA 6 – RADIATION

20 Nuclear radiation 330
- *The nature of alpha (α), beta (β) and gamma (γ) radiation* 332
- *Ionistaion* 333
- *Detecting radiation* 334
- *Activity* 338
- *Dangers associated with ionising radiations* 340
- *Equivalent dose (rate)* 344
- *Background radiation* 345
- *Applications of nuclear radiation* 347
- *Applications of radiation* 347
- *Half-life* 349
- *Fission* 354
- *Fusion* 358

Exam-style questions **361**

Answers to Exercises https://collins.co.uk/pages/scottish-curriculum-free-resources

Answers to Exam-style questions https://collins.co.uk/pages/scottish-curriculum-free-resources

Relationship sheet

$E_p = mgh$

$E_k = \frac{1}{2}mv^2$

$Q = It$

$V = IR$

$R_T = R_1 + R_2 + \ldots$

$\frac{1}{R_T} = \frac{1}{R_1} + \frac{1}{R_2} + \ldots$

$V_2 = \left(\frac{R_2}{R_1 + R_2}\right)V_s$

$\frac{V_1}{V_2} = \frac{R_1}{R_2}$

$P = \frac{E}{t}$

$P = IV$

$P = I^2 R$

$P = \frac{V^2}{R}$

$E_h = cm\Delta T$

$p = \frac{F}{A}$

$\frac{pV}{T} = \text{constant}$

$p_1V_1 = p_2V_2$

$\frac{p_1}{T_1} = \frac{p_2}{T_2}$

$\frac{V_1}{T_1} = \frac{V_2}{T_2}$

$d = vt$

$v = f\lambda$

$T = \frac{1}{f}$

$A = \frac{N}{t}$

$D = \frac{E}{m}$

$H = Dw_R$

$\dot{H} = \frac{H}{t}$

$s = vt$

$d = \bar{v}t$

$s = \bar{v}t$

$a = \frac{v - u}{t}$

$W = mg$

$F = ma$

$E_w = Fd$

$E_h = ml$

$f = \frac{N}{t}$

Data sheet

Speed of light in materials

Material	Speed in m s^{-1}
Air	$3{\cdot}0 \times 10^8$
Carbon dioxide	$3{\cdot}0 \times 10^8$
Diamond	$1{\cdot}2 \times 10^8$
Glass	$2{\cdot}0 \times 10^8$
Glycerol	$2{\cdot}1 \times 10^8$
Water	$2{\cdot}3 \times 10^8$

Gravitational field strengths

	Gravitational field strength on the surface in N kg^{-1}
Earth	9·8
Jupiter	23
Mars	3·7
Mercury	3·7
Moon	1·6
Neptune	11
Saturn	9·0
Sun	270
Uranus	8·7
Venus	8·9

Specific latent heat of fusion of materials

Material	Specific latent heat of fusion in J kg^{-1}
Alcohol	$0{\cdot}99 \times 10^5$
Aluminium	$3{\cdot}95 \times 10^5$
Carbon Dioxide	$1{\cdot}80 \times 10^5$
Copper	$2{\cdot}05 \times 10^5$
Iron	$2{\cdot}67 \times 10^5$
Lead	$0{\cdot}25 \times 10^5$
Water	$3{\cdot}34 \times 10^5$

Specific latent heat of vaporisation of materials

Material	Specific latent heat of vaporisation in J kg^{-1}
Alcohol	$11{\cdot}2 \times 10^5$
Carbon Dioxide	$3{\cdot}77 \times 10^5$
Glycerol	$8{\cdot}30 \times 10^5$
Turpentine	$2{\cdot}90 \times 10^5$
Water	$22{\cdot}6 \times 10^5$

Speed of sound in materials

Material	Speed in m s^{-1}
Aluminium	5200
Air	340
Bone	4100
Carbon dioxide	270
Glycerol	1900
Muscle	1600
Steel	5200
Tissue	1500
Water	1500

Specific heat capacity of materials

Material	Specific heat capacity in J kg^{-1} °C^{-1}
Alcohol	2350
Aluminium	902
Copper	386
Glass	500
Ice	2100
Iron	480
Lead	128
Oil	2130
Water	4180

Melting and boiling points of materials

Material	Melting point in °C	Boiling point in °C
Alcohol	−98	65
Aluminium	660	2470
Copper	1077	2567
Glycerol	18	290
Lead	328	1737
Iron	1537	2737

Radiation weighting factors

Type of radiation	Radiation weighting factor
alpha	20
beta	1
fast neutrons	10
gamma	1
slow neutrons	3

Introduction

About this book

This book provides a resource to practise and assess your understanding of the physics covered for the National 5 qualification. The book has been organised to reflect the course specifications and is packed with examples, experiments, exercises and features to deepen your understanding of Physics and help you prepare for the assessments and final exam.

Features

IN THIS CHAPTER

There is a list of the topics covered at the start of each chapter.

In this chapter you will learn about:

- Electric charge in terms of positive and negative charge.
- Electrons and protons.
- Static electricity and some of its applications.

EXPERIMENT

Topics are usually introduced by way of an experiment that aims to familiarise you with the apparatus and methods that you are likely to encounter. Each includes a list of things to do to perform the experiment, as well as questions that the experiments will raise.

GO! Experiment: Measuring power and energy transfer

This experiment compares the amount of energy transferred from electricity to heat and light by low and high power bulbs.

You will need:

- Bulbs of various power ratings
- Power supply (DC)
- Wires

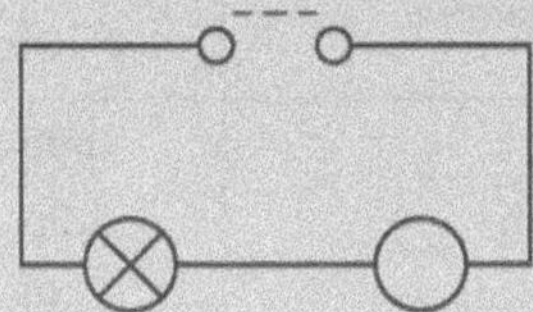

Fig 1.5.1

WORKED EXAMPLES

New topics involving calculations are introduced with at least one Worked Example, which shows how to go about tackling the questions and activities. Each Example breaks the question and solution down into steps, so you can see what calculations are involved, what kind of rearrangements are needed and how to work out the best way of answering the question.

Worked example	An electric fire converts 25 000 J of electrical energy into heat and light energy in a time of 30 s. Calculate the power of the electric fire.

TECHNIQUE

Examples are often complemented by features that focus on the techniques needed to understand a particular calculation or concept.

Technique: Power, energy and time

1. Write down what you know from the question, leaving blank what you are trying to find:

STEP BACK IN TIME

Some chapters have Step Back in Time features that add context to the topics. They help to show the history of the physics and the people involved in the major discoveries and developments in the science.

STEP BACK IN TIME: HORSEPOWER

Power was initially measured in horsepower (hp). The term was adopted in the late 18th century by Scottish engineer James Watt to compare the output of steam engines with the power of draft horses. It was later expanded to include the output power of piston engines, turbines, electric motors and other machinery. 1 hp = 746 watts

KEY POINTS AND WORD BANKS

Where appropriate, Key point features provide useful summaries of critical information. Word banks help to secure the terms and phrases used in the physics course to help you to become familiar with them and their usage.

Key point

In an isolated system the total amount of energy remains the same. Energy cannot be created or destroyed.

Word bank

- **force**

an action, either internal or external, that can cause a change. This might be a change to shape, direction, structure or something else. Force is measured in newtons.

MAKE THE LINK

Physics is not a subject that exists in isolation! Where appropriate, links to other subject areas show the links that exist between different disciplines. Make the Link also shows how different aspects of physics link to each other.

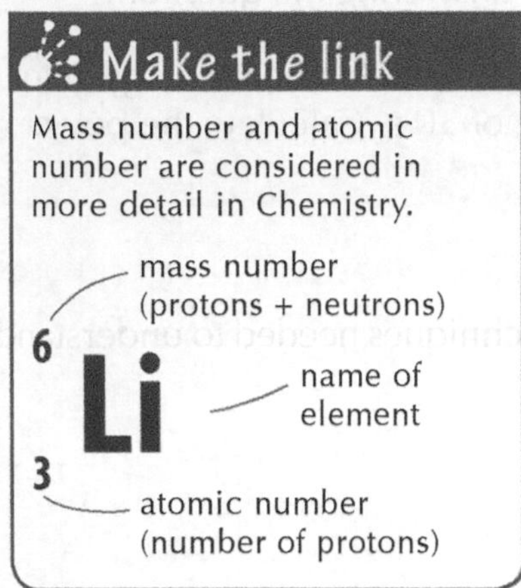

LEARNING CHECKLIST

Each chapter closes with a summary of learning statements showing what you should be able to do when you complete the chapter. You can use the Learning checklist to check you have a good understanding of the topics covered in the chapter. Traffic light icons can be used for your own self-assessment.

Learning checklist

In this chapter you have learned:

- How to state the principle of energy conservation.
- How to apply the principle of energy conservation to solve problems.

ASSESSMENTS

In each chapter, exercises provide topic-linked questions. At the end of the book, assessments are provided for each of the areas. These assessments contain a number of questions and cover the minimum competence for the area content.

Answers to exercises and to the assessment questions at the end of the book are available online at: https://collins.co.uk/pages/scottish-curriculum-free-resources

1 **Scalars and vectors**

- Definition and identification of scalar and vector quantities
- The resultant of two vector quantities
- Distance and displacement
- Speed and velocity
- Experiments to measure average and instantaneous speed

2 **Velocity-time graphs**

- Drawing velocity-time and speed-time graphs
- Interpreting velocity-time graphs
- Displacement from a velocity-time graph

3 **Acceleration**

- Definition of acceleration
- Acceleration from a velocity-time graph
- Experiment to measure acceleration

4 **Newton's laws**

- Balanced forces and Newton's first law
- Unbalanced forces and Newton's second law
- Newton's third law
- Free-fall and terminal velocity

5 **Energy**

- Energy conservation, conversion and transfer
- Gravitational potential energy
- Kinetic energy

6 **Projectile motion**

- Horizontally launched projectiles
- Explaining satellite orbits
- Benefits of satellite technology
- Geostationary satellites

AREA 1

Dynamics

1 Scalars & Vectors

In this chapter you will learn about:

- Vectors – how to describe what a vector quantity is, and how to combine vector quantities.
- Identifying scalar and vector quantities.
- Speed – how to measure average and instantaneous speed.

Scalars and vectors

All quantities in Physics can be placed into one of two categories. They can be either regarded as 'scalar' quantities or 'vector' quantities.

A scalar quantity is one which has a magnitude (size) only.

Speed is one example of a scalar quantity and we could say that, "A car is travelling with a speed of 30 ms^{-1}." This statement provides no indication of the direction that the car is travelling, only how fast it is moving.

Some examples of scalar quantities are shown in the table opposite.

Scalar	Vector
Time	Force
Distance	Displacement
Speed	Velocity
Mass	Acceleration
Energy	

Table 1.1.1

Note, identification of these scalars is a requirement for the examination.

Adding scalar quantities is very simple since it uses the simple rules of arithmetic:

i.e. 5 s + 10 s = 15 s

The only real requirement is that the quantities added must be the same type, i.e. **do not** try adding a time of 5 s to a speed of 8 ms^{-1} as this would not make sense.

A vector quantity is one which has both a magnitude (size) **and** a direction. In this case we might say that, "A car is travelling with a velocity of 30 ms^{-1} due south". The addition of the direction turns the speed (mentioned above) into a velocity. The addition of direction provides more information about the motion of an object but also means that adding two vectors together requires more skill than adding two scalar quantities together (see later). As with scalar quantities, only identical quantities can be added together, i.e. **do not** try adding a displacement of 2 m south to a velocity of 5 ms^{-1} south as this would not make sense either.

Note, like scalars, identification of the vector quantities given in the table is a requirement for the examination.

Distance and displacement

Consider walking from point A to point B, as shown opposite. You follow the path which is shown in green. The distance travelled does not have a direction; it is just the total distance that you walk in going from A to B.

Displacement, *s*, is a vector quantity – it has both size and direction. Your displacement is the straight line from the start to the end point (shown in red). It does not depend on the route taken; it only depends on the start and end point. When stating displacement, you must always state direction! For example, a displacement of 5 miles on a bearing of 045°.

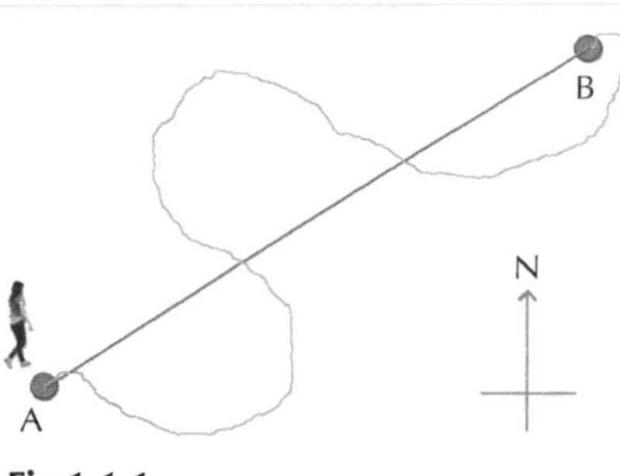

Fig 1.1.1

Experiment: Orienteering challenge

This experiment makes use of displacements to navigate around your school playing field. You will need:

- A trundle wheel
- A compass

Plan a route in terms of displacements which can be given to a partner. Your partner should follow this route using a compass and a trundle wheel.

Finding displacements by scale drawing

The resultant displacement of a journey can be found using a scale diagram. Remember that displacement is the straight line from the start point to the end point. Consider the journey below, where a walker has left the car park and climbed Schiehallion, a mountain in Highland Perthshire. The route is shown by the path on the map. The displacement is shown by the straight line from the car park to the peak of the mountain.

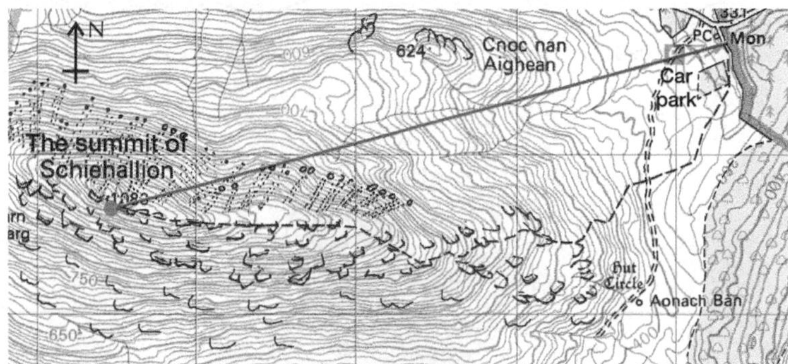

Fig 1.1.2: *OS map showing Schiehallion*

Measuring the length of the line and comparing with the scale will give us the magnitude of the displacement vector.

If the scale of the above map is 1 cm = 500 m

and the length of the line is 7·5 cm

then the size of the displacement vector is therefore:

$$s = 7{\cdot}5 \times 500$$

$$s = 3750\ m$$

(Note: s = displacement and d = distance)

The direction of the displacement can be found by using a protractor to find the angle. Directions can be quoted as bearings. A bearing is always measured clockwise from north. Therefore north would be a bearing of 000. East would be a bearing of 090. West would be a bearing of 270. In the above example, north is marked by the arrow at the top left. Using a 360° protractor, the angle from north is found to be approximately 250.

We can now quote the displacement as:

$$s = 3750 \text{ m on a bearing of } 250$$

Technique: Using three figure bearings

A bearing is always quoted as the angle you have to go through moving clockwise from north. So, you would always start at north and then see how many degrees you move round to get to the direction of your vector. Some examples of this are shown below.

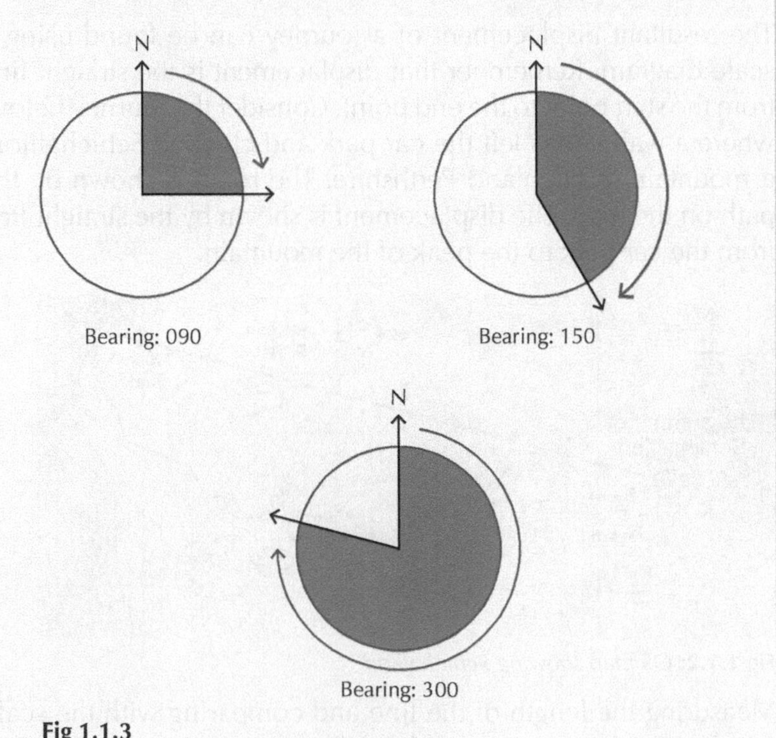

Fig 1.1.3

Key point

Distance is a scalar quantity which depends on the journey taken. Displacement is a vector quantity and is the straight line from the start to the end point. Displacement has both magnitude and direction.

Exercise 1.1.1 Distance and displacement

1 Shown below are displacement vectors for walking journeys assuming a scale of 1 cm = 100 m. North is shown on the diagram.

For each of the displacement vectors, find the displacement. Remember, you must quote both magnitude and direction.

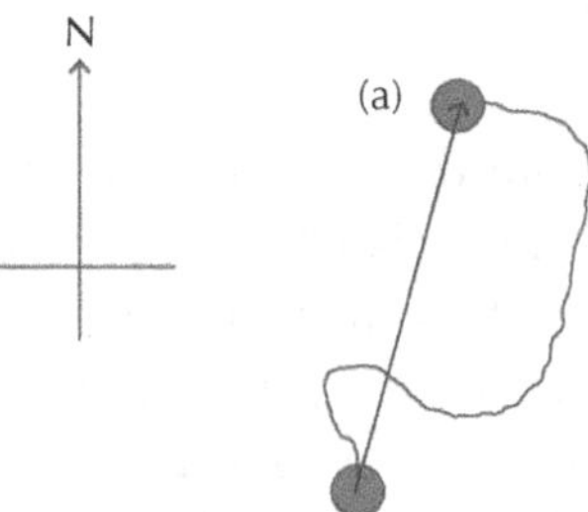

Fig 1.1.4

2 Shown below are four points on an orienteering course. The scale of the map is 1 cm = 500 m. North is shown.

a) One competitor navigates from P to S, then from S to R. Find the total distance travelled and the resultant displacement.

b) A second competitor navigates from Q to R, then from R to P, then P to Q. Find the total distance travelled and the resultant displacement.

c) A third competitor navigates directly from P to S. Find the total distance travelled and the resultant displacement.

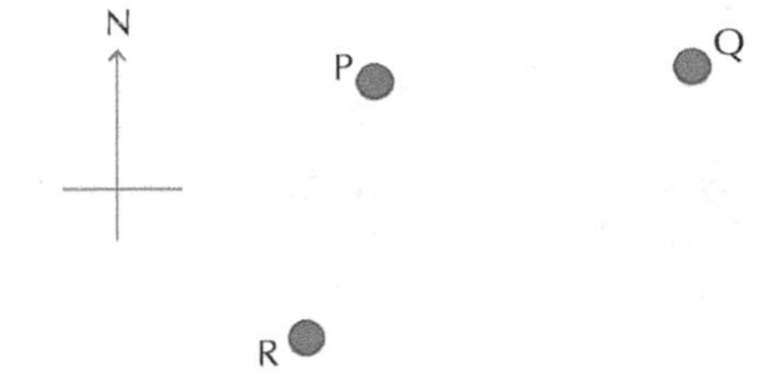

Fig 1.1.5

3 A walking group wants to climb Ben Lawers. They have the following OS map. Using a scale of 1·5 cm = 1 km, and north as shown on the map, find the displacement of the walkers when they go from the car park (V on the map) to the summit of Ben Lawers.

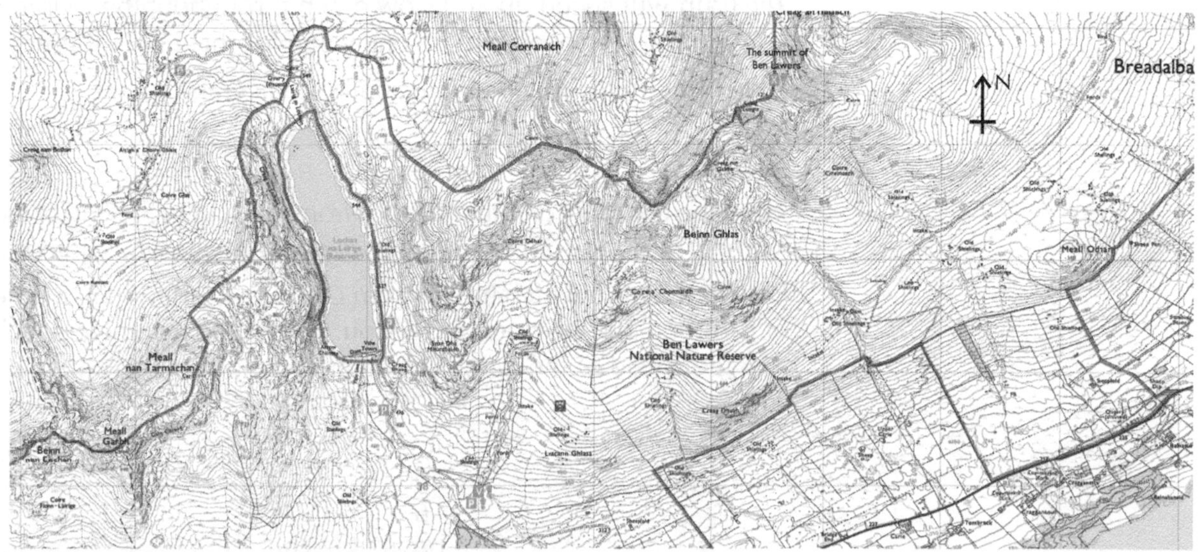

Fig 1.1.6: *OS map showing Ben Lawers*

Speed

The speed of an object is defined as the rate of change of distance in a given time. The faster an object is moving, the greater the distance it will cover in a given time. This leads us to the familiar equation that links speed (v), distance (d) and time (t):

$$v = \frac{d}{t}$$

This equation can be used to work out the speed of an object if the time taken to cover a certain distance is known. Typically in physics we measure distances in metres and time in seconds which gives us the unit for speed as metres per second (ms^{-1}). However, for some applications (such as the average speed of a car), measuring distance in miles or kilometres and time in hours is more appropriate. In these cases, the speed will be measured in either miles per hour (mph) or kilometres per hour (km/hr).

Worked example

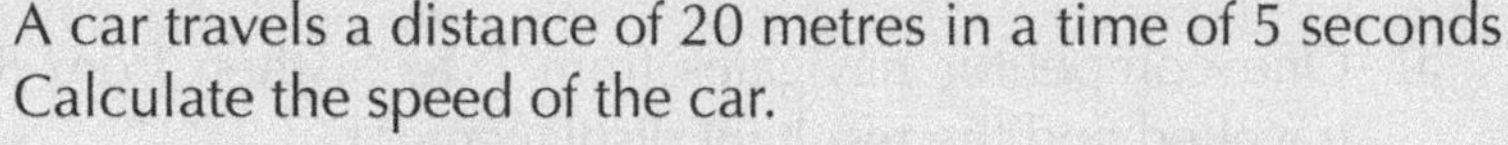
A car travels a distance of 20 metres in a time of 5 seconds. Calculate the speed of the car.

Fig 1.1.7: *Speed of a car*

$$v = \frac{d}{t}$$

$$v = \frac{20}{5}$$

$$v = 4\ ms^{-1}$$

Average speed

The average speed of an object is the speed measured over a large distance and a long time. We talk of the average speed being measured over the whole journey — for example, the average speed of a train travelling from Dundee to Edinburgh. The train will speed up and slow down throughout the journey, but the average speed will be one value for the whole journey.

The equation for speed above applies to average speed:

$$v = \frac{d}{t}$$

Here, the distance, d, is the distance from the start to the end of the journey. The time, t, is the time taken to complete the whole journey.

Consider a train which travels from Dundee to London in a time of 6 hours. The distance of the journey is 700 kilometers. The average speed of the train is given by:

$$v = \frac{d}{t}$$

$$v = \frac{700}{6}$$

$$v = 116{\cdot}67\ km/hr$$

Experiment: Measuring average speed

This experiment measures the average speed of an object.
You will need:

- A ramp
- A dynamics cart
- A metre stick
- A stop clock
- Ramp holder (or books)

Instructions

1 Use a ramp holder or books to prop one end of the ramp up. Mark start and finish points on the ramp.
2 Use the metre stick to measure the distance between the start and finish points.
3 Release the dynamics cart from the starting point and let it roll down to the end point.
4 Use the stop clock to measure the time taken for the cart to travel between the start and finish points.
5 Use the following equation to calculate the average speed of the cart:

$$v = \frac{d}{t}$$

Average speed camera

Average speed cameras (known as SPECS cameras) measure the average speed of a vehicle as it travels between two points. Using infrared technology, so that they can work both day and night, these cameras record a car's number plate at two fixed sites. The time taken for the car to travel between these sites is then measured. As the distance between the sites is known, the average speed can be calculated using the equation:

$$v = \frac{d}{t}$$

Fig 1.1.8: *SPECS average speed cameras*

For example, if a car passes the first SPECS camera at 0900 hours and then passes the second SPECS camera, a distance of 2 miles away, at 0905 hours, the average speed of the car between the cameras would be:

$$v = \frac{d}{t}$$

$$v = \frac{2}{0{\cdot}083}$$

$$v = 24\ mph$$

Key point

The average speed is one value for the distance between two given points on the journey. To find the average speed, you need to know the distance travelled and the time taken to travel that distance.

Technique: Calculations involving average speed

1. Write down what you know from the question, leaving blank what you are trying to find:

   Distance = ____

   Time = ____

   Average speed = ____

2. Write down the equation linking the quantities above as you see it on the formula sheet:

   $$d = vt$$

3. Substitute into the equation what you know.
4. Solve for the unknown.

Worked example

Fig 1.1.9: *Bus*

A bus does a journey at an average speed of 40 km/hr. The journey was 30 km. Calculate the time taken to complete the journey.

Write down what we know and what we are looking for:

Distance = 30 km

Time = ?

Average speed = 40 km/hr

We are using the equation that links distance, speed and time:

$$d = vt$$

Substitute what we know:

$$30 = 40t$$

Solve for time:

$$t = \frac{30}{40}$$

$$t = 0{\cdot}75 \text{ hours}$$

$$t = 45 \text{ min}$$

Exercise 1.1.2 Average speed

1 Use the equation for average speed to fill in the missing values in the table below:

Distance (m)	Time (s)	Speed (ms^{-1})
25	5	(a)
40	4	(b)
(c)	6	12
150	(d)	9
2500	(e)	33·50

Table 1.1.2

2 An oil rig supply vessel leaves port in Edinburgh and sails directly to an oil rig 75 miles away. The ship leaves at 1400 hours and arrives at 1730 hours. Calculate the average speed of the vessel in miles per hour.

Fig 1.1.10: *Oil rig supply vessel*

3 A car can cover a distance of 3000 metres in a time of 120 seconds. Calculate the average speed of the car in metres per second.

4 In the 2012 London Olympic games, the men's 100 m sprint final was won by Usain Bolt. He ran the distance of 100 m in a time of 9·63 s. Calculate Bolt's time in:

a) metres per second

b) kilometres per hour

5 An aeroplane can maintain an average speed of 250 ms^{-1} when at its cruising altitude. How long will it take the plane to cover a distance of 6000 km on a transatlantic flight?

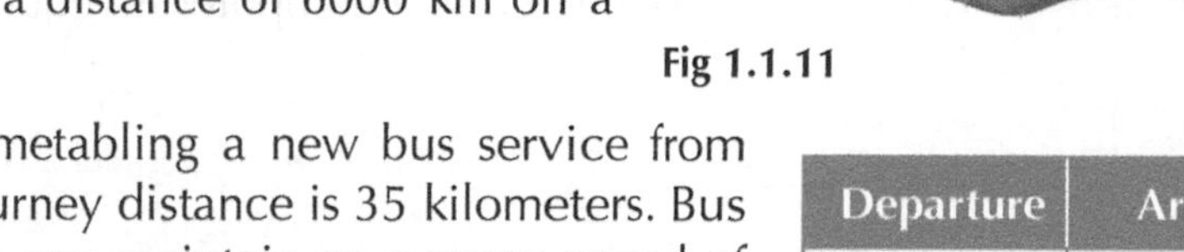

Fig 1.1.11

6 You are in charge of timetabling a new bus service from Dundee to Perth. The journey distance is 35 kilometers. Bus drivers tell you that they can maintain an average speed of 50 km/hr on the journey between Dundee and Perth. On the timetable shown, fill in the corresponding arrival or departure times based on the average speed and distance given.

Departure	Arrival
0950	(a)
(b)	1145
(c)	1235
1330	(d)

Table 1.1.3

(continued)

7 At your school's sports day you have been asked to calculate the average speed of runners during the 100 m, 200 m and 400 m races. Describe how you would measure the average speed of a runner. Your answer should include the measurements that you would make and how you would make them, and any calculations you would do.

8 You are on a school trip to Paris to investigate the physics of rollercoasters at Euro Disney. On the return journey you have to catch a ferry which leaves Calais at 1930 hours. You leave Paris at 1600 hours with a distance of 200 miles to travel.

   a) During the first 100 miles of the journey, an average speed of 50 mph is maintained. Calculate the time taken for the first stage of the journey.

   b) The coach has a maximum speed of 62 mph. Is it possible to complete the second stage of the journey and make it to the ferry on time?

9 A football is kicked from the penalty spot on a football pitch. The ball travels a distance of 9 m in a time of 2 s. Calculate the average speed of the football.

Fig 1.1.12: *Speedometer*

Instantaneous speed

The instantaneous speed of an object is its speed at any given instant in time. Unlike the average speed of a whole journey, the instantaneous speed will change throughout a journey as the moving object speeds up and slows down. The instantaneous speed is the speed that is recorded on a car's speedometer at that very moment.

In order to measure the instantaneous speed, we need to measure the very short distance that an object covers in a very small period of time. The instantaneous speed is found using the familiar equation linking speed, distance and time:

$$v = \frac{d}{t}$$

Measuring the instantaneous speed of an object is more difficult than measuring the average speed of a journey. As the time is very short, it is not possible to measure it accurately manually with a stop clock. Instead, a timing device such as a light gate connected to a computer is required.

Experiment: Measuring instantaneous speed

This experiment measures the instantaneous speed of an object as it rolls down a slope.

You will need:

- A ramp
- A dynamics cart
- A mask
- A light gate and timer
- Ramp holder (or books)

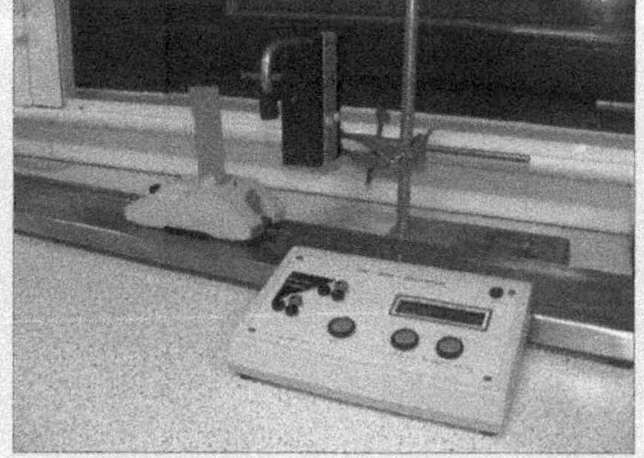

Fig 1.1.13: *Measuring instantaneous speed*

Instructions

1. Use a ramp holder or books to prop the ramp up at one end.
2. Place your light gate across the ramp so that the cart can travel freely through it (see photograph above).
3. Attach a mask to the cart which has a known length, *l*. Ensure that this mask passes between the light source and the light sensor of the gate.
4. Let the cart roll freely down the ramp.
5. Use the light gate connected to a timer to measure the time taken, *t*, for the cart to pass through the gate.
6. The instantaneous speed of the cart can then be calculated using the equation for speed, distance, time:

$$v = \frac{d}{t} = \frac{l}{t}$$

where *l* is the length of the mask.

Gatso speed cameras

Gatso speed cameras are designed to measure the instantaneous speed of a car as it passes by. A radar beam is emitted from the back of the camera towards oncoming cars. This is used to measure the speed of the car which is approaching. If the car is travelling faster than a set speed (the speed limit on the road), it *triggers* the camera to take two photographs of the car as it passes by.

The photographs are separated by a known time, *t*. The car will travel a set distance between the two photographs. This allows the instantaneous speed to be calculated using:

$$v = \frac{d}{t}$$

where *d* is the distance the car travels between the photographs. The distance is found in practice by using lines on the road. The example below highlights how the speed is calculated.

Fig 1.1.14: *Gatso speed camera*

Worked example

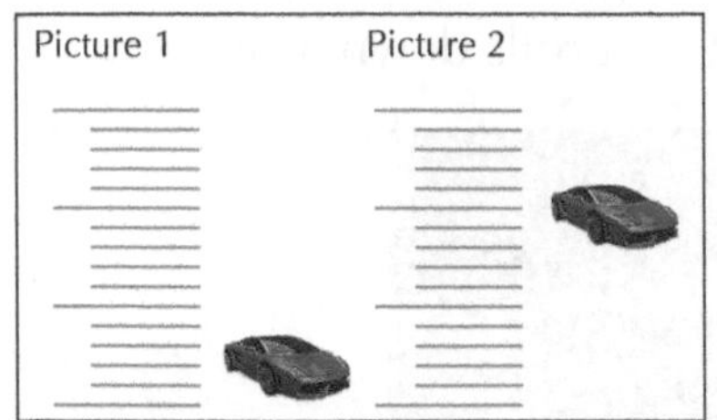

Fig 1.1.15

A speed camera takes two photographs of a passing car which are shown below. The photographs are separated by a time of 0.2 seconds. Each of the lines on the road are 1 metre apart. Calculate the speed of the passing car.

First of all, calculate the distance the car has travelled. It has passed 8 lines in between the photographs so the distance is:

$$d = 8 \times 1$$

$$d = 8\ m$$

The speed can then be calculated using the speed, distance, time equation:

$$d = vt$$

Substitute what we know:

$$8 = v \times 0{\cdot}2$$

Then solve for speed:

$$v = \frac{8}{0{\cdot}2}$$

$$v = 40\ ms^{-1}$$

Key point

The instantaneous speed changes over the course of a journey. To find the instantaneous speed, you need to know the mask or object length and the time taken to pass a set point. As the times and distances are so small, electronic timing equipment is required to make the measurements. However, the same equation applies as for average speed:

$$v = \frac{d}{t}$$

Experiment: The speed camera

This experiment calculates the instantaneous speed of a cart using a digital camera set up to work like a speed camera.

You will need:

- A dynamics cart
- Digital camera
- A1 sheet of paper

Instructions

1 Mark lines on the sheet of paper at set distances apart (for example, every 10 cm).

2 Set your digital camera to take multiple photographs at high speed (e.g. one every 0·1 second – different models of camera will vary). The time between photographs gives you the time, *t*, for calculating speed.

3 As the cart moves over the lines, use the camera to take two (or more) photographs of the cart passing.

4 Compare two photographs of the cart to work out the distance, *d*, travelled between the pictures.

5 Use the equation below to find the instantaneous speed of the cart:

$$v = \frac{d}{t}$$

Exercise 1.1.3 Instantaneous speed

1 The experiment shown in the photograph below is used to work out the instantaneous speed of a cart as it passes through a light gate.

Different carts with different mask lengths, *l*, were passed through the gate. The mask lengths and times taken are shown in the table below. For each one, find the instantaneous speed of the cart.

Fig 1.1.16

Mask length (cm)	Time (s)	Speed (ms^{-1})
5	0·20	
5	0·05	
10	0·5	
7·5	2×10^{-3}	

Table 1.1.4

2 Describe an experiment that you would carry out to calculate the instantaneous speed of a bobsleigh at different points on a track. Your answer should include the apparatus you would use, the measurements you would make and how you would make them, and any calculations you would do.

3 A speed camera is used to measure the instantaneous speed of cars passing on a dangerous stretch of road. Two photographs, taken 0·3 s apart, of a white car passing are shown. The distance between the lines on the road is 1 m. The speed limit on the road is 25 ms^{-1}.

Determine whether or not the car was speeding when it passed the camera.

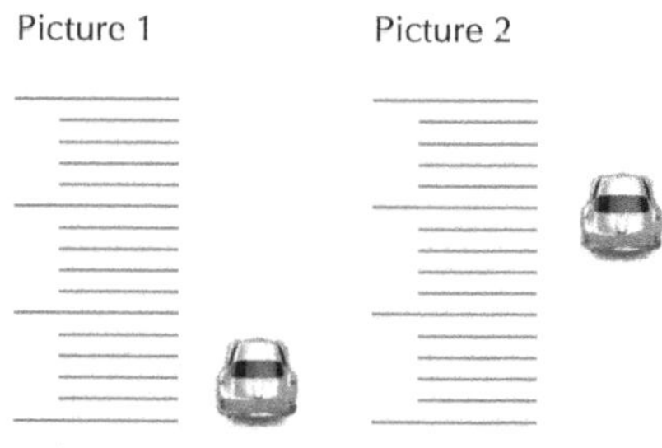

Fig 1.1.17

4 A stone of length 14 cm is dropped from a height through a light gate. The stone breaks the beam of the light gate for a time of 25 ms. Calculate the instantaneous speed of the stone in:

a) metres per second

b) kilometres per hour

Speed and velocity

Previously we defined speed as the rate of change of distance in a given time. Speed is a scalar quantity; speed has no direction. Velocity is the rate of change of displacement in a given time. Velocity has both a magnitude and a direction; it is a vector quantity.

When calculating the average speed (v) of an object, we use the distance (d) and the time (t):

$$v = \frac{d}{t}$$

When calculating the average velocity (v) of an object, we use the displacement (s) and the time (t):

$$v = \frac{s}{t}$$

The direction of the velocity is the same as the direction of the displacement used in the calculation.

Technique: Calculating the average velocity

1. Find the displacement of the object, remembering to find both the magnitude and direction.
2. Write down what you know and what you are trying to find out:

   Displacement = ______ m on a bearing of ______ °

   Time = ______ s

   Average velocity = ______ ms^{-1} on a bearing of ______ °
3. Use the following equation to work out the magnitude of the average velocity:

$$v = \frac{s}{t}$$

4. Substitute into the equation what you know.
5. Solve for the unknown.
6. The direction of the average velocity is the same as the direction for the displacement.

SPOTLIGHT ON INDUSTRY: AIR TRAFFIC CONTROL

When guiding aircraft, air traffic controllers issue pilots with a vector for their velocity. With so many aircraft in the air, air traffic control is a high pressured and very challenging job! It is essential that both the speed and the direction are known so that it can be calculated where an aircraft is at any given time and where it will be after a set time.

Vectoring in air traffic control is where a plane is issued with a specific heading – a bearing which they must follow. The direction is essential to ensure that the plane makes it to the correct location and does not come too close to other aircraft in the air.

Fig 1.1.18: *Air traffic controllers*

Worked example

A ship leaves port A and sails to an oil rig at point B as shown in the diagram. The journey takes 4 hours to complete.

The scale for the diagram is 1 cm = 10 km, and north is shown.

a) Calculate the displacement of the ship.

b) What is the average velocity of the ship?

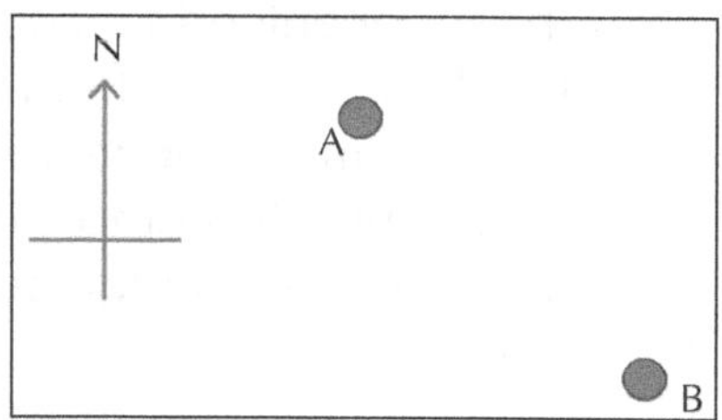

Fig 1.1.19

Working:

a) *Using a ruler, the length of the line was measured to be approximately 2·5 cm. This means that the magnitude of the displacement is:*

$$s = 2{\cdot}5 \times 10$$

$$s = 25\ km$$

Using a protractor, the angle is measured to be 43° south of east, which is a bearing of 133°. This gives a displacement of:

s = 33 km on a bearing of 133°

b) *To work out the average velocity, write down what we know and what we are trying to find out:*

- *Displacement = 25 km on a bearing of 133°*
- *Time = 5 hours*
- *Average velocity = ?*

We are using the equation which links velocity, displacement and time:

$$s = vt$$

Substitute what we know (magnitude only):

$$25 = v \times 5$$

(continued)

Solve for average velocity:

$$v = \frac{25}{5}$$

$$v = 5\ km/hr$$

The direction of the average velocity will be the same as that for the displacement – on a bearing of 133°. This gives for the average velocity:

$$v = 5\ km/hr \text{ on a bearing of } 133°$$

Exercise 1.1.4 Speed and velocity

1 A ship sails from port to an oil rig. It sails 95 miles due west in a time of 6 hours. Calculate the average velocity of the ship.

2 A plane is flying across the Atlantic Ocean. Air traffic control vector the plane to fly with a heading of 280°. The plane maintains this heading with a speed of 550 mph for a time of 45 minutes. Calculate the resultant displacement of the plane.

3 In an orienteering competition, competitors need to navigate from the start point to the finish point as shown in the diagram. If it takes 12 minutes to complete the course, calculate the average velocity of the competitor in metres per second.

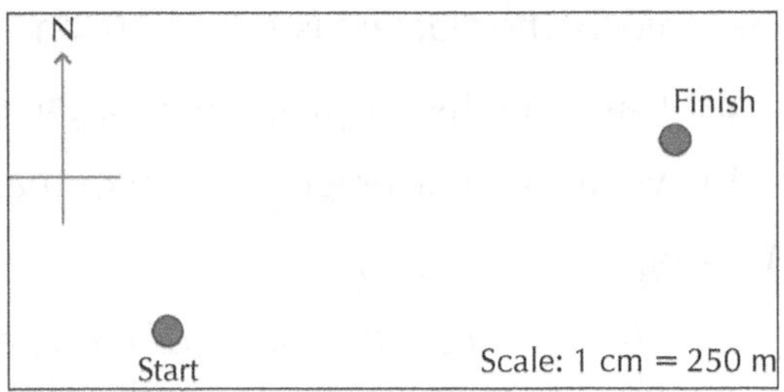

Fig 1.1.20

4 A walker walks from East Lomond to West Lomond, between the two points marked with an X on the map below.

Assume the scale of the map is 1·5 cm = 1 km.

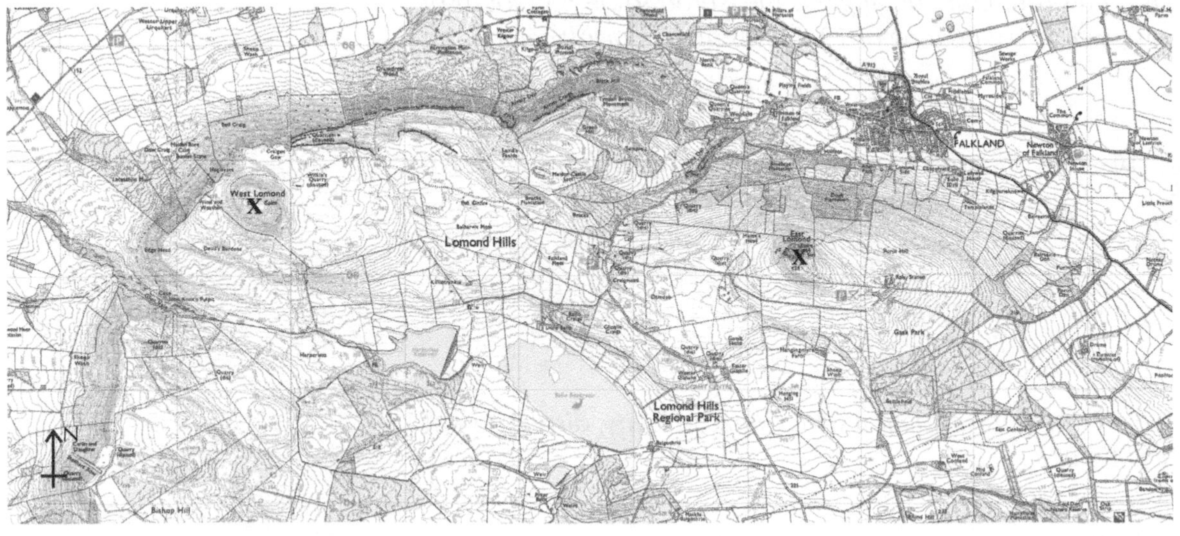

Fig 1.1.21

(continued)

a) Find the resultant displacement of the walker in kilometres.

b) If the walk is completed in 2 hours, find the average velocity of the walker in km/hr.

c) Explain why the average speed of the walker is likely to be different from the average velocity.

Addition of vectors

So far we have looked at individual vectors to represent displacement and velocity of objects. Vectors can be added together to give a resultant vector. The resultant vector must take both the magnitude and the direction of the vectors into account. To achieve this, we add vectors 'tip-to-tail' on a diagram, with the resultant vector being the single straight line which goes from the start point to the end point. We have already seen something similar when looking at displacements – the displacement is the straight line from the start to the end point.

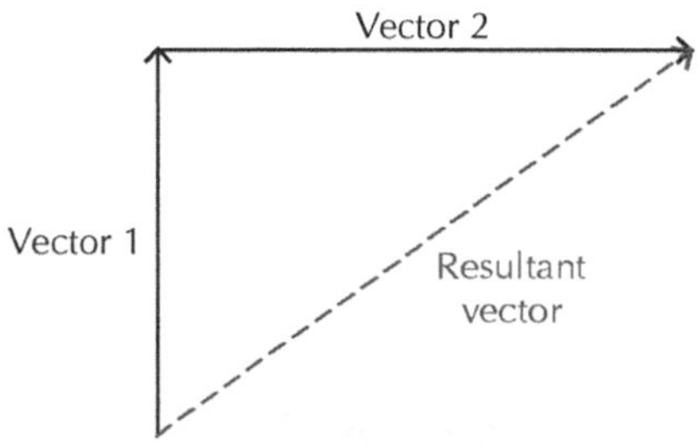

Vector 2

Vector 1

Resultant vector

Fig 1.1.22: *Addition of vectors tip-to-tail*

One dimension

In one dimension, we will consider vectors acting either right to left or up and down. The direction of a vector in one dimension can be represented using positive or negative numbers. Conventionally:

- Vectors acting to the right are positive and to the left are negative.
- Vectors acting upwards are positive and downwards are negative.

It is not necessary to stick to convention so long as you clearly state which direction you are considering to be positive and which to be negative.

Consider walking due north for 10 m, then turning and walking due south for 5 m. Each part of the journey can be represented by a vector (shown blue on the diagram opposite). These vectors are added tip-to-tail to give the resultant vector (shown red). The resultant vector is the straight line from the start point to the finish point.

Adding the vectors together, where due north is positive and due south is negative, we get:

$$s = 10 + (-5)$$
$$s = +5 \text{ m}$$

The resultant displacement that you will have walked is 5 m due north.

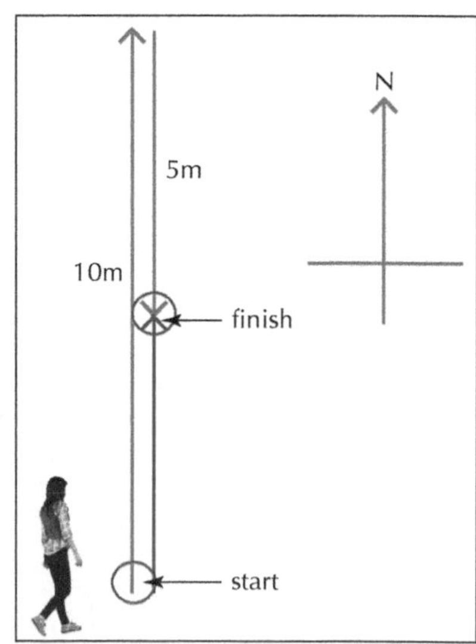

Fig 1.1.23

Technique: Adding vectors in one dimension

1. Define a positive direction – for example, *upwards is positive*.
2. Draw a diagram showing the vectors that you are adding together, remembering:
   - to mark the start position
   - to add vectors tip-to-tail
   - that the resultant vector goes from start point to end point
3. Find the magnitude and direction (positive or negative) through vector addition.

Worked example

In an orienteering competition, a competitor runs due west for a distance of 800 m. Realising he has made a mistake, he turns and runs due east for a distance of 1500 m to the end point. Calculate the resultant displacement of the competitor.

First of all, define a positive direction and then draw a diagram of the vectors, including the resultant vector which we are trying to find.

Define positive direction as due east.

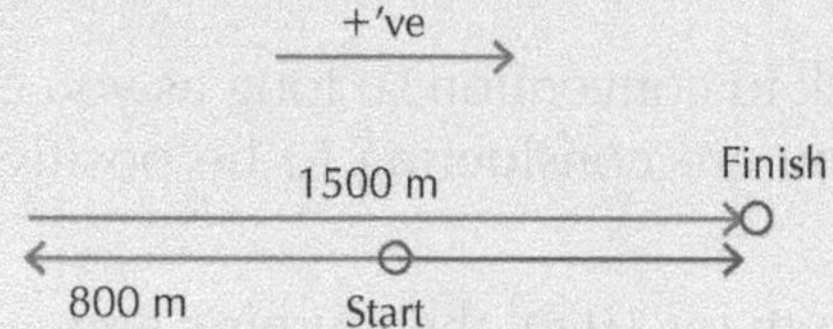

Fig 1.1.24 A

Fig 1.1.24 B: *Orienteering combines map reading skills and racing*

Resultant displacement is given by:

$$s = -800 + 1500$$
$$s = +700 \text{ m}$$

Positive displacement means that the resultant is due east, so:

$$s = 700 \text{ m due east}$$

Two dimensions

In two dimensions, we will consider vectors which can act on any bearing. The same rules that we considered for one dimension work in two dimensions. The vectors are added tip-to-tail with the resultant being the straight line that goes from the start point to the end point.

One of the main differences when working in two dimensions is that using a positive or negative sign on its own is no longer suitable for representing direction. Instead, direction must be quoted as an angle – this can be a bearing, or an angle from a specific direction.

GO! Experiment: Orienteering in two dimensions

This experiment highlights how the resultant of two vectors acting at right angles to each other can be found.

You will need:

- A trundle wheel
- A compass
- Markers

Instructions

1 Find a wide open space such as the school playing fields.
2 Mark a starting position using a marker.
3 Using the compass, identify due north and then, using the trundle wheel to measure distance, walk for a distance of 4 metres.
4 Using the compass, identify due east and then, using the trundle wheel to measure distance, walk for a distance of 3 metres.
5 Place a marker at the finishing point.
6 Return to the starting point and, using the compass, identify the direction of the straight line from the start to finish point.
7 Use the trundle wheel to measure the distance in metres from the start point to the finishing point.
8 Note down the resultant displacement.
9 Repeat with different distances and directions, ensuring the two vectors you are adding are at right angles to each other.

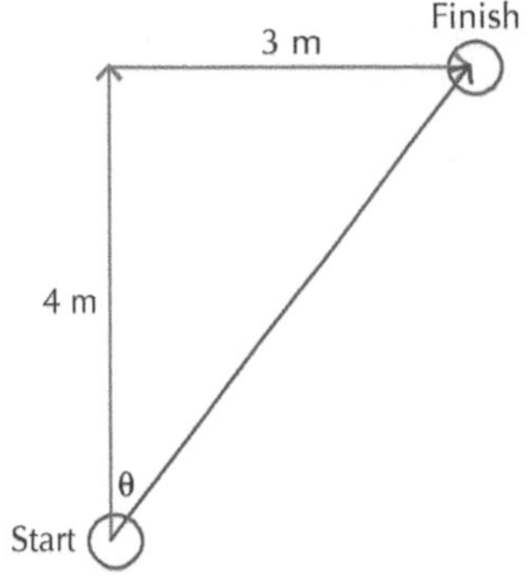

Fig 1.1.25

Consider the example from the above experiment of walking 4 metres due north and then walking 3 metres due east. These vectors are added tip-to-tail to give the resultant vector as shown on the diagram. This forms a right-angled triangle.

The length (size) of the resultant vector can be found using Pythagoras' theorem:

$$h^2 = o^2 + a^2$$

The direction, angle θ, can be found using trigonometry for right-angled triangles, SOH-CAH-TOA:

$$\theta = \sin^{-1}\frac{o}{h} \quad \theta = \cos^{-1}\frac{a}{h} \quad \theta = \tan^{-1}\frac{o}{a}$$

Usually we use the equation involving $\tan^{-1}$ because we are given both the opposite and adjacent lines (the two given vectors) so we are using numbers given in the question rather than numbers calculated which may contain an error.

Key point

Vectors must always be added tip-to-tail as shown on the diagram. Pythagoras' theorem is used in two dimensions to find the magnitude of the resultant vector. Trigonometry is used to find the direction in terms of an angle which can be converted to a bearing.

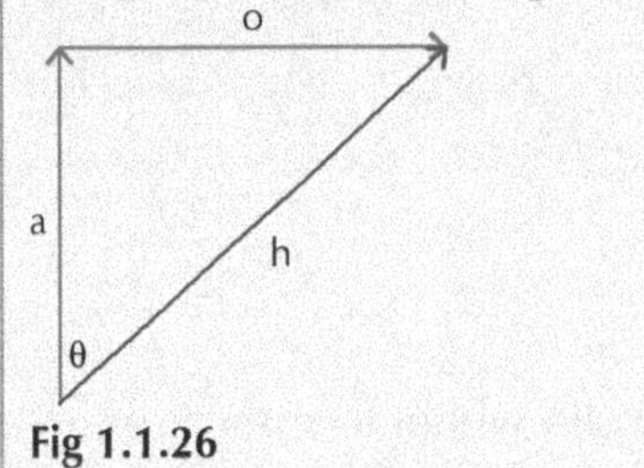

Fig 1.1.26

Technique: Adding vectors in two dimensions

1. Draw a diagram showing the vectors that you are adding together, remembering:
   - add vectors tip-to-tail
   - resultant vector goes from start point to end point
2. Find the magnitude of the resultant vector using Pythagoras' theorem:
$$h^2 = o^2 + a^2$$
3. Find the direction of the resultant vector using trigonometry:
$$\theta = \tan^{-1}\frac{o}{a}$$

Worked example

An aeroplane flies with a velocity of 550 mph due west for a time of 2 hours before turning and flying with a velocity of 490 mph due north for a time of 3 hours. Find:

a) The total distance flown by the plane.

b) The resultant displacement of the plane.

a) *To find the total distance flown by the plane, we need to find the distance flown during each stage and then add them together. We are using the equation linking speed, distance and time:*

$$d = vt$$

(continued)

For the first stage, flying at 550 mph for 2 hours:

$$d = 550 \times 2$$
$$d = 1100 \text{ miles}$$

For the second stage, flying at 490 mph for 3 hours:

$$d = 490 \times 3$$
$$d = 1470 \text{ miles}$$

The total distance flown is:

$$d = 1100 + 1470$$
$$d = 2570 \text{ miles}$$

b) *To find the resultant displacement, we must consider both magnitude and direction because displacement is a vector. We add vectors together using vector addition. Start by drawing a diagram:*

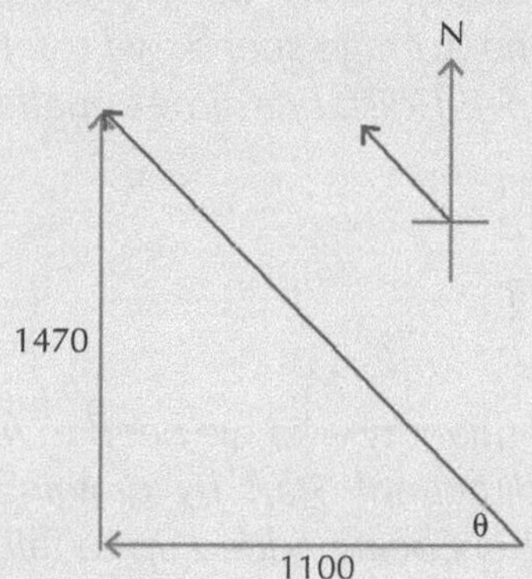

Fig 1.1.27

Use Pythagoras' theorem to work out the magnitude of the resultant vector:

$$h^2 = o^2 + a^2$$
$$h^2 = 1100^2 + 1470^2$$
$$h^2 = 3370900$$
$$h = 1836 \text{ miles}$$

The angle,θ, is found by using trigonometry:

$$\theta = \tan^{-1}\frac{o}{h}$$
$$\theta = \tan^{-1}\frac{1470}{1100}$$
$$\theta = 53{\cdot}2°$$

The resultant vector is included on the compass points on the diagram to make calculating the bearing clearer – the angle 53·2 ° is above west, so the bearing is:

$$270 + 53{\cdot}2 = 323{\cdot}2°$$

Hence the resultant displacement of the aeroplane is:

$$s = 1836 \text{ miles on a bearing of } 323{\cdot}2°$$

(continued)

Worked example

A walker walks 3 km north, then turns and walks 4 km west in a time of 2 hours. Find:

a) the distance walked

b) the average speed

c) the displacement of the walker

d) the average velocity

a) *Distance is a scalar quantity. So to find the total distance walked, you just need to add together the individual distances:*

$$d = 3 + 4$$

$$d = 7 \text{ km}$$

b) *Speed is also a scalar quantity so we do not need to quote direction. To find the average speed, we use the distance travelled in the speed, distance, time equation:*

$$d = vt$$

$$v = \frac{7}{2}$$

$$v = 3{\cdot}5 \text{ km/hr}$$

c) *Displacement is a vector quantity – so we need to find both the size and the direction! Start by drawing a diagram showing the vectors being added tip-to-tail:*

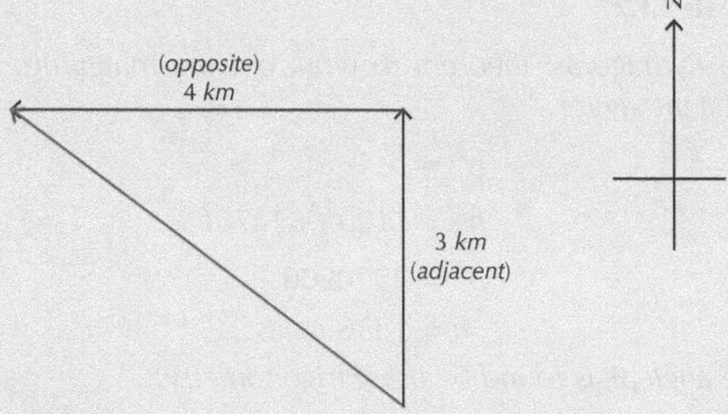

Fig 1.1.28

To find the size of the displacement, use Pythagoras' theorem:

$$h^2 = o^2 + a^2$$

$$h^2 = 3^2 + 4^2$$

$$h = \sqrt{25}$$

$$h = 5 \text{ km}$$

(continued)

To find the direction, use trigonometry:

$$\tan\theta = \frac{\text{opposite}}{\text{adjucent}}$$

$$\tan\theta = \frac{4}{3}$$

$$\theta = \tan^{-1}\frac{4}{3}$$

$$\theta = 53°$$

Using the compass points, this angle is west of north so the direction is 53° west of north, or a bearing of 307°. So:

s = 5 km on a bearing of 307°

d) *Velocity is also a vector quantity so we need to find both the size and direction. To find the average velocity, we use the displacement. Start with the size of the velocity:*

$$s = vt$$

$$5 = v \times 2$$

$$v = \frac{5}{2}$$

$$v = 2{\cdot}5 \text{ km/hr}$$

The direction of the average velocity is the same as the direction of the displacement, so we can say:

v = 2·5 km/hr on a bearing of 307°

Relative velocities

An important application of addition of vectors is in the calculation of relative velocity. Imagine standing on a station platform when a train passes you at 40 ms^{-1}.

Fig 1.1.29: *A train travelling at speed through a station*

Now, consider a passenger on that train walking *in the direction the train is travelling* at 3 ms^{-1}. The person will appear to you to be moving with a velocity of 43 ms^{-1}. Their velocity relative to you is that of the train plus their own walking velocity. In other words, we must add together the two vectors to find the resultant.

The above one dimensional example is illustrated here:

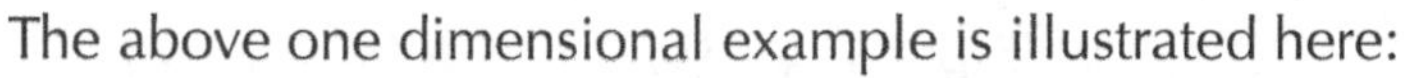

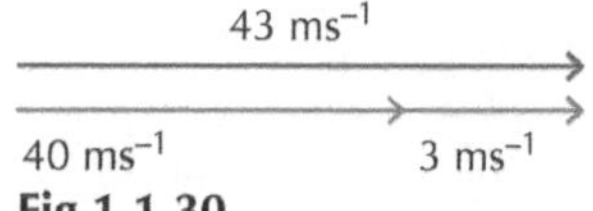

Fig 1.1.30

We can also apply this idea in two dimensions. Consider swimming across a river from one side to the other. You would swim in the direction south to north as shown on the diagram below. However, the current of the river would be flowing west to east. Therefore you will have a resultant

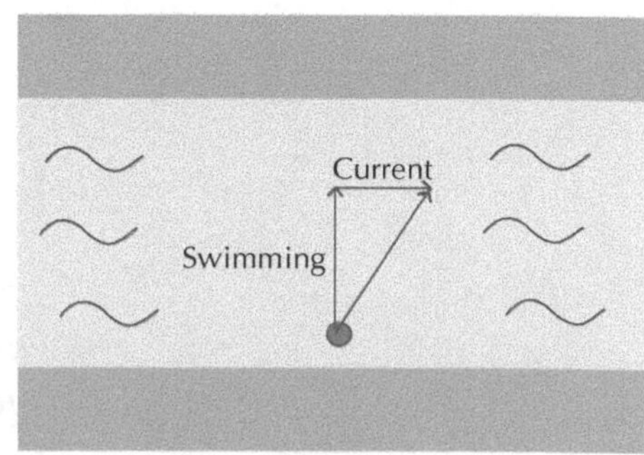

Fig 1.1.31: *Swimming across a river*

velocity which is given by adding your swimming velocity to the river current velocity.

This means that the river current sweeps you off your intended course.

Worked example

An aircraft is flying due north with a speed of 300 mph. There is a wind blowing from west to east with a speed of 50 mph. Find the resultant velocity of the aircraft.

To find the resultant velocity of the aircraft, we must add the two velocities together using vector addition. Start by drawing a diagram showing the two vectors being added tip-to-tail:

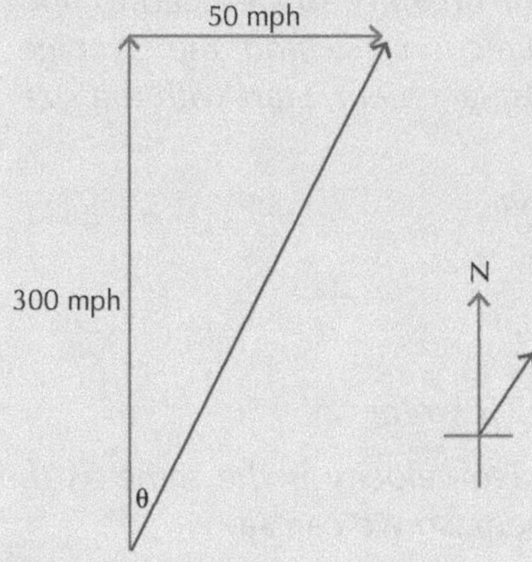

Fig 1.1.32

Use Pythagoras' theorem to work out the magnitude of the resultant vector:

$$h^2 = o^2 + a^2$$
$$h^2 = 300^2 + 50^2$$
$$h^2 = 92500$$
$$h = 304{\cdot}1 \text{ mph}$$

The angle, θ, is found by using trigonometry:

$$\theta = \tan^{-1} \frac{o}{h}$$
$$\theta = \tan^{-1} \frac{50}{300}$$
$$\theta = 9{\cdot}5°$$

The resultant vector is shown on the compass points and the angle marked is equal to the bearing so we can say that the resultant velocity is:

v = 304 mph on a bearing of 009·5°

SPOTLIGHT ON INDUSTRY: AUTOPILOT AND NAVIGATION

Modern aircraft have navigation systems that automatically fly the plane in the specified direction, or automatically fly the plane between set waypoints. This system must take into account headwinds, crosswinds and tailwinds, and ensure that the resultant velocity of the aircraft gets it to the next waypoint on schedule.

Exercise 1.1.5 Addition of vectors

1 A student wishes to add together two vectors to find his resultant displacement when walking from the school canteen to the school office. He draws the diagram shown. Assess this vector diagram and explain why this will not give the correct answer for the resultant displacement. Redraw the diagram to show how the vectors should be added correctly.

Resultant

Fig 1.1.33

2 A boat sails 14 km in a northerly direction before turning and sailing back south for 10 km. Calculate the distance sailed and the resultant displacement of the ship.

3 In an orienteering race, a team hikes 400 metres east before realising they have gone the wrong way. They turn round and hike 800 metres west. What is the distance they have hiked? What was their resultant displacement?

4 A plane flies on a bearing of 045° a distance of 30 km. It turns and flies on a bearing of 225° a distance of 15 km. Find the total distance flown and the resultant displacement of the plane.

5 A walker walks at 4 ms^{-1} for a time of 60 seconds in a northerly direction. She turns and walks due south at 2 ms^{-1} for 100 seconds. What total distance did the walker cover? What was her displacement?

6 A ship sails due north for 100 km before turning and sailing due east for 50 km. Calculate the distance sailed and the resultant displacement of the ship.

7 A runner runs due south for 300 metres, then runs due west for 800 metres. Calculate the distance and resultant displacement of the runner.

8 A plane flying from London Heathrow to Dyce takes off due west and flies for 40 miles in a westerly direction before turning and flying due north for 500 miles. Calculate the total distance flown and the displacement between Heathrow and Dyce airports.

9 An aircraft is flying north with a velocity of 450 mph into a headwind of velocity 60 mph. What is the resultant velocity of the aircraft?

Fig. 1.1.34: *A runner*

10 A man is walking in the direction of travel along a train with a speed of 2 ms^{-1}. The train passes through a station at 40 ms^{-1}. What is the speed of the man relative to people standing on the station platform?

(*continued*)

11 A river flows from west to east at a speed of 14 ms^{-1}. A boat leaves point A on the south bank of the river and sails with a velocity of 9 ms^{-1} north. Calculate the resultant velocity of the boat.

12 A river flows from south to north with a velocity of 11 ms^{-1}. A speedboat leaves the west bank of the river and sails with a velocity of 30 ms^{-1} east. The boat sails for a time of 300 seconds. Calculate:

a) the resultant velocity of the boat

b) the displacement of the boat crossing the river

Learning checklist

In this section you will have learned:

- The definition of scalar and vector quantities.
- The identity of scalar and vector quantities.
- How to calculate the resultant of two vector quantities in one dimension or at right angles.
- How to determine the resultant distance and/or displacement using a scale diagram or calculation.
- How to determine the resultant speed and/or velocity using a scale diagram or calculation.
- How to use appropriate relationships to solve problems involving distance, displacement, speed, velocity and time.
- How to describe experiments to measure average and instantaneous speeds.

2 Velocity-time graphs

In this chapter you will learn about:

- Velocity-time graphs – how to plot a velocity-time graph, and how to find velocity, acceleration, displacement or the motion of an object from a velocity-time graph.

Velocity-time graphs

The motion of an object can be represented on a velocity-time graph (or speed-time graph) which shows the speed of an object at any given time. The velocity-time graph is a way to represent motion which is easy to read at a glance. For example, an acceleration is shown as a positive gradient on a velocity-time graph, and the steeper the gradient the greater the acceleration. Being able to see this at a glance is very useful.

Plotting a velocity-time graph

A velocity-time graph is a line graph which is used to represent the motion of an object. We will only deal with constant accelerations so we will always have straight lines between the different points on the graph.

Technique: Manually plotting a velocity-time graph

1. Draw and label the axes on your sheet of graph paper (time on the horizontal axis, speed on the vertical axis).
2. Choose an appropriate scale for the axes: look at the number of boxes on the axis and the minimum and maximum values to be plotted; choose a scale which uses as many of the boxes as possible to make best use of the available paper.
3. For each time and speed in your data set, mark an X on your graph.

(continued)

4 Join the points with straight lines.

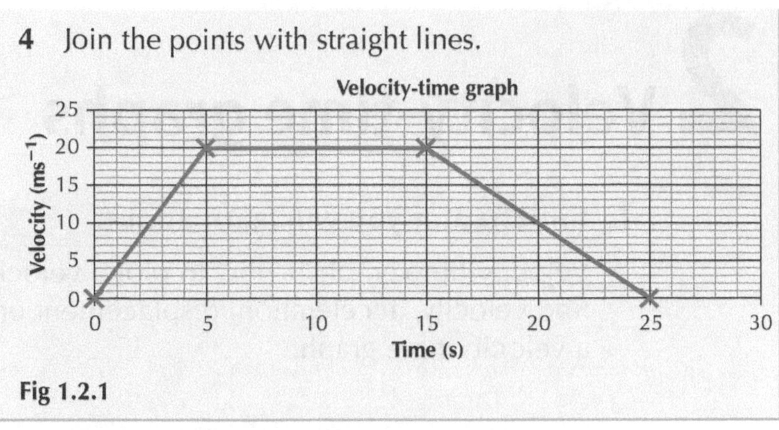

Fig 1.2.1

Worked example

The speed of a car as it drove along a road was recorded at different points. These speeds are shown in the table below.

Time (s)	0	5	15	20	25	45	55	60
Velocity (ms^{-1})	0	10	10	0	15	15	20	0

Table 1.2.1

Plot a velocity-time graph for the data.

Examine the times and velocities and choose suitable axes scales for the graph to make the best use of the graph paper.

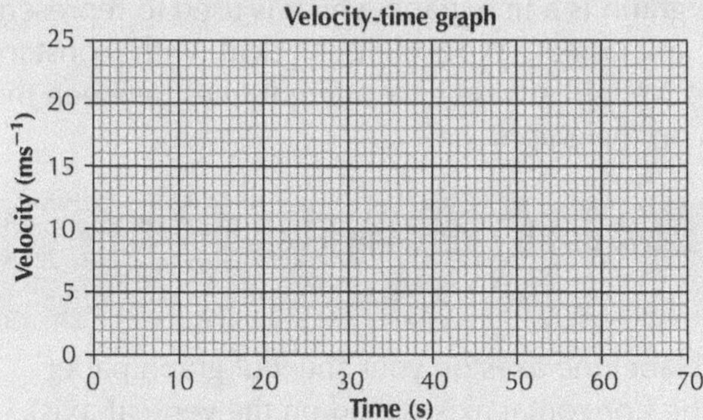

Fig 1.2.2

Plot each of the points on the graph by marking them with an X.

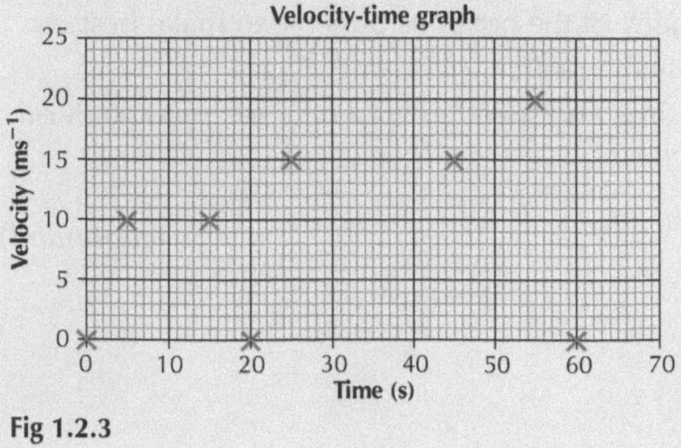

Fig 1.2.3

(continued)

Join the points with straight lines.

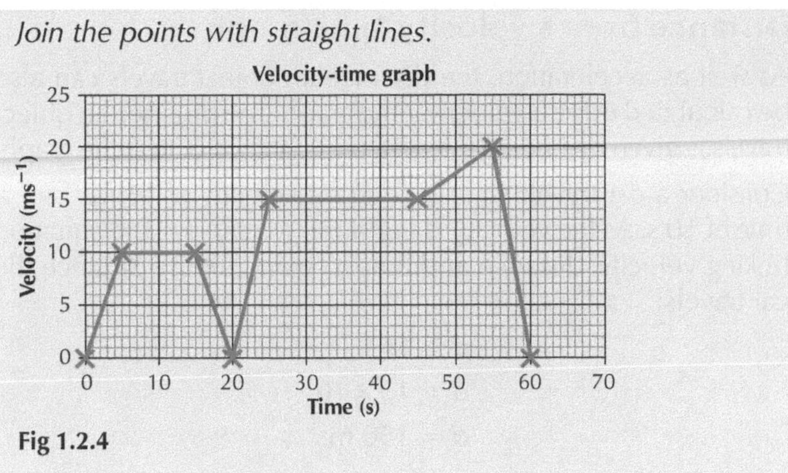

Fig 1.2.4

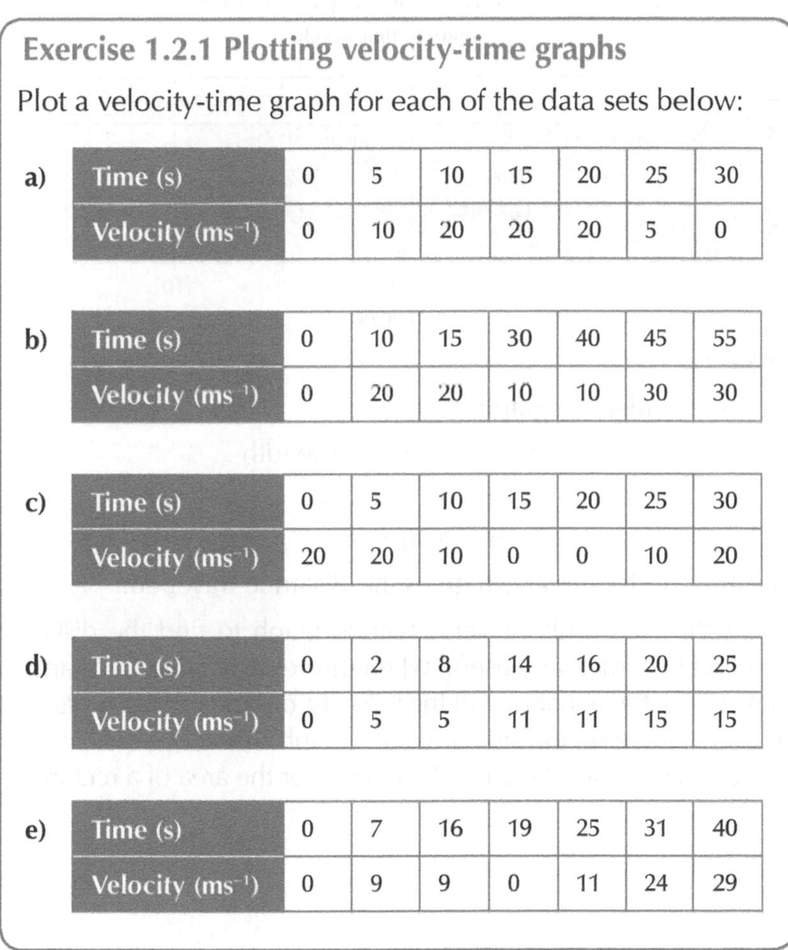

Exercise 1.2.1 Plotting velocity-time graphs

Plot a velocity-time graph for each of the data sets below:

a)

Time (s)	0	5	10	15	20	25	30
Velocity (ms^{-1})	0	10	20	20	20	5	0

b)

Time (s)	0	10	15	30	40	45	55
Velocity (ms^{-1})	0	20	20	10	10	30	30

c)

Time (s)	0	5	10	15	20	25	30
Velocity (ms^{-1})	20	20	10	0	0	10	20

d)

Time (s)	0	2	8	14	16	20	25
Velocity (ms^{-1})	0	5	5	11	11	15	15

e)

Time (s)	0	7	16	19	25	31	40
Velocity (ms^{-1})	0	9	9	0	11	24	29

Distance from a velocity-time graph

As well as acceleration, the distance an object travels can also be calculated using a velocity-time graph. The distance an object travels is given by finding the area under the velocity-time graph.

Consider a car travelling at a constant velocity of 15 ms^{-1} for a time of 10 s. As the velocity is constant, we can use the equation linking velocity, distance and time to work out the distance the car travels:

$$d = vt$$
$$d = 15 \times 10$$
$$d = 150 \text{ m}$$

We can also plot a velocity-time graph for the car above:

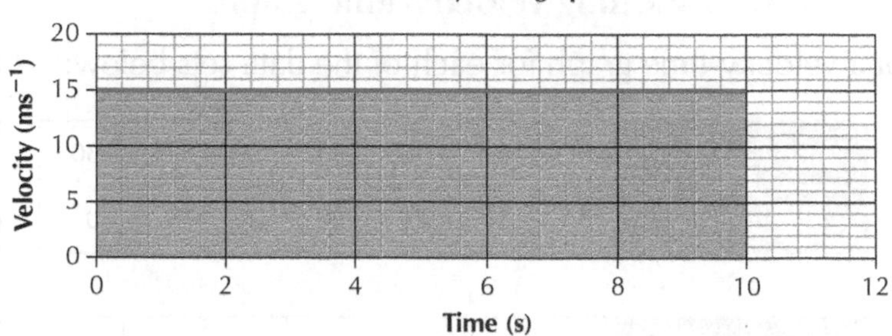

Fig 1.2.5

The area under the graph is given by:

$$Area = \text{length} \times \text{breadth}$$
$$Area = 10 \times 15$$
$$Area = 150 \text{ m}$$

The area under the graph gives the distance travelled.

Using the area under a velocity-time graph to find the distance travelled becomes very useful when the velocity is not constant. In these cases, the equation linking velocity, distance and time cannot be used. However, the area under the graph can still be easily found using equations for the area of a triangle or the area of a rectangle.

Technique: Find the distance from a velocity-time graph

1. Divide the area under the graph into triangles and rectangles:

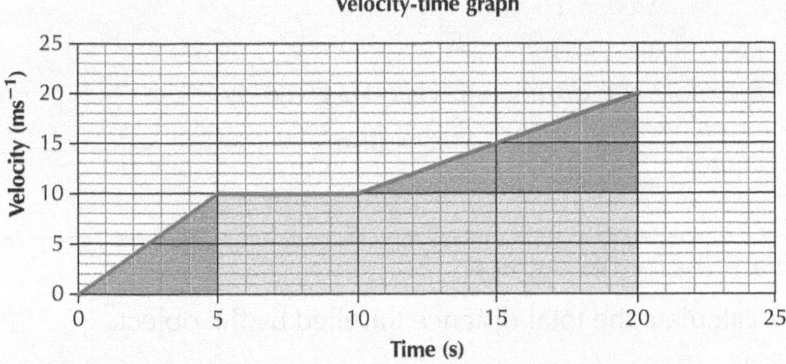

Fig 1.2.6

2 Work out the area of each of the shapes on your graph using:

$$\textit{Area of rectangle} = \textit{length} \times \textit{breadth} \quad \text{and} \quad \textit{Area of triangle} = \frac{1}{2} \times \textit{breadth} \times \textit{height}$$

3 Add the area of each individual shape together to find the total distance travelled.

Key point

The distance of an object is given by finding the area under a velocity-time graph. The area can be divided into triangles or rectangles and the size of the areas found using the following equations:

rectangle: $area = l \times b$

triangle: $area = \frac{1}{2} \times b \times h$

Worked example

The velocity-time graph for an object travelling across the lab floor is shown below. Calculate the total distance travelled by the object.

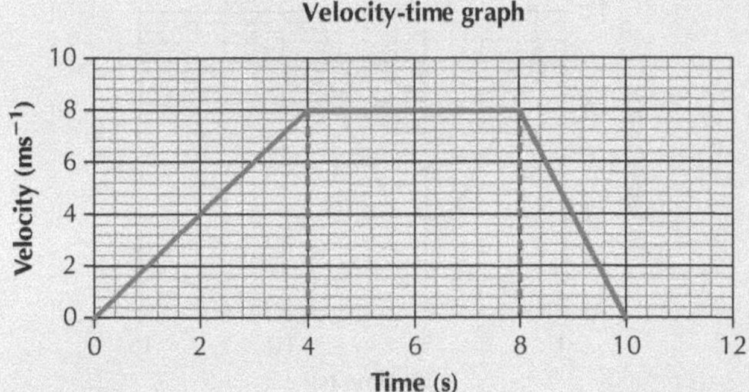

Fig 1.2.7

The distance travelled will be given by the area under the velocity-time graph. Divide the area into shapes to allow the area to be easily calculated (dashed lines).

(continued)

The total area under the graph is given by:

$$Area = \frac{1}{2}bh + lb + \frac{1}{2}bh$$

$$Area = \left(\frac{1}{2} \times 4 \times 8\right) + (4 \times 8) + \left(\frac{1}{2} \times 2 \times 8\right)$$

$$Area = 16 + 32 + 8$$

$$Area = 56$$

Therefore the distance travelled is given by:

$$d = 56m$$

Exercise 1.2.2 Distance from a velocity-time graph

1 For each of the graphs below, calculate the total distance travelled by the object.

a)

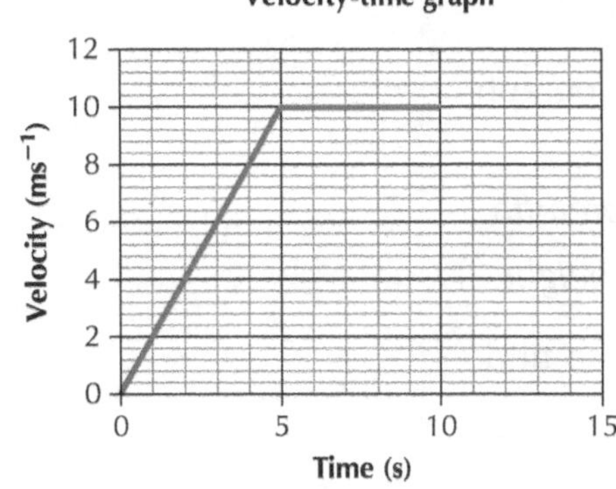

Fig 1.2.8

b)

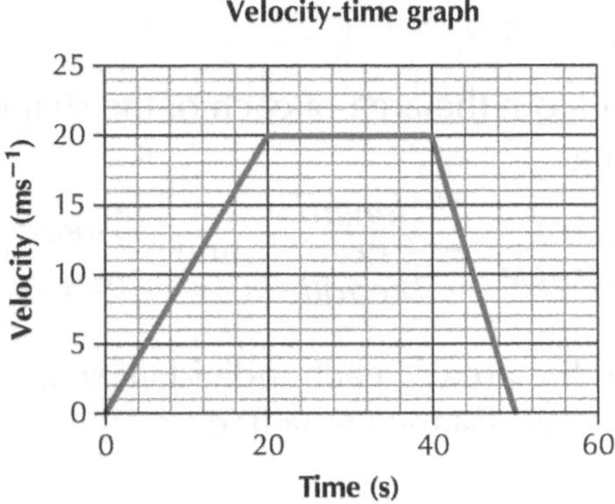

Fig 1.2.9

c)

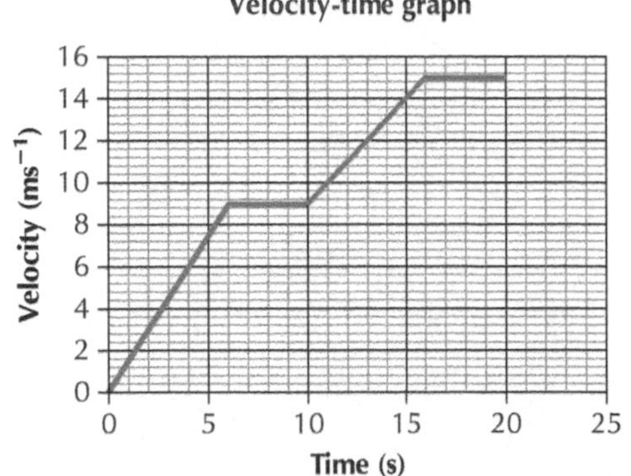

Fig 1.2.10

d)

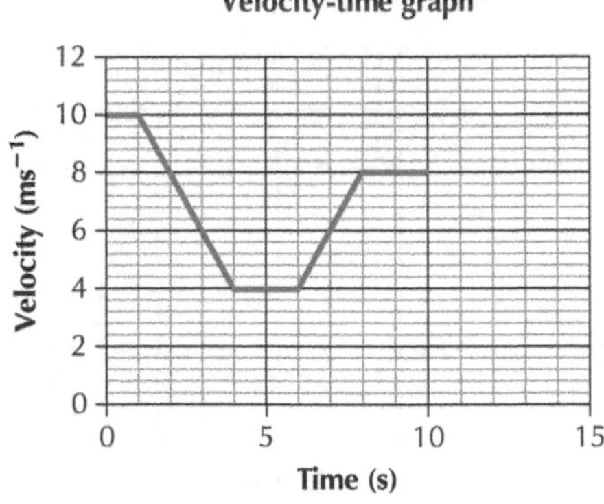

Fig 1.2.11

(continued)

2 A ball is thrown vertically upwards with an initial velocity of 25 ms^{-1}. It leaves the thrower's hand and is subject to acceleration due to gravity of $-9{\cdot}8\ ms^{-2}$. This downwards acceleration causes the ball to slow down to rest. The velocity-time graph for the motion of the ball as it reaches its maximum height is shown below.

Calculate the maximum height that the ball reaches.

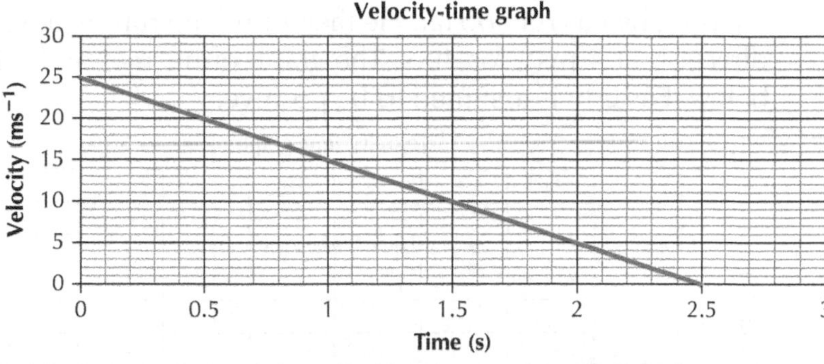

Fig 1.2.12

3 A car is travelling along the road at 20 ms^{-1} when the driver sees a hazard and brakes to a halt. The graph for the motion of the car is shown below.

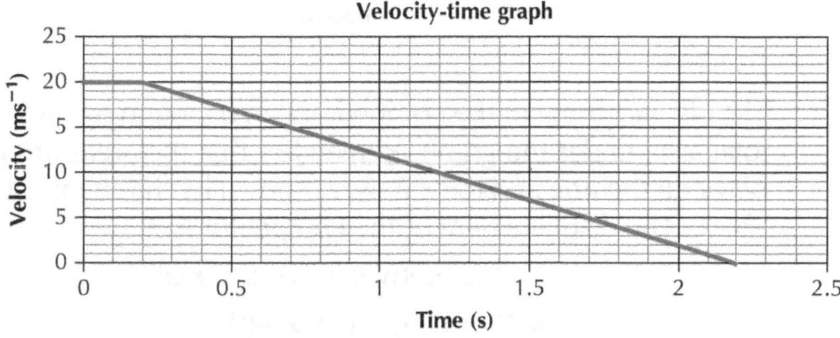

Fig 1.2.13

a) At what time did the driver hit the brake pedal?

b) Calculate the total stopping distance of the car.

Motion graphs and vectors

We have already looked at the concept of vector quantities – a vector is a quantity with both magnitude and direction. A velocity-time graph must therefore have the ability to represent direction as well as magnitude to give a true representation of the vector.

In one dimension we can represent the direction of a vector using positive or negative signs. We define the positive direction, for example, upwards, and any motion upwards is positive and any motion downwards is negative. We can represent positive and negative velocities on a graph. Therefore, in one dimension the sign of the velocity from the graph is very important; it tells us which way the object is moving. This will also lead to positive or negative displacements. Any area above the x-axis is positive; any area below the x-axis is negative. This gives the direction of the displacement.

Consider a competitor in an orienteering competition. Say he runs in an easterly direction at 5 ms^{-1} for a time of 100 s before realising he has run too far. He then turns and runs in a westerly direction at 4 ms^{-1} for 40 s. The graph of his motion is shown below. Easterly is assumed to be positive.

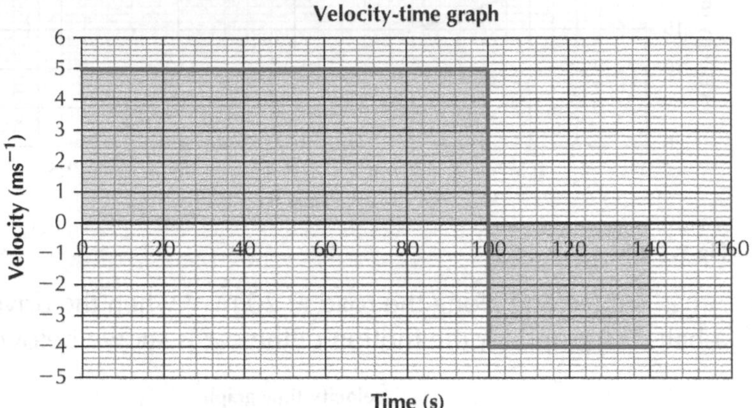

Fig 1.2.14

The displacement of the competitor can be found by finding the total area under the graph. Remember that the areas are found between the line and the x-axis, as shown by the shapes shaded above. The displacement of the competitor is

$$s = \textit{Area between line and x-axis}$$

$$s = (5 \times 100) + (-4 \times 40)$$

$$s = 500 - 160$$

$$s = 340 \text{ m}$$

A positive displacement indicates that the direction is easterly.

Learning checklist

In this section you will have learned:

- How to draw or sketch speed-time and velocity-time graphs from data.
- How to interpret a velocity-time graph to describe the motion of an object.
- How to determine the displacement from a velocity-time graph.

3 Acceleration

In this chapter you will learn about:

- Acceleration – how to measure acceleration and calculations involving acceleration.

Acceleration

When an object changes speed, we say that it accelerates. In everyday language, when an object is speeding up we say that it is accelerating and when an object is slowing down we say it is decelerating. However, in our later study of vectors we will speak only of acceleration: the change in speed is positive (getting faster) or negative (getting slower).

Acceleration is defined as the change in speed in a given time. The greater the change of speed in a set time, the greater the acceleration of the object. Acceleration is given by:

$$a = \frac{\Delta v}{t}$$

where Δv is the change in speed and t is the time. We can also define acceleration as:

$$a = \frac{v - u}{t}$$

where u is the initial speed, v is the final speed and t is the time. The units for acceleration are metres per second per second (ms^{-2}). Notice that u and v are in alphabetical order, u being the initial speed and v being the final speed.

In order to measure the acceleration we need to measure the instantaneous speed at the starting point and at the end point, and the time taken to travel between these points. This can be done as an average acceleration over a large distance or an instantaneous acceleration over a small distance. Either method will require the use of electronic timing equipment to measure the instantaneous speed.

Experiment: Measuring acceleration

This experiment measures the instantaneous acceleration of a cart as it passes a single point.

You will need:

- A ramp
- A dynamics cart
- A double mask
- A light gate and computer
- A ramp holder (or books)

Fig 1.3.1

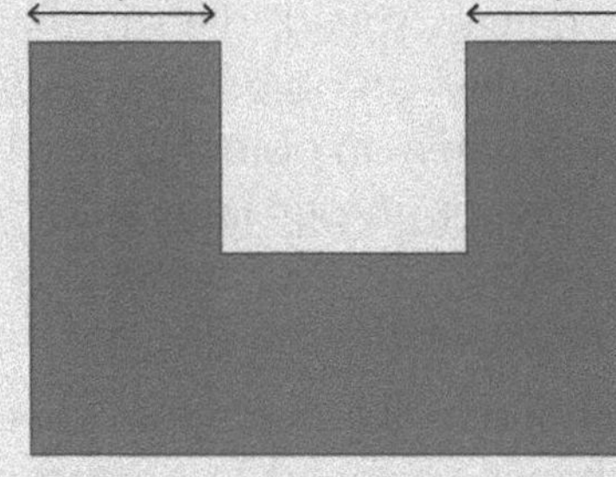

Fig 1.3.2

Instructions

1. Use a ramp holder or books to prop the ramp up at one end.
2. Place your light gate across the ramp so that the cart can travel freely through it.
3. Attach a double mask to the cart with known lengths, l, with a separation between them as shown. Ensure that this mask passes between the light source and the light sensor of the gate.
4. Let the cart roll freely down the ramp.
5. Use the timer connected to a computer to measure the following times.
   a) Time for mask one to pass through the gate – this gives the initial speed, u.
   b) Time for mask two to pass through the gate – this gives the final speed, v.
   c) Time between the end of the first mask and the start of the second – this gives the time, t.
6. The acceleration can then be calculated using the equation:

$$a = \frac{v - u}{t}$$

Technique: Calculations involving acceleration

1. Write down what you know from the question, leaving blank what you are trying to find:

   Initial speed, u = ____ ms^{-1}

   Final speed, v = ____ ms^{-1}

   Time, t = ____ s

   Acceleration, a = _____ ms^{-2}

2 Write down the equation linking the quantities above as you see it on the formula sheet:

$$a = \frac{v - u}{t}$$

3 Substitute into the equation what you know.

4 Solve for the unknown.

Worked example

A lorry accelerates from 10 ms^{-1} to 25 ms^{-1} in a time of 50 s. Calculate the acceleration of the lorry.

Write down what we know and what we are looking for:

Initial speed, u = 10 ms^{-1}

Final speed, v = 25 ms^{-1}

Time, t = 50 s

Acceleration = ? ms^{-2}

We are using an equation which links initial speed, final speed, time and acceleration:

$$a = \frac{v - u}{t}$$

Substitute what we know:

$$a = \frac{25 - 10}{50}$$

Solve for acceleration:

$$a = \frac{15}{50}$$

$$a = 0{\cdot}3\ ms^{-2}$$

Fig 1.3.3: *Lorry*

Worked example

Fig 1.3.4: *Hot air balloon*

A hot air balloon ascends at a constant speed. The occupants of the balloon throw some sandbags overboard and the balloon begins to accelerate upwards with a constant acceleration of 0·1 ms^{-2} for 1 minute. If the balloon reached a speed of 7 ms^{-1} after this minute, what was its initial speed?

Write down what we know and what we are trying to find out:

- *Initial speed, u = ? ms^{-1}*
- *Final speed, v = 7 ms^{-1}*
- *Time, t = 1 minute = 60 s*
- *Acceleration, a = 0·1 ms^{-2}*

We are using the equation which links initial speed, final speed, time and acceleration:

$$a = \frac{v - u}{t}$$

Substitute what we know:

$$0{\cdot}1 = \frac{7 - u}{60}$$

Solve for initial speed:

$$0{\cdot}1 \times 60 = 7 - u$$

$$7 - u = 6$$

$$u = 1\ ms^{-1}$$

Exercise 1.3.1 Acceleration

1 Use the equation for acceleration to fill in the missing values in the table below.

Initial speed (ms^{-1})	Final speed (ms^{-1})	Time (s)	Acceleration (ms^{-2})
0	20	10	(a)
7	24	6·5	(b)
11	25	(c)	3
24	(d)	5	–4
(e)	12	1·8	4·2

Table 1.3.1

2 A small rocket is launched from rest with an acceleration of 14 ms^{-2}. It accelerates for a time of 9 s. What is the rocket's final speed?

(*continued*)

3 Car manufacturers often quote the 0–100 km/hr time for their cars. The 0–100 km/hr time for three cars is shown below:

- car A: 7·8 s
- car B: 11·8 s
- car C: 4·6 s

a) Convert 100 km/hr into metres per second.

b) Work out the acceleration of each of the cars in metres per second per second.

4 A child sledges down a hill. He starts from rest at the top of the hill and reaches a speed of 15 ms^{-1} at the bottom of the hill. He takes 10 s to reach the bottom of the hill. After reaching the bottom of the hill, he decelerates back to rest again in a time of 15 s.

a) Calculate the acceleration of the child when going down the hill.

b) Calculate the deceleration of the child at the bottom of the hill.

Fig 1.3.5: *Child sledging down a hill*

5 In a drag race, a car accelerates from rest to a speed of 50 ms^{-1} in a time of 4·7 s. Calculate the acceleration of the car.

6 A plane accelerates down the runway before taking off. In order to take off, it must reach a speed of 55 ms^{-1}. If the plane starts from rest and accelerates with an acceleration of 7 ms^{-2}, how long will it take for the plane to reach its take off speed?

7 A sprinter accelerates from rest for 0·7 s with an acceleration of 9 ms^{-2}. Calculate the final speed of the sprinter.

8 An experiment with two light gates was set up to measure the acceleration of a cart travelling down an inclined plane.

The cart has a mask of length 5 cm fitted to it. The times measured by the two light gates are shown below:

gate 1 = 0·42 s

gate 2 = 0·08 s

The time for the cart to travel between the gates was 6 s.

Fig 1.3.6

a) Calculate the speed of the cart through each of the light gates.

b) Find the acceleration of the cart down the inclined plane.

9 Two pupils are studying acceleration due to gravity on Earth. They set up an experiment that involves two light gates separated by a distance of 1 metre. They drop a block with a known length through the light gates and measure the following speeds:

top gate = 2·0 ms^{-1}

bottom gate = 21·5 ms^{-1}

(continued)

The time taken for the block to travel between the gates is 2·0 s.

a) Calculate the acceleration of the block due to gravity.

b) Suggest an alternative experiment which could be conducted that uses only one light gate.

10 A brick is dropped out of a window. It takes the brick a time of 1·6 seconds to reach the ground. It accelerates with an acceleration of 9·8 ms^{-2}. Find the speed of the brick when it hits the ground assuming it is dropped from rest.

11 A bus is travelling at 25 ms^{-1}. The driver sees people waiting for the bus at the stop in front of her so applies the brakes. The bus slows down with a steady deceleration of –3 ms^{-2}. How long does it take for the bus to slow down to rest?

Fig 1.3.7: *Old London routemaster bus*

Acceleration from a velocity-time graph

We can tell at a glance at a velocity-time graph whether an object is speeding up (accelerating), slowing down (decelerating) or travelling at a constant speed. If the graph is increasing, then the object is accelerating. If the graph is decreasing, then the object is decelerating. A graph which is horizontal represents a constant speed.

Speeding up

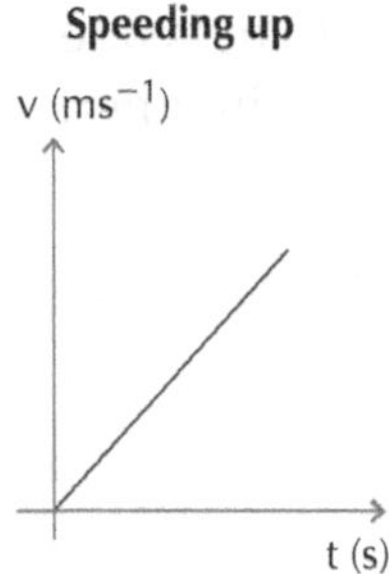

Constant speed

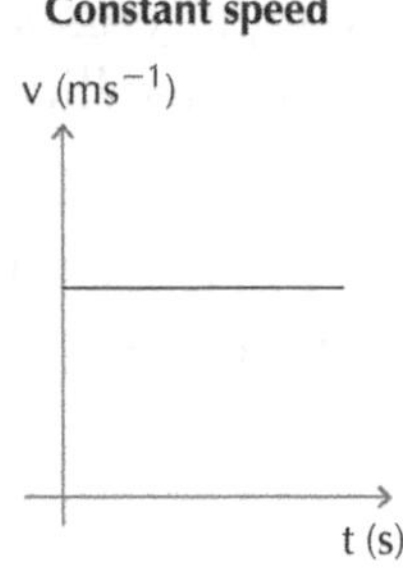

Slowing down

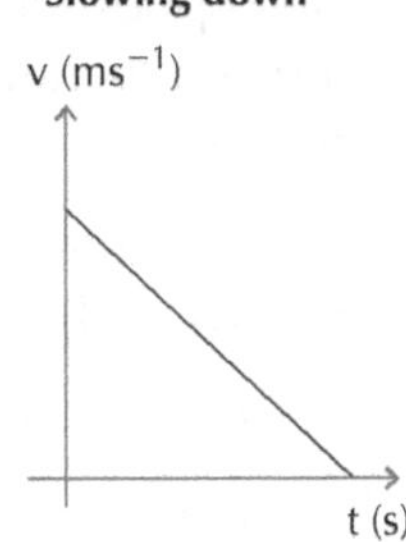

Fig 1.3.8

By using the equation for acceleration,

$$a = \frac{v - u}{t}$$

we can work out the acceleration of an object using the velocity-time graph. The initial and final velocities can be read from the graph, as can the time between these velocities. This is highlighted by the worked example below.

Worked example

The velocity-time graph for the motion of a car driving along a road is shown below.

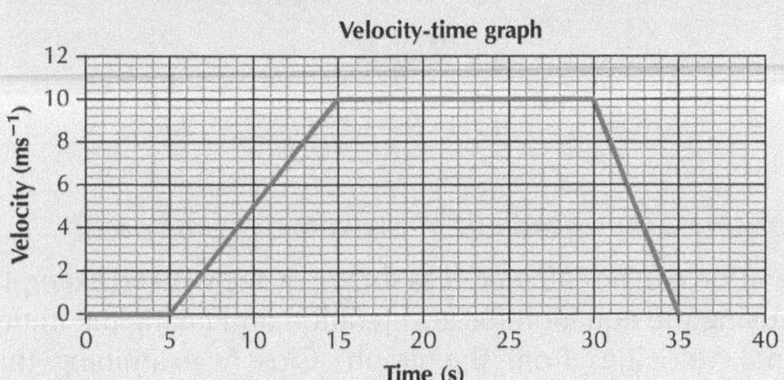

Fig 1.3.9

Calculate the acceleration and deceleration of the car during the journey.

Acceleration is where the graph is increasing – between 5 s and 15 s. Use the graph to write down what we know and what we are looking for:

Initial velocity, $u = 0\ ms^{-1}$

Final velocity, $v = 10\ ms^{-1}$

Time, $t = 10$ s

Acceleration = ? ms^{-2}

We are using the equation for acceleration:

$$a = \frac{v - u}{t}$$

Substitute what we know and then solve for acceleration:

$$a = \frac{10 - 0}{10}$$

$$a = 1\ ms^{-2}$$

Deceleration is where the graph is decreasing – between 30 s and 35 s. Use the graph to write down what we know and what we are looking for:

Initial velocity, $u = 10\ ms^{-1}$

Final velocity, $v = 0\ ms^{-1}$

Time, $t = 5$ s

Acceleration, a = ? ms^{-2}

We are using the equation for acceleration:

$$a = \frac{v - u}{t}$$

(continued)

Substitute what we know and solve for acceleration (which will be negative because the car is slowing down):

$$a = \frac{0 - 10}{5}$$

$$a = -2\ ms^{-2}$$

Key point

The acceleration of an object is given by the gradient of a velocity-time graph.

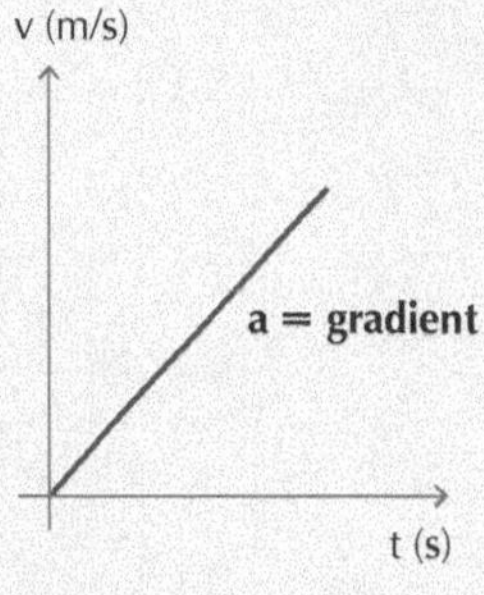

Fig 1.3.10

$$a = \frac{v - u}{t} = \frac{y_2 - y_1}{x_2 - x_1} = m$$

Acceleration as the gradient of the velocity-time graph

Acceleration can be calculated as shown by the worked example above using the equation for acceleration and taking the initial and final velocities from the graph. Closely examining this equation will show that it is the same as the equation for the gradient of a straight line:

$$m = \frac{y_2 - y_1}{x_2 - x_1}$$

We have the following:

- y_2 = *final velocity* = v
- y_1 = *initial velocity* = u
- $x_2 - x_1$ = *time between velocities* = t

Exercise 1.3.2 Acceleration from a velocity-time graph

1 For each of the graphs below, calculate the acceleration and deceleration of the object.

a)

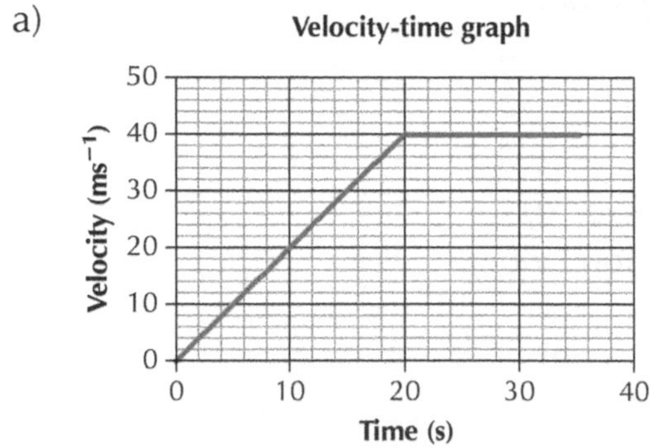

Fig 1.3.11

b)

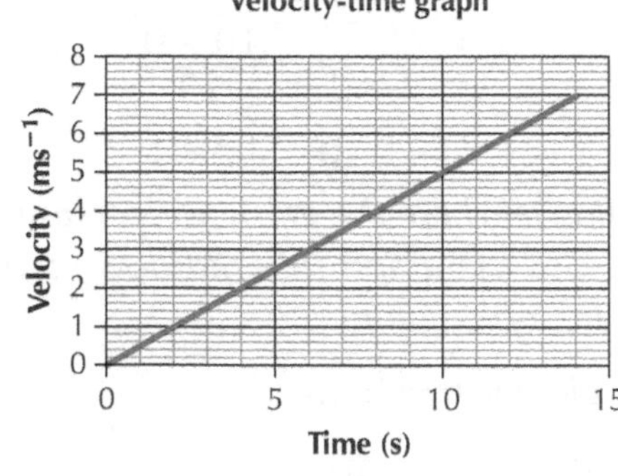

Fig 1.3.12

(continued)

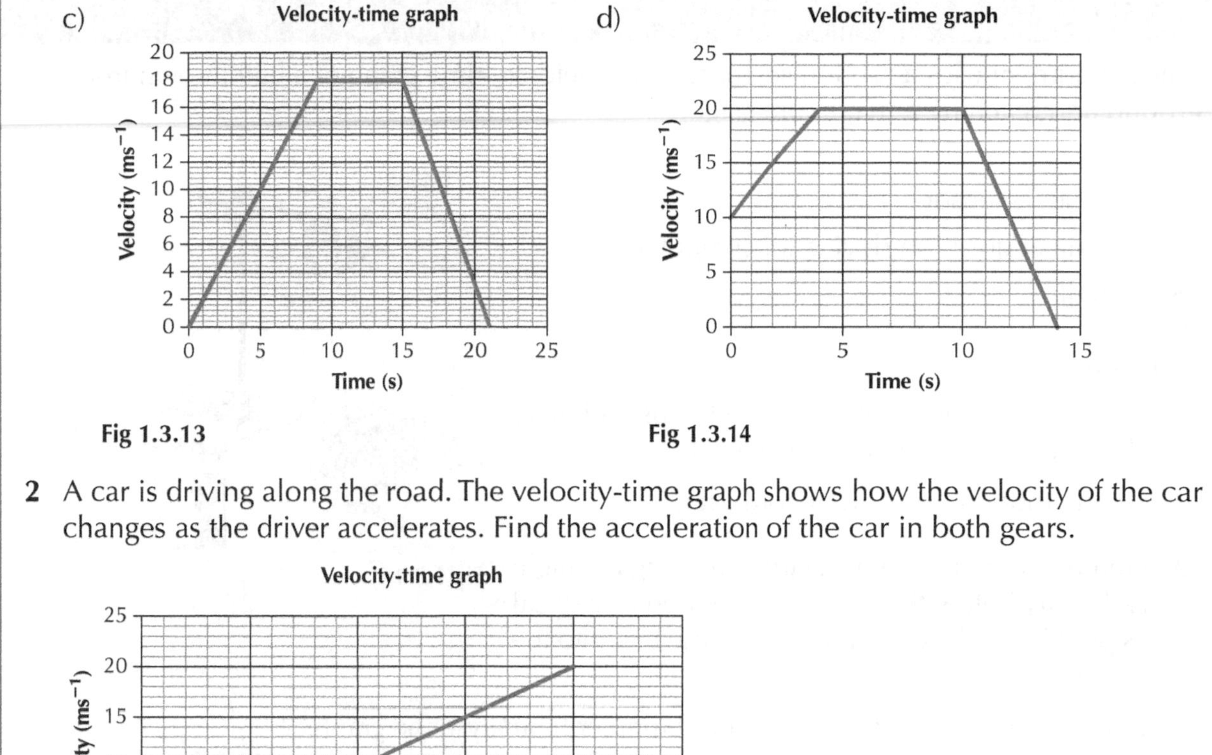

Fig 1.3.13

Fig 1.3.14

2 A car is driving along the road. The velocity-time graph shows how the velocity of the car changes as the driver accelerates. Find the acceleration of the car in both gears.

Velocity-time graph

Fig 1.3.15

Velocity-time graphs from experiment

Velocity-time graphs can be plotted using data-capturing devices connected to motion sensors. These can be used to plot the velocity of an object in real time, leading to the production of a velocity-time graph. We will consider two main examples here: the velocity-time graph for a ball thrown vertically into the air and the velocity-time graph for a bouncing ball.

Experiment: Thrown ball velocity-time graph

This experiment uses a motion sensor to plot a velocity-time graph of a ball thrown from the ground and reaching a maximum height.

You will need:

- A motion sensor
- A data acquisition device (or computer)
- A ball

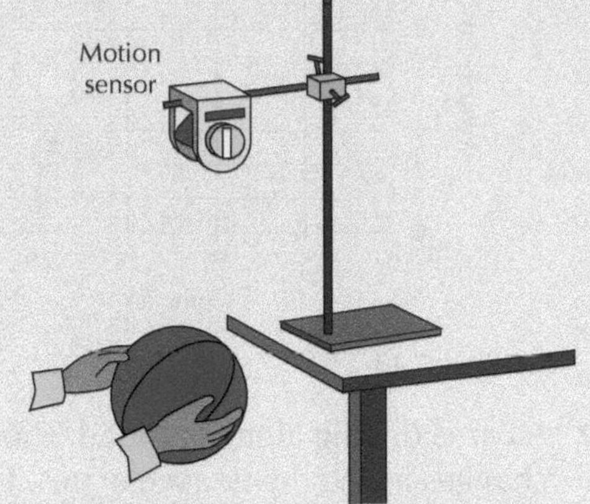

Fig 1.3.16

Instructions

1 Position the motion sensor on a clampstand, held above where you will throw the ball.

2 Set the motion sensor to record data.

3 Gently throw the ball upwards, ensuring the flight is directly below the motion sensor and as vertical as possible. Take care not to throw the ball into the sensor!

4 Use the data acquisition device or computer to analyse the graph.

5 Use the graph to find:

a) the time at which the maximum height is reached

b) the maximum height of the ball

Thrown ball

Consider throwing a ball vertically from the ground (as in the experiment above). It will be subject to a constant acceleration due to gravity of 9·8 ms^{-2} downwards. If we assume that upwards is positive, the acceleration of the ball will be negative:

$$a = g = -9{\cdot}8ms^{-2}$$

The initial velocity of the ball will be positive. If we throw the ball with an initial velocity of 19·6 ms^{-1}:

- the ball will decelerate to 0 ms^{-1} at the maximum height,
- the ball will then begin to accelerate back down to the ground.

The velocity-time graph for the motion of this ball is shown below.

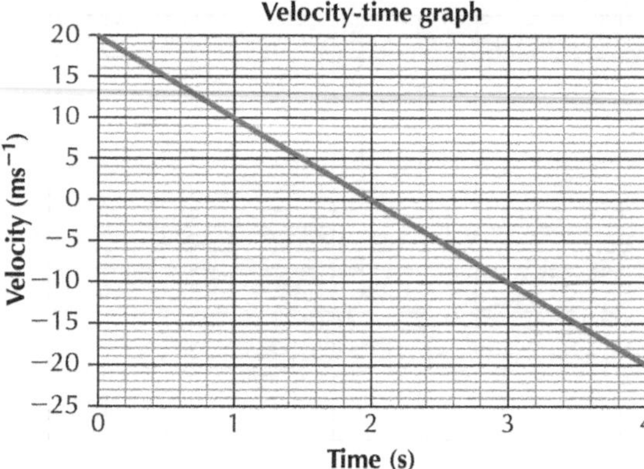

Fig 1.3.17

From the graph, we can work out the acceleration by finding the gradient of the line – this acceleration is constant throughout the flight as we would expect, because the ball is always being accelerated due to gravity. Calculating the acceleration:

$$a = \frac{v - u}{t}$$

$$a = \frac{-19{\cdot}6 - 19{\cdot}6}{4}$$

$$a = \frac{-39{\cdot}2}{4}$$

$$a = -9{\cdot}8 \text{ ms}^{-2}$$

The velocity of the ball is positive to begin with which indicates that the ball is travelling upwards. This velocity decreases to zero after 2 seconds – at this stage the ball has reached its maximum height. By finding the area under the graph up to this point we can work out the maximum height of the ball. The area is:

$$A = \frac{1}{2}bh$$

$$A = \frac{1}{2} \times 2 \times 19{\cdot}6$$

$$A = 19{\cdot}6$$

Hence the maximum height of the ball,

$$h_{max} = 19{\cdot}6 \text{ m}$$

After this point, the velocity of the ball becomes negative – this indicates that the ball is moving back downwards. After 4 seconds, the ball stops moving as it has returned to the thrower's hand. The diagram below shows the position of the ball in flight at different stages and the corresponding points on the graph.

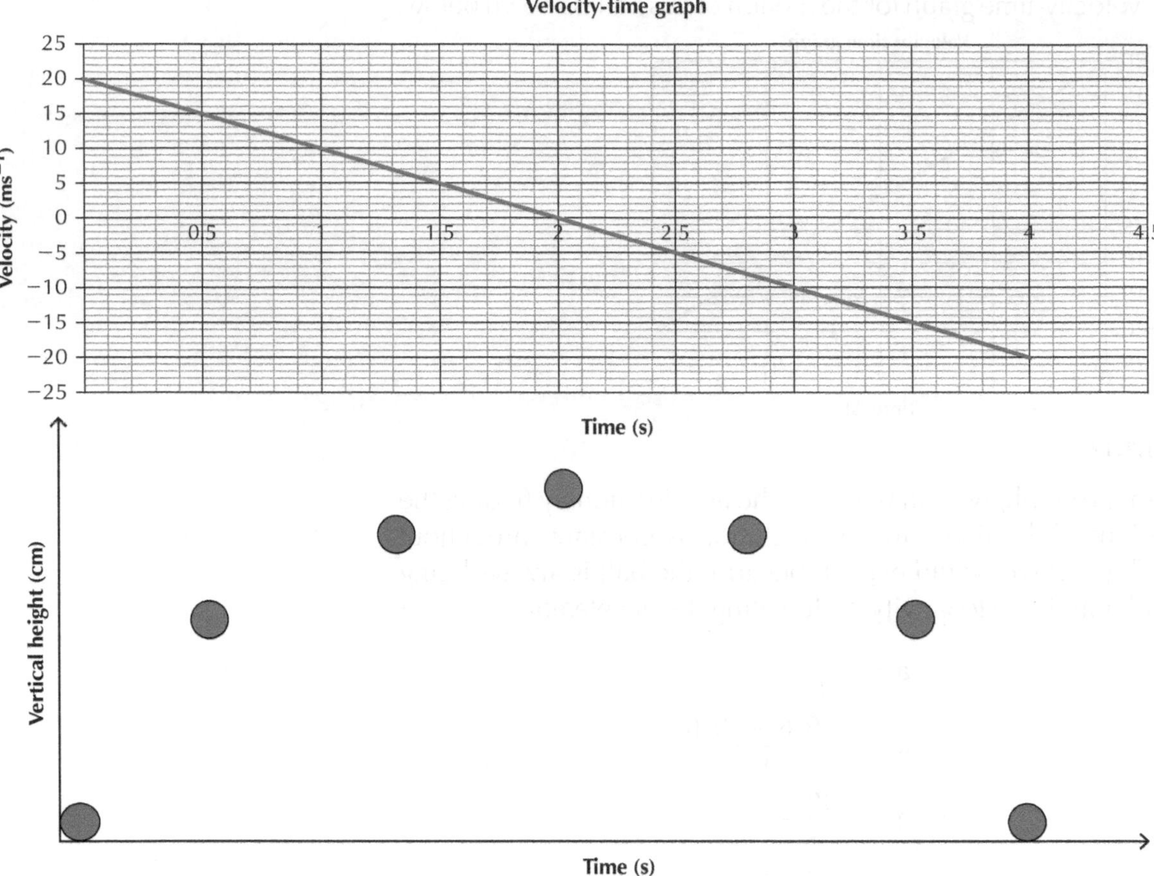

Fig 1.3.18

Experiment: The bouncing ball

This experiment uses a motion sensor to plot the velocity-time graph for a bouncing ball.

You will need:

- A motion sensor
- A data acquisition device (or computer)
- A ball

Instructions

1. Position the motion sensor on a clampstand, held above where you will drop the ball.
2. Set the motion sensor to record data.
3. Drop the ball from just underneath the motion sensor and allow it to bounce off the ground and return to the sensor. You may wish to let the ball bounce more than once.
4. Use the data acquisition device or computer to analyse the graph.

(*continued*)

5. Use the graph to find:
   a) the time at which the ball hits the ground
   b) the height from which the ball was dropped
   c) the height to which the ball returns after the first bounce*

*By the theory of energy conservation, this height should be the same as the height from which the ball was dropped. However, frictional losses during the flight and bounce will cause a loss in energy, and the height will be reduced.

The bouncing ball

Consider dropping a ball from a maximum height and letting it bounce off the ground (as in the experiment above). It will be subject to a constant acceleration due to gravity of 9·8 ms^{-2} downwards. If we assume that upwards is positive, the acceleration of the ball will be negative:

$$a = g = -9{\cdot}8\ ms^2$$

As the ball is dropped from rest, the initial velocity of the ball will be zero. It then falls to the ground, so the velocity of the ball will be negative. After bouncing, the ball will rebound upwards, so will have a positive velocity that will decrease back to zero as it is being decelerated by gravity. The velocity-time graph for the motion of the ball is shown below:

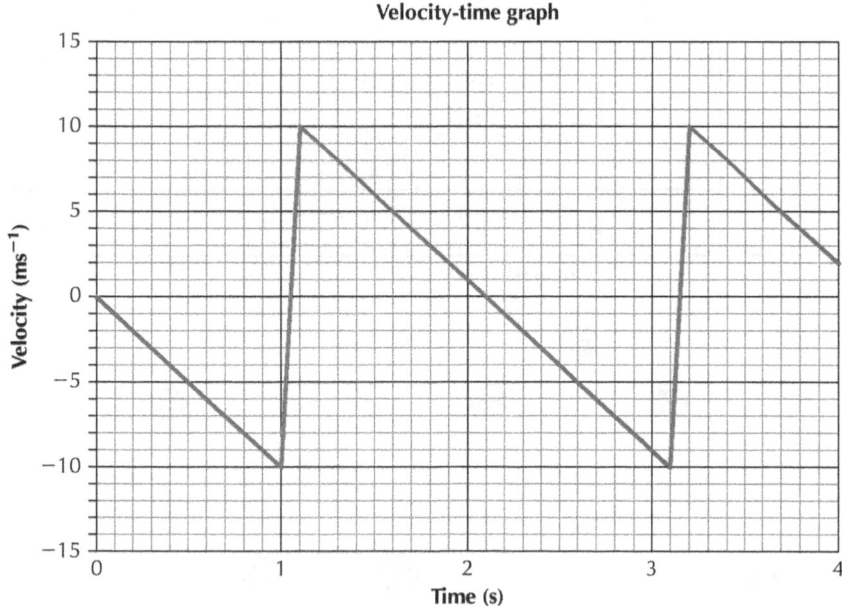

Fig 1.3.19

As above for the thrown ball, the acceleration can be found by calculating the gradient of the graph. As the ball is accelerating due to gravity, this acceleration will be constant and equal to −9·8 ms^{-2}. Notice the negative gradient on the graph which highlights that the acceleration is acting downwards.

The ball's velocity changes rapidly from negative to positive at a time of 1 s into the flight. This is the point at which the ball bounces off the ground. The downwards velocity is changed to an upwards velocity.

The height from which the ball was dropped can be found by working out the area under the graph between 0 s and 1 s:

$$A = \frac{1}{2}bh$$

$$A = \frac{1}{2} \times 1 \times -9{\cdot}8$$

$$A = -4{\cdot}9$$

The negative number here simply means that the ball covered a negative displacement, in other words it was travelling downwards. The height from which the ball was dropped was:

$$h = 4{\cdot}9 \text{ m}$$

The diagram below shows the vertical position of the flight of the ball corresponding to different points on the graph.

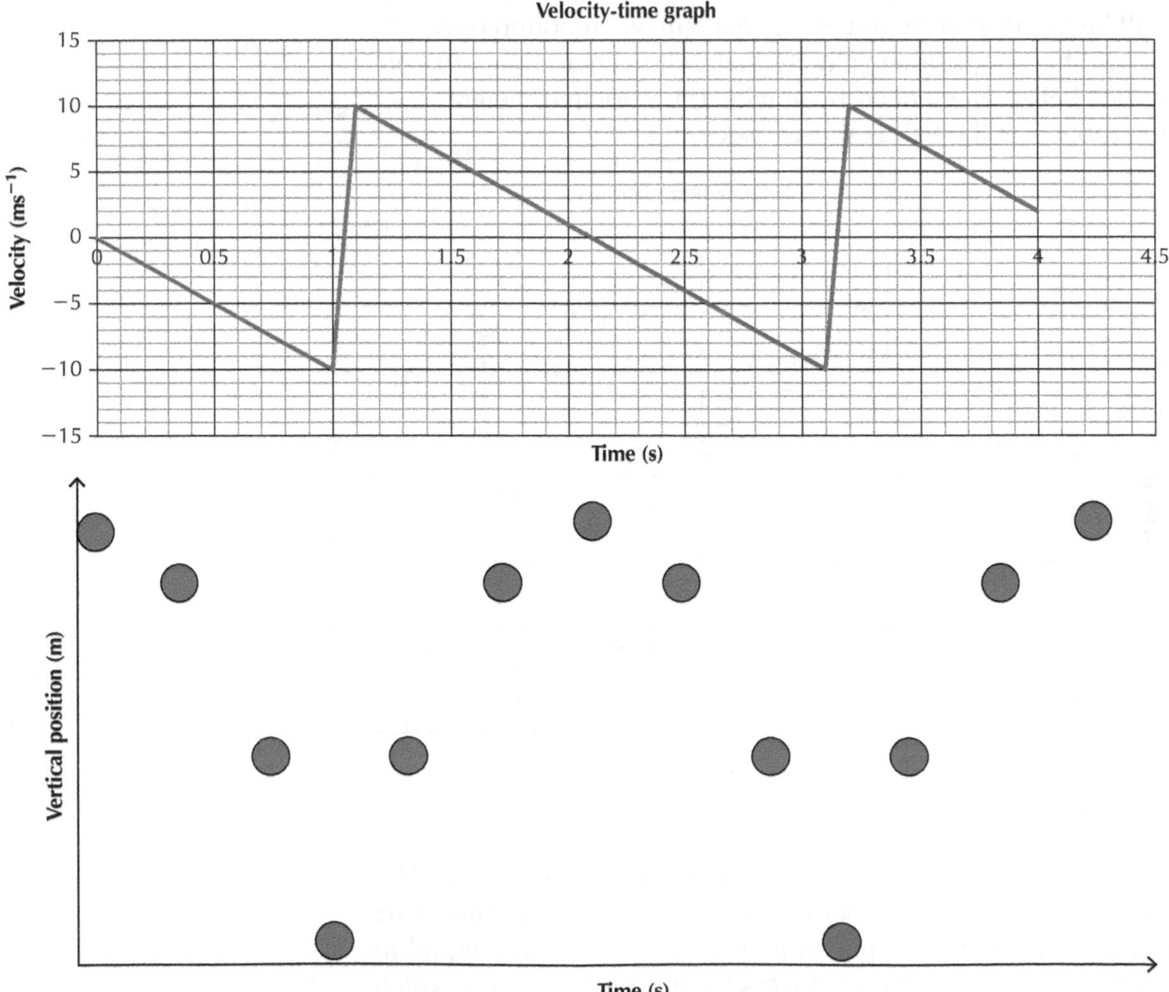

Fig 1.3.20

Exercise 1.3.3 Analysing velocity-time graphs

1 A ball is thrown vertically into the air from the ground with an initial speed of 29·4 ms^{-1}. It reaches a maximum height and then falls back down to the thrower's hand. The velocity-time graph for the motion of the ball is shown below.

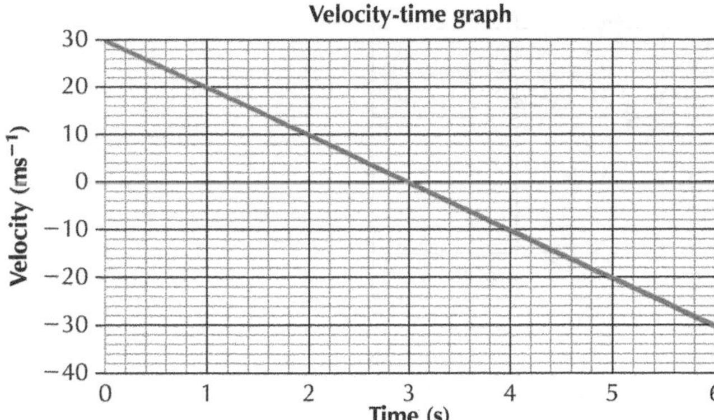

Fig 1.3.21

a) How long does it take for the ball to reach its maximum height?

b) What is the maximum height that the ball reaches?

c) Use the graph to find the acceleration of the ball throughout its flight.

2 A ball is dropped from a height and allowed to fall to the ground. The ball is dropped from rest and falls to the ground in a time of 1·5 seconds. The ball falls under gravity with a constant acceleration of 9·8 ms^{-2} downwards.

a) Calculate the velocity of the ball at the point when it hits the ground (remember if you define upwards as positive, this velocity will be negative).

b) Plot a velocity-time graph to represent the motion of the ball as it falls to the ground.

c) Use your velocity-time graph to calculate the height from which the ball was dropped.

3 A ball is thrown vertically upwards from the ground. After 0·8 seconds, the ball reaches its maximum height where its velocity is zero. The ball then returns to the thrower's hand.

a) Calculate the initial velocity that the ball was thrown with, assuming acceleration due to gravity of 9·8 ms^{-2} downwards.

b) Plot a velocity-time graph to represent the motion of the ball as it reaches its maximum height and then returns to the thrower's hand.

c) Use your velocity-time graph to calculate the height above the thrower's hand that the ball reaches.

Fig 1.3.22

4 In an experiment to investigate the motion of a bouncing ball, a ball is dropped from under a motion sensor and allowed to bounce.

(continued)

The velocity-time graph for the motion of the ball is shown below.

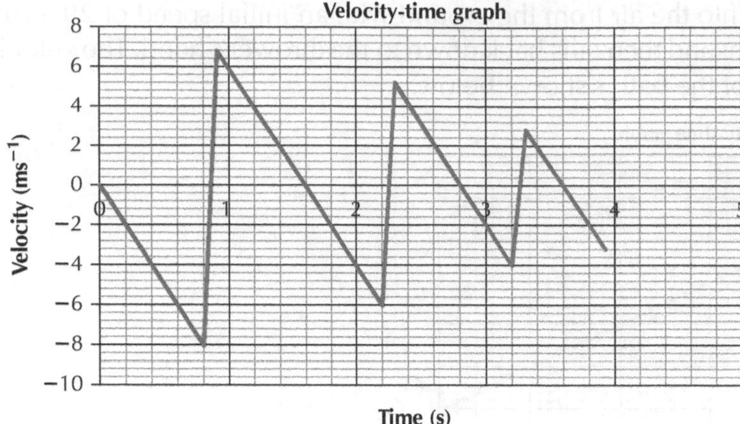

Fig 1.3.23

a) How long does it take for the ball to fall to the ground?

b) What is the velocity of the ball when it hits the ground?

c) From what height was the ball dropped?

d) What is the rebound velocity of the ball after the first bounce?

e) What height does the ball reach after the first bounce?

f) Account for the difference in your answers to part (c) and (e).

5 A ball is dropped from a height and allowed to bounce off the ground. The ball accelerates under gravity. The ball takes 1·2 seconds to reach the ground and after the first bounce takes 1·1 seconds to reach its maximum height. The bounce lasts just 0·1 seconds.

a) Calculate the velocity of the ball immediately before it hits the ground.

b) Calculate the velocity of the ball immediately after the bounce.

c) Plot a velocity-time graph for the motion of the ball for the 2·4 seconds described above.

d) Use your velocity-time graph to find:

i) the height from which the ball was dropped

ii) the height to which the ball rebounded

Learning checklist

In this section you will have learned:

- To define acceleration in terms of initial velocity, final velocity and time.
- How to use an appropriate relationship to solve problems involving acceleration, initial velocity (or speed), final velocity (or speed) and time.
- How to determine the acceleration from a velocity-time graph.
- How to describe an experiment to measure acceleration.

4 Newton's laws

In this chapter you will learn about:

- The application of Newton's laws to everyday life.
- Newton's first law – the effects of balanced and unbalanced forces on the motion of an object.
- Newton's second law – the link between unbalanced force and the acceleration of an object.
- Newton's third law – the concept of Newton's pairs of forces and application of this law to rockets and the jet engine.

Fig 1.4.1: *Sir Isaac Newton*

Think about it

A book sitting on a table does not move. Are there any forces acting on the book? Discuss this in small groups and write a short summary of your thoughts.

Introduction to Newton's laws

Isaac Newton (1642–1727) is considered one of the greatest and most influential scientists who ever lived! In 1687 he described three laws of motion. These laws are so important they are still applied today in aeronautical engineering, space travel and motor sport to name just a few examples.

Newton's three laws of motion are:

- **Newton's first law:** an object will remain at rest or continue moving with a constant speed in a straight line unless acted on by an unbalanced force.
- **Newton's second law:** the acceleration of an object is directly proportional to the unbalanced force and inversely proportional to the mass of the object.
- **Newton's third law:** every action force has an equal and opposite reaction force.

Newton's first law

We all know Newton's first law as a part of everyday life. Think about a book on your desk. It doesn't move anywhere unless you push it or pick it up. That is Newton's first law of motion – the book will stay still unless a force acts on it.

The book is on planet Earth, so it must be subject to gravity which pulls the book downwards. The table is applying a force to the book to stop it from falling downwards to the ground. We say that these forces are balanced because the overall force acting on the book is zero: gravity accelerating the book downwards is cancelled out by the force of the table acting upwards. When the forces are balanced, Newton's first law states that the book must remain still.

Adding forces in one dimension

It is important to note that when we discussed the book above we considered the overall force being exerted on the book. Force is a vector quantity – it has both size and direction. In one dimension, the direction of the force is given by either a positive or negative sign. We call the overall force the **resultant force**. To work it out, you have to add or subtract the individual forces. It is usually easiest to draw a diagram showing the individual forces acting – such a diagram is called a **free-body diagram**. If forces act in opposite directions then they can cancel each other out. The overall force is equal to zero when these forces are balanced.

Word bank

- **resultant force**
the overall force acting when more than one force is combined (added or subtracted)

Word bank

- **free-body diagram**
a simplified diagram showing the forces acting on an object

A book on a table

On a diagram, the forces acting on a book sitting on a table are represented by arrows that point in the direction of the forces, as shown opposite.

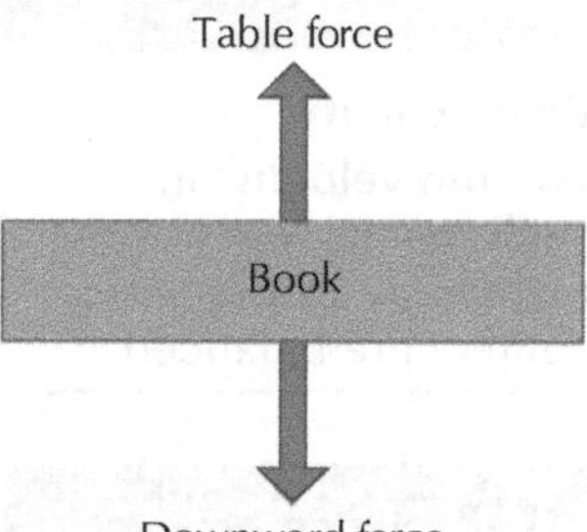

Fig 1.4.2: *Free-body diagram showing forces acting on a book*

As discussed in the section on vectors, we must define a positive and a negative direction. Sticking with convention, upwards is positive. From the diagram, we can see that the resultant force is given by:

Resultant force = table force – downward force

The table force and the downward force are equal so the resultant force is zero. According to Newton's first law, when the resultant force is zero (forces are balanced), the book does not move.

Key point

Force is a vector quantity. In one dimension, the direction of the force is represented by a positive or negative sign. You can choose which direction you wish to be positive. The opposite direction will be the negative direction. See the three diagrams in Fig 1.4.3.

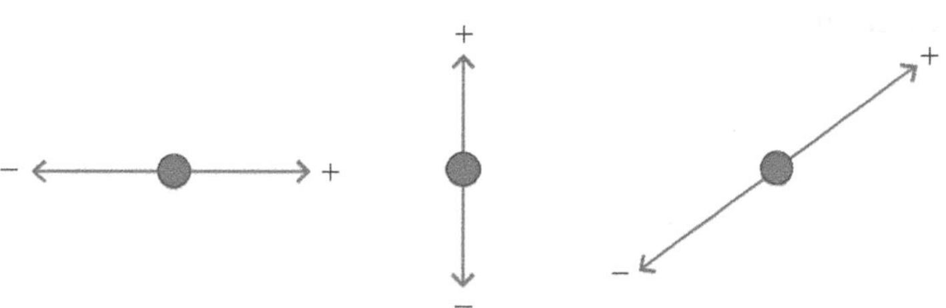

Fig 1.4.3

Tug of war

During a tug of war, two teams hold on to a single rope and pull in opposite directions.

Fig 1.4.4: *Tug of war*

When the two teams apply forces of equal size in opposite directions the rope stays in the same place. This is because the forces are balanced. The diagram below shows the forces from the two teams.

Fig 1.4.5: *Forces acting during a tug of war*

Terminal velocity

Previously we stated that if the forces acting on an object are balanced, the object will remain still. However, Newton's first law also states that an object may continue moving at a constant speed (velocity) if the forces acting on it are balanced.

Car manufacturers state the maximum speed that a car can reach. A car will reach its maximum speed, in other words its **terminal velocity**, when the frictional forces such as air resistance and rolling resistance become equal to the maximum force the engine can produce. In other words, the maximum speed is reached when the forces acting on the car are balanced. As a car moves more quickly, the frictional forces acting against it get bigger. The engine must produce a greater force to make the car move more quickly. This is why sports cars have big, powerful engines!

Think about it

In the example of the book on the table, the resultant force was expressed as an equation. Can you express the resultant force in the *tug of war* example as an equation?

Word bank

- **terminal velocity**
  the maximum velocity an object will reach, which occurs when the forces acting on the object are balanced

Think about it

The terminal velocity of a car will depend on the maximum force that the engine can produce. What other factors about the car's design will affect its maximum speed?

Key point

An object will reach its terminal velocity when the propulsion force (e.g. car engine) is equal to the resistive force (e.g. air resistance).

Fig 1.4.6: *Streamlined design to reduce air resistance*

Fig 1.4.7: *Skydivers*

Skydiver's safe landing – the parachute

During the fall, skydivers can reach speeds of 100 miles per hour. Once the skydiver falls to a certain altitude they must deploy a parachute to slow them down. Skydivers are often referred to as experiencing free fall. However, in terms of Newton's laws, free fall is any motion of a body where its weight is the only force acting upon it.

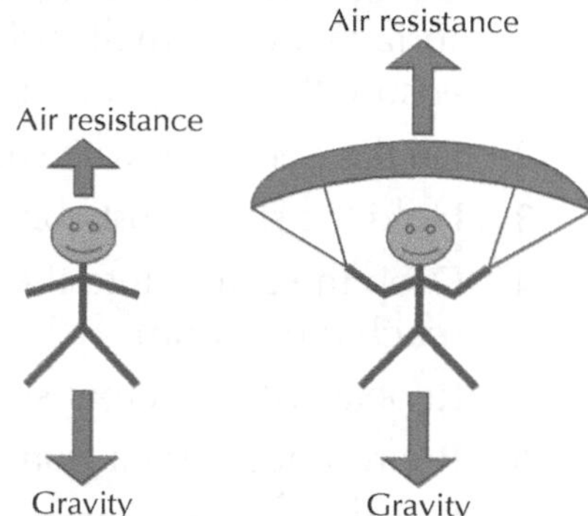

Fig 1.4.8: *Forces acting on a parachutist*

The forces acting on the skydiver with and without a parachute are shown in the diagram below. As the speed of the skydiver increases, the air resistance acting against him also increases. The parachute increases the air resistance because it has a large surface area. Therefore, with a parachute, the downward force is balanced at a much lower speed – in other words, the terminal velocity of the skydiver is lower with a parachute. This allows the skydiver to land safely.

> **Make the link**
>
> Falling objects and the relationship between potential and kinetic energy are considered in Chapter1: Conservation of energy.

Air resistance

Air resistance is one of the main forces that moving objects are subject to. Consider riding a bike – as your speed increases, you have to work harder against the air pushing against you. This is air resistance. The greater the air resistance acting, the harder it is to go faster. If you can reduce the air resistance acting against you, you will be able to reach a greater terminal velocity for the same propulsion force.

Consider objects designed to go quickly. They have a streamlined shape which helps to reduce air resistance. The bobsleigh shown in the picture has a tear drop shape which makes it streamlined. By reducing the air resistance, the bobsleigh can reach greater speeds during a run.

Fig 1.4.9: *The bobsleigh's streamlined shape helps it reach greater speeds*

Experiment: Streamlining

This experiment investigates the effect of shape on the resistive forces acting on an object.
You will need:

- A tall watertight container
- Water
- Shampoo
- Plasticine
- Stop clock

Instructions

1. Using plasticine, make models of different shapes to test – these shapes can have large surface areas, small surface areas, tear drop shapes, etc. To ensure a fair test, you must ensure that the same mass of plasticine is used in each.
2. Fill the container with shampoo – the thicker the shampoo, the greater the resistance force.
3. Hold the shape just above the surface of the shampoo in the container.
4. Drop the shape into the shampoo and use the stop clock to time how long it takes to reach the bottom.
5. Repeat with different shapes.
6. Represent the data from your experiment in an appropriate format.
7. Use the results of your experiment to answer the following questions.
   a) Of the shapes you tested, which one experienced the lowest resistance?
   b) Of the shapes you tested, which one experienced the greatest resistance?
   c) Which shape had the greatest terminal velocity?
   d) How does terminal velocity depend on the size of the resistance?
   e) How could you change your experiment to represent conditions on a windy day?
   f) Explain, based on the results of your experiment, how a parachute acts to ensure a safe landing for a skydiver.
   g) Consider the shape of the bicycle helmet shown here. Explain, based on the results of your experiment, why this shape is good for ensuring the cyclist can go as fast as possible.

Fig 1.4.10: *Streamlined design*

Key point

Friction is a force which acts to oppose the motion of an object. Friction always acts in the opposite direction to the direction of travel.

Friction

Air resistance is a type of friction which acts on an object. It is sometimes referred to as air friction. Friction is a term used to describe a force which acts to oppose the motion of an object. However, friction is also vital to allow objects to move. Consider a car – without friction, the tyres would simply spin on the ground with no grip. Friction prevents the tyres from spinning and allows them to grip the road to move the car forwards.

Experiment: Static friction

This experiment investigates the effects of static friction acting on an object and investigates methods that can be used to reduce static friction.

You will need:

- A ramp
- Blocks of various materials
- Straws
- A ruler

Instructions

1. Place a block on the ramp with the ramp lying horizontal on the bench.
2. Carefully lift the ramp at one end until the block begins to move.
3. Hold the ramp in position at the point where the block starts to move. Be careful to catch the block at the bottom of the ramp.
4. Measure the height of the ramp above the ground.
5. Repeat this experiment a further two or three times and find the average height at which the block moved.
6. Repeat with blocks made from different materials to compare the static friction acting on different objects – the greater the height of the ramp, the greater the static friction acting on the block.
7. Present the results from your experiment in an appropriate format.
8. Lay straws onto the ramp at 90 degrees to the direction of the incline. Choose one of the blocks you have used above and place it on top of the straws which will act like rollers. Repeat the above experiment and compare the height of the ramp with and without straws.
9. Use the results of your experiment to write a short report about which materials experience greater static friction, and the methods that could be used to reduce the static friction acting.

Experiment: Dynamic friction

This experiment investigates dynamic friction acting on moving objects.

You will need:

- A ramp
- Dynamics cart(s)
- Block(s)
- Toy car(s)
- Ruler/metre stick
- Books
- Straws
- Paper

Instructions

1. Use books or a ramp holder to raise one end of the ramp into the air to make an inclined plane.

(*continued*)

2 Check that at the bottom of the ramp the dynamics cart can run freely onto the ground without catching or scraping.
3 Mark a starting position near to the top of the ramp – all experiments will be started from this point.
4 Hold the cart at the start position and then let it go without pushing it.
5 The cart will roll down the ramp and onto the ground. On the ramp and on the ground, dynamic friction will act against the motion of the cart to slow it down.
6 Use the ruler or metre stick to measure the distance taken for the cart to stop.
7 Now make a sail using a straw and a sheet of paper and attach this to the dynamics cart. Repeat steps 4 to 6.
8 Write a short report describing your findings on dynamic friction from the above experiment.
9 Repeat the above experiment using different toy cars. Present your results in an appropriate format and write a short report discussing which car experienced the least dynamic friction and which car experienced the most.

Key point

Static friction acts on stationary objects to oppose motion – you must overcome static friction to get an object to move. Dynamic friction acts on moving objects to oppose motion.

Balanced forces

Newton's first law tells us that an object will remain at rest or continue moving at a constant speed if the forces acting on it are balanced. In such a situation, the resultant force acting on the object is zero. In describing the examples above, we have considered balanced forces such as:

- Air resistance versus weight due to gravity for falling objects.
- Air resistance versus engine driving force for cars.

If these pairs of forces are equal to each other we say that they are balanced, and therefore according to Newton's first law the object keeps moving at a constant speed (or remains at rest). This principle leads to the idea of terminal velocity – the object does not move any faster because the forces acting on it are balanced!

As force is a vector quantity, individual forces must be added together using vector addition (see the section on vectors). An example is given below.

Worked example

Fig 1.4.11: *Cyclist*

A cyclist is cycling along a road at a constant speed of 12 ms^{-1}. She pedals with a force of 470 N. What is the size of the frictional force acting against her?

The cyclist is moving with a constant speed so by Newton's first law the forces acting on her must be balanced. Therefore:

Friction = Pedal force = 470 N

Exercise 1.4.1 Balanced forces

1 Look at the free-body diagrams shown below. Which ones show balanced forces?

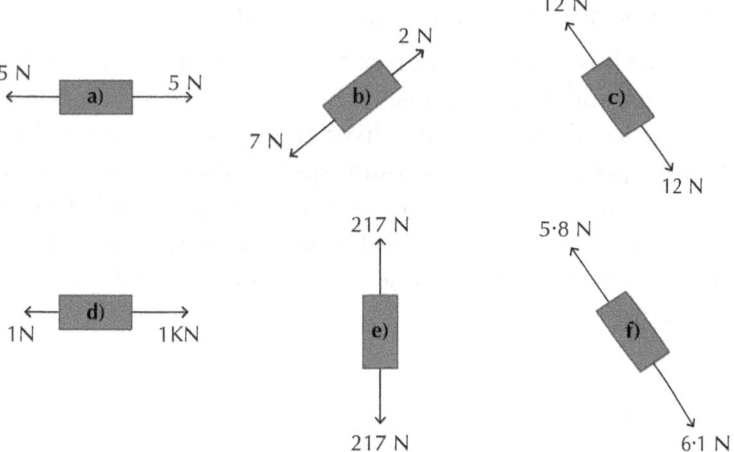

Fig 1.4.12

2 The weight (downwards force) of a shopping bag is 24 N. What is the size of the upwards force required to keep the bag at a constant height?

3 A plane is cruising at a constant altitude of 36 000 feet. The upwards force provided by the wings is 40 000 N. What is the weight of the aircraft (downward force)?

4 A helicopter's rotors provide a constant upwards force of 18 000 N. The weight of the helicopter is 18 000 N downwards. What can be said about the change in altitude of the helicopter?

Fig 1.4.13: *Helicopter*

5 A parachutist jumps out of an aircraft. He opens his parachute during the descent in order to slow down for a safe landing. The velocity-time graph for the motion of the parachutist is shown below.

a) During what time(s) of the descent are the forces acting on the parachutist balanced?

b) At what time does the parachutist open his parachute?

c) Explain, in terms of the forces acting, the difference in terminal velocities between the parachutist falling with, and without, his parachute.

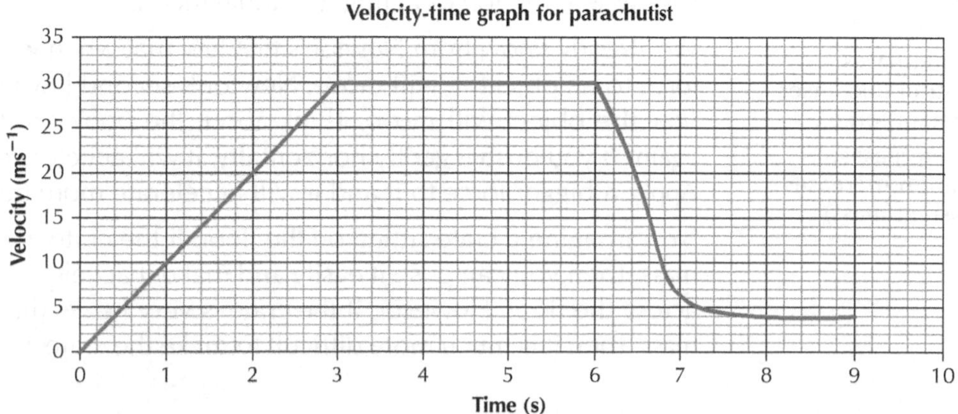

Fig 1.4.14

Fig 1.4.15: *The law requires you to wear a seatbelt when travelling in a car*

Applying Newton's first law to car safety

You must wear a seatbelt when you are travelling in a car – it is the law! The seatbelt is a crucial safety device which keeps you in your seat in the event of an accident.

A seatbelt is an application of Newton's first law. Consider a car stopping suddenly because it hits a wall. The driver's seat will stop suddenly because it is fixed to the car. However, the driver will keep moving at a constant speed unless there is something to apply an unbalanced force to him. In a car, it is the seatbelt that applies the force needed to keep you in your seat. Without it, you would keep moving out of your seat with possibly fatal consequences!

Exercise 1.4.2 Applying Newton's first law

1 State Newton's first law.

2 During a tug of war, both teams are applying equal forces in opposite directions.

a) Describe the motion of a small marker attached to the rope half way between the two teams.

b) One of the team on the left slips on a patch of oil – describe what happens to the marker on the rope in this case.

3 Dragsters like the one shown often use parachutes to slow them down at the end of a race. Explain, using Newton's first law, how a parachute reduces the terminal velocity of the dragster.

Fig 1.4.16

Key point

The acceleration of an object depends on the unbalanced force acting on the object and the object's mass.

- As mass increases, acceleration decreases (for constant force).
- As force increases, acceleration increases (for constant mass).

Newton's second law

So far we have only considered what happens in the case of forces being balanced. Newton's second law focuses on the case where the forces acting are unbalanced.

We already know from Newton's first law that when we apply a force to an object, it will move. Newton's second law describes *how* the object will move. Think about the book on the table again. If you push it gently horizontally, it will accelerate slowly. Apply a bigger force to it and it will accelerate more quickly!

Now imagine a crate sitting on the floor. If the crate has a small mass, then you can push it easily and make it accelerate quickly across the floor. However, if the crate is very heavy (has a large mass) then it is much more difficult to make the crate accelerate.

Fig 1.4.17: *Man pushing a heavy box*

Experiment: Discovering Newton's second law

This experiment demonstrates Newton's second law through investigating the effects of force on acceleration for a constant mass.

You will need:

- A trolley with low-friction bearings
- A ramp (optional)
- A light gate and suitable timing device
- A pulley
- Hanging masses
- String

Fig 1.4.18

Instructions

1. Set up the experiment as shown in the photograph above. Attach the hanging masses to the trolley using a piece of string fed over a pulley.
2. Attach a double mask to the trolley for the measurement of acceleration. The light gate must be connected to a suitable timing device.
3. Copy the table below to record your results:

Mass (kg)	Applied force (N)	Acceleration result (ms^{-2})	Acceleration result (ms^{-2})	Acceleration result (ms^{-2})	Average acceleration (ms^{-2})
0·1					
0·2					
0·3					
0·4					
0·5					

Table 1.4.1

(*continued*)

4 For each hanging mass, calculate the applied force using the formula: $W = mg$. For the purposes of this experiment, friction should be ignored.
5 Attach a 100 g mass to the end of the string, holding the trolley in place.
6 Set up the light gate and timer so that they are ready to measure the acceleration.
7 Release the trolley and the hanging mass – ensure the mask passes through the light gate.
8 Note the acceleration in the first results column of your table.
9 Repeat the experiment with the same hanging mass two more times.
10 Find the average acceleration and note it in the average acceleration column of the table.
11 Repeat the experiment with different hanging masses.
12 Plot a graph of the applied force versus the acceleration.
13 Find the gradient of the straight line on the graph. Compare this gradient to the mass of the cart used in the experiment.

Extra

You can carry out a similar experiment to the one above to investigate the effects of mass on acceleration for a constant force. This time the hanging mass is kept constant and the mass of the cart is changed. A graph of acceleration versus mass can be plotted to analyse the results.

The results of the experiments above should highlight the following points:

- If the size of the unbalanced force increases, the size of the acceleration increases: F up $\rightarrow$ a up
- If the mass increases, the size of the acceleration decreases: m up $\rightarrow$ a down

This relationship can be summarised by the following important equation:

$$F_{un} = ma$$

where F_{un} is the unbalanced force (N), m is the mass (kg) and a is the acceleration (ms^{-2}).

Deriving relationships from graphs

Above we have described Newton's second law, and written it in the form of an equation,

$$F = ma$$

It is possible to get this equation from our experiment. Remember from mathematics that the equation for a straight line is:

$$y = mx + c$$

where m is the gradient of the line and c is the y-intercept. Our results do not plot y against x, but instead we plot F against a. As shown below, this gives us a straight line through the origin:

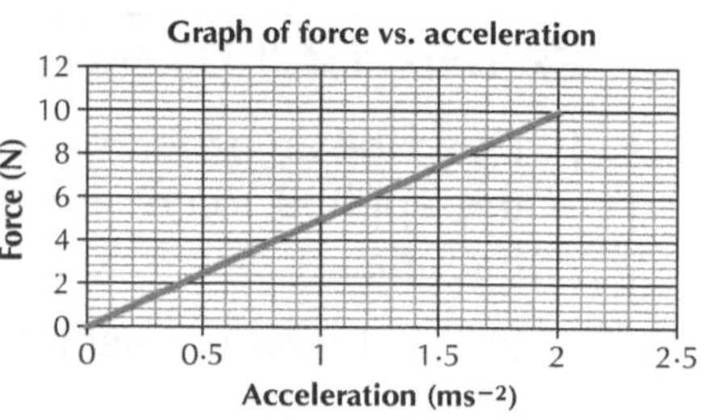

Fig 1.4.19

Comparing to the equation of a straight line ($y = F$) and ($a = x$), we have:

$$F = ma + c$$

As the straight line goes through the origin,

$$c = 0$$

The gradient of the line is the mass of the object. We get the equation for Newton's second law:

$$F = ma$$

Technique: Using Newton's second law

1. Write down what you know from the question, leaving blank what you are trying to find:
   Force, F = _______ N
   Mass, m = _______ kg
   Acceleration, a = _______ ms^{-2}
2. Write down the equation linking the quantities above as you see it on the formula sheet:
   $$F = ma$$
3. Substitute into the equation what you know.
4. Solve for the unknown.

Make the link

Calculations involving the relationship W = Fd between work done, unbalanced force and distance/displacement are looked at in Chapter 1: Conservation of energy.

Worked example

A force of 25 N is applied to a crate of mass 10 kg which is sitting on a frictionless surface. Calculate the acceleration of the crate.

Write down what we know and what we are looking for:

Force, F = 25 N

Mass, m = 10 kg

Acceleration, a = ? ms⁻²

We are using the equation which links force, mass and acceleration:

$$F = ma$$

Substitute what we know:

$$25 = 10 \times a$$

Then solve for the unknown:

$$\frac{25}{10} = a$$

$$a = 2{\cdot}5\ ms^{-2}$$

Fig 1.4.20: Crate

Exercise 1.4.3 Discovering and using Newton's second law

1 Describe an experiment that you would carry out to investigate the relationship between the applied force and the acceleration of an object. In your description, you should include:

a) What measurements you would make and how you would make them.

b) What variables you would keep constant and how you would keep them constant.

c) How you would analyse the data to find Newton's second law.

2 Use the equation for Newton's second law to fill in the missing values in the table below:

Unbalanced force (N)	Mass (kg)	Acceleration (ms^{-2})
(a)	2	5
(b)	1·5	6·2
12	2	(c)
18·9	11	(d)
13·4	(e)	2·7

Table 1.4.2

3 An unbalanced force of 1200 N is applied to a mass of 340 kg. Calculate the acceleration.

4 What size of unbalanced force is required to accelerate an object with a mass of 12 kg with an acceleration of 4 ms^{-2}?

5 A crate with a mass of 14 kg is being towed across a frictionless surface. It is found to have an acceleration of 3·5 ms^{-2}. Find the size of the towing force.

6 A van of mass 3000 kg is travelling along a stretch of road when the driver sees an accident ahead. She applies the brakes and decelerates at a rate of 2·5 ms^{-2}. Calculate the braking force applied to the van.

Fig 1.4.21: *3000 kg van*

7 A sports car has a mass of 1100 kg. The manufacturer specifies that it must have the ability to decelerate at 10 ms^{-2}. Find the minimum braking force required to meet the manufacturer's specification.

8 A car accelerates from rest to a velocity of 15 ms^{-1} in a time of 4 s. The mass of the car is 900 kg.

a) Calculate the acceleration of the car.

b) What is the size of the unbalanced force required to produce this acceleration?

9 A car engine can produce a force of 14 kN. The mass of the car is 1600 kg.

a) Calculate the maximum acceleration of the car.

b) Assuming the car accelerates from rest and frictional forces are negligible, what will be the speed of the car after it accelerates for 8 seconds?

Unbalanced forces

Newton's second law describes the effects of an unbalanced force on the motion of an object. The greater the unbalanced force, the greater the acceleration. In order to work out the unbalanced force, we use a free-body diagram of the situation, drawing on all of the forces which are acting.

Free-body diagrams

A free-body diagram is used to represent all of the forces that are acting on an object. Drawing a free-body diagram gives a visual representation of the forces which makes it easier to calculate the unbalanced force.

Consider a car which is driving along the road. It is propelled by the engine, so there will be an engine force in the direction of the motion. There will be resistive forces acting against the motion. Two of these are air resistance and rolling friction. A free-body diagram to represent these forces would look like this:

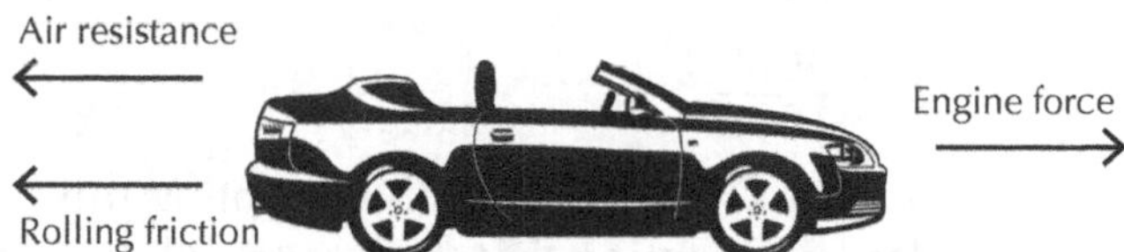

Fig 1.4.22: *Free-body diagram of forces on a car*

The direction of the arrows shows the direction of the force which is important because force is a vector quantity. These forces can be added together in the same way as displacements to find the resultant. Here we call the resultant force the unbalanced force.

Worked example

Calculate the unbalanced force acting on a car. The car's engine produces a force of 10 000 N. Rolling friction is 2000 N and air resistance at 70 mph is 3000 N.

Draw a free-body diagram to show the forces acting:

Fig 1.4.23

Then calculate the unbalanced force:

Unbalanced force = 10 000 – 2000 – 3000

Unbalanced force = 5000 N in the direction of the engine force

Worked example A car's engine can produce a force of 12 000 N. The mass of the car is 1500 kg. If the resistive forces acting on the car are 8000 N, calculate the acceleration of the car.

First of all, find the unbalanced force acting on the car. Draw a simple free-body diagram showing all forces:

Fig 1.4.24

The unbalanced force acting:

Unbalanced force = 12 000 – 8000

Unbalanced force = 4000 N to the right

Then use Newton's second law to find the acceleration:

$$F = ma$$

$$4000 = 1500 \times a$$

$$a = \frac{4000}{1500}$$

$a = 2{\cdot}67\,ms^{-2}$ to the right (adopting the standard convention that the positive direction is to the right)

Exercise 1.4.4 Unbalanced forces and Newton's second law

1 A lorry of mass 18 000 kg has an engine capable of providing a force of 40 kN. Frictional forces acting against the lorry amount to 25 000 N. Calculate the acceleration of the lorry.

2 a) Calculate the resultant forces acting on each of the blocks below.

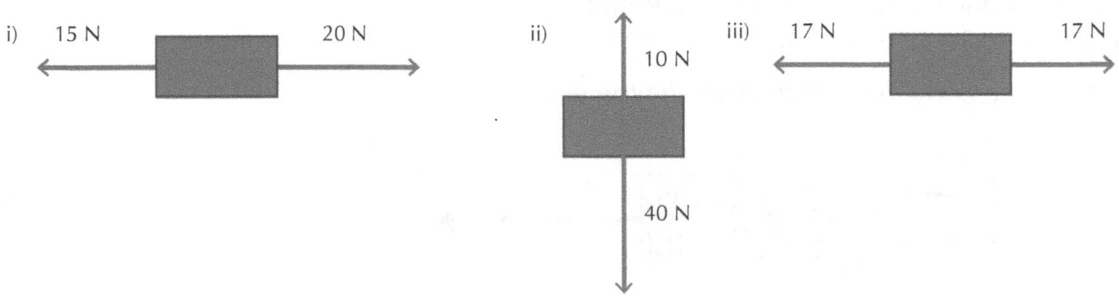

Fig 1.4.25

b) If each box has a mass of 5 kg, calculate its acceleration in each case.

3 A car's engine produces a force of 26 000 N. The mass of the car is 1250 kg. The force of air resistance acting on the car is 7500 N and the force of rolling resistance is 1900 N.

(continued)

a) Calculate the unbalanced force acting on the car.

b) Calculate the acceleration of the car.

4 A box is pulled along a rough surface with a constant force of 160 N. If the box has a mass of 20 kg and is accelerating at 2 ms^{-2}, calculate:

a) the unbalanced force required to produce the acceleration

b) the force of friction acting on the box

5 A train is subject to the forces shown in the diagram below.

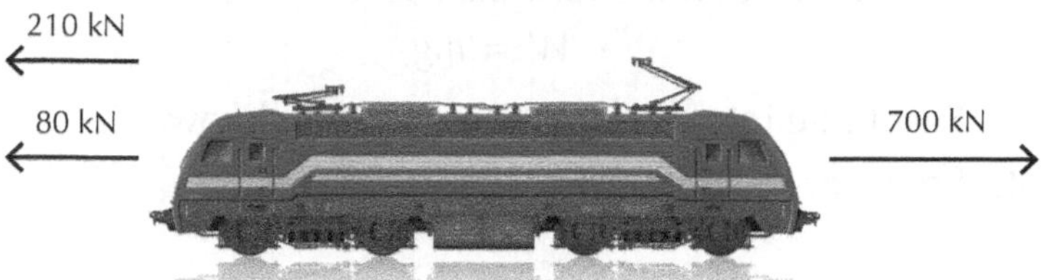

Fig 1.4.26

a) Calculate the unbalanced force acting on the train.

b) If the train has a mass of 120 000 kg calculate the acceleration of the train.

c) Assuming the acceleration calculated in part (b) is constant, how long will it take the train to accelerate from rest to 40 ms^{-1}?

6. A cyclist and her bike have a total mass of 75 kg. She accelerates from rest to 8 ms^{-1} in a time of 2·6 s. A constant frictional force of 80 N acts against her.

a) Calculate the unbalanced force acting on the cyclist to produce the acceleration.

b) Calculate the force produced by the cyclist.

Weight and mass

In physics there is a key difference between weight and mass.

The mass of an object is constant regardless of where the object is, and depends on the amount of matter contained in the object. Mass is measured in kilograms (kg).

The weight of an object is the downward force that acts on an object due to gravity. It depends on the size of the gravitational field, g, and as it is a force, it is measured in newtons.

Weight and mass are linked by the following equation:

$$W = mg$$

This equation is strongly linked to Newton's second law. Weight is a force, and g is defined as the acceleration due to gravity – substituting this into the above equation gives us Newton's second law:

$$F_{un} = ma$$

> **Key point**
>
> Weight is a force which depends on the gravitational field strength. It is a vector quantity.
>
> Mass depends on the amount and type of matter in an object. It is a scalar quantity.

Key point

The gravitational field strength, g, varies from planet to planet. The size of the gravitational field on other planets can be found in the physics data booklet.

On Earth, we take: $g = 9.8$ N/kg *(note: the unit N/kg is equivalent to ms^{-2})*

Technique: Weight and mass

1. Write down what you know from the question, leaving blank what you are trying to find:

   Weight, W = ______ N

   Mass, m = ______ kg

   Gravitational field strength, g = ______ ms^{-2}

2. Write down the equation linking the quantities above as you see it on the formula sheet:

   $$W = mg$$

3. Substitute into the equation what you know.
4. Solve for the unknown.

Worked example

A footballer has a mass of 90 kg. Calculate his weight on Earth.

Fig 1.4.27

The weight of the footballer will depend on his mass and the gravitational field strength. We use the equation,

$$W = mg$$

On Earth, the gravitational field strength is 9·8 N/kg. Substitute what we know,

$$W = 90 \times 9{\cdot}8$$

Solve for weight:

$$W = 882\,N$$

Exercise 1.4.5 Weight and mass

1 Use the equation linking weight, mass and gravitational field strength to fill in the missing entries in the table below.

Weight (N)	Mass (kg)	Gravitational field strength (N/kg)
	12	10
	124	1·6
120		11
17·5		0·98
240	16	

Table 1.4.3

(continued)

2 On the moon, the gravitational field strength is 1·6 N/kg. Calculate the weight of the following objects on both the moon and Earth:

a) A boy of mass 35 kg

b) A ball of mass 1·5 kg

c) A car of mass 900 kg

d) A packet of sweets of mass 35 g (careful with units!)

3 A man has a weight of 890 N on Earth. What will his weight be on the planet Jupiter where the gravitational field strength is 25·95 N/kg?

Fig 1.4.28: *Jupiter*

Applying Newton's second law: lifting problems

So far we have looked at applying Newton's second law to finding the acceleration of objects on level ground. However, Newton's second law can also be applied to calculate the acceleration of objects which are being lifted (or powered in the case of a rocket!) into the air. In these cases, one of the forces acting to oppose the motion is the weight of the object which acts downwards due to gravity.

Consider lifting a shopping bag into the air. If the bag has a mass of 2 kg, then it has a weight as follows:

$$W = mg$$
$$W = 2 \times 9{\cdot}8$$
$$W = 19{\cdot}6N$$

Our lifting force must counteract this weight. A free-body diagram is shown.

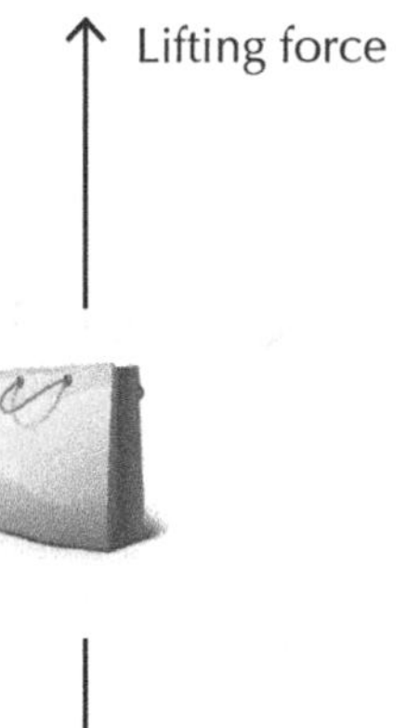

Fig 1.4.29

The unbalanced force, F_{un}, that produces the acceleration is given by:

$$F_{un} = \text{Lifting force} - \text{Weight}$$

This can then be used to find the acceleration of the shopping bag using Newton's second law:

$$F_{un} = ma$$

The same method can also be used to find the acceleration of a rocket leaving the surface of a planet. If the gravitational field strength of the planet and mass of the rocket are known, then the weight of the rocket can be found. The thrust produced by the engines must be greater than this weight for the rocket to lift off. The difference between the thrust and the weight is the unbalanced force which produces the upwards acceleration of the rocket. We will consider rockets in more detail in the section on Newton's third law.

Fig 1.4.30: *Rocket taking off*

Technique: Lifting problems

1. Draw a free-body diagram of the problem showing all forces acting (weight downwards; lift upwards).
2. Calculate the weight of the object using:

$$W = mg$$

3. Find the unbalanced force acting on the object using:

$$F_{un} = \text{Lift} - \text{Weight}$$

4. Use Newton's second law to find the acceleration of the object. Write down what we know from the calculations above:

   Unbalanced force, F_{un} = _______ N

   Mass, m = _______ kg

   Acceleration = _______ ms^{-2}

5. Write down the equation that links unbalanced force, mass and acceleration:

$$F_{un} = ma$$

6. Substitute into the equation what we know.
7. Solve for the unknown.

Worked example

A crate has a mass of 250 kg. If a crane can produce an upwards force of 5000 N, calculate the upwards acceleration of the crate. For the purposes of this question you should assume air resistance is negligible.

Start by drawing a free-body diagram of the forces acting on the crate. Calculate the weight of the crate using,

$$W = mg$$

Work out the unbalanced force acting on the crate:

$$F_{un} = Lift - Weight$$

$$F_{un} = 5000 - 2450$$

$$F_{un} = 2550\,N$$

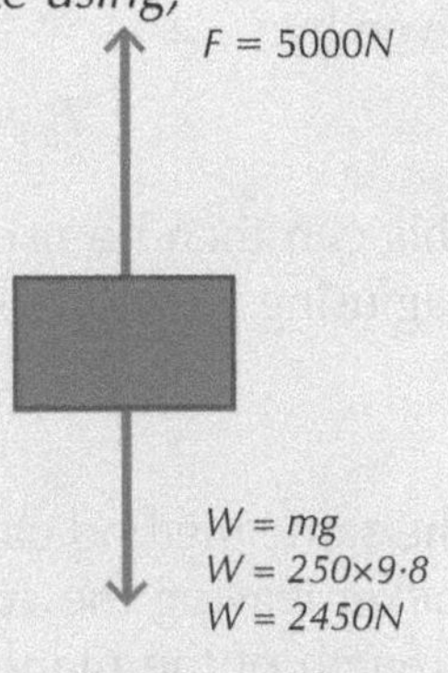

Fig 1.4.31

Then use Newton's second law to work out the acceleration of the crate:

$$F_{un} = ma$$

$$2550 = 250a$$

$$a = \frac{2550}{250}$$

$$a = 10{\cdot}2\,ms^{-2}$$

Exercise 1.4.6 Lifting problems

1 A rocket has a mass of 400 kg. Its engines can produce a thrust of 8000 N.

a) Calculate the weight of the rocket on Earth.

b) What is the maximum upwards acceleration of the rocket from Earth?

c) Assuming the rocket maintains its maximum upwards acceleration, how long does it take for the rocket to reach a maximum speed of 200 ms^{-1} if it starts from rest?

2 A mass of 4500 kg is to be lifted upwards by a helicopter winch with an acceleration of 3 ms^{-2}. Air resistance of 2500 N acts against the mass.

a) Calculate the weight of the mass.

b) What is the unbalanced force required to accelerate the mass upwards?

c) Find the lifting force required from the helicopter winch.

3 A crane is designed to lift crates off a ship in a harbour. The crates that are being lifted by the crane have a mass of up to 1400 kg. The crane must be able to accelerate the crates upwards with a minimum acceleration of 1·5 ms^{-2}. Air resistance can be ignored.

a) What is the maximum weight of crate that the crane can lift?

b) What is the minimum unbalanced force required to lift the heaviest crate with the desired acceleration?

c) Calculate the minimum force the crane must be able to provide when lifting the heaviest crate.

Fig 1.4.32: *Crane for lifting crates off a ship*

4 A person can exert a maximum upwards force of 150 N using their arms. Calculate the maximum acceleration they can provide to a box which has a mass of 10 kg. Assume air resistance is negligible.

Applying Newton's second law: car safety

We have seen that unbalanced force and acceleration are linked – the greater the unbalanced force, the greater the acceleration. This can be applied to car safety – if the deceleration of the car and its occupants can be reduced, then the forces acting on them can be reduced. This might result in less severe injuries.

Fig 1.4.33: *Modern car after an accident*

Crumple zones

Modern cars are designed with crumple zones at the front and rear. These are regions of the car which are designed to deform in the event of an accident. When a car strikes a solid object, the crumple zones cause the car to take longer to slow down because it takes time for the crumple zone to be squashed. The deceleration is therefore less. According to Newton's second law, if the deceleration is less, then the average force is lower.

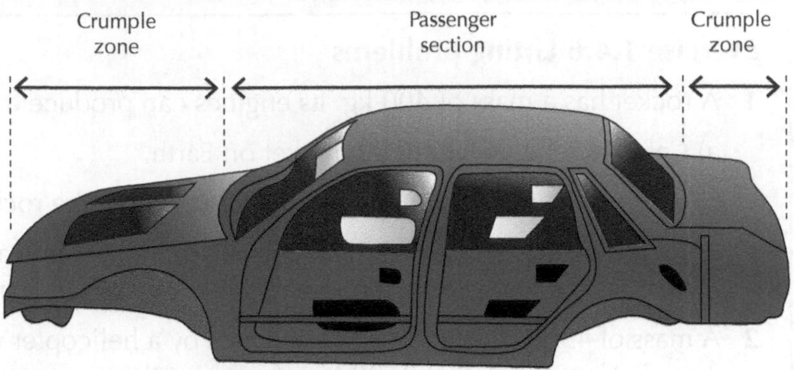

Fig 1.4.34: *Car body showing crumple zones*

Airbags

A car airbag works on a similar principle to a crumple zone. By being softer than the steering wheel, the length of time taken for the driver's head to come to rest is increased. This means the average force acting on the driver's head is reduced, which in turn reduces the potential for serious injury.

SPOTLIGHT ON INDUSTRY: CAR SAFETY

Safety is a key consideration in the design of modern cars. Every car on the market must undergo numerous crash tests to test the safety of occupants in different conditions. Newton's laws are applied in car safety – reducing the deceleration of the vehicle and its occupants in the event of a crash will reduce the forces acting on them. This is a key consideration for car designers.

Experiment: Crumple zones

This experiment investigates the properties of crumple zones on the forces acting on a car during a collision.

You will need:

- A dynamics cart (an old one is preferable!)
- Materials to act as crumple zone (e.g. paper, bubble wrap)
- A ramp
- A solid object to crash the cart into
- A motion sensor

The challenge

Your task is to design a crumple zone for the cart which will minimise the forces acting on it during a head-on collision. The size of the forces acting will be directly linked to the deceleration of the cart which is measured using a motion sensor. Be careful – some materials may seem squishy but it is deformation that is the key to reducing the force!

(*continued*)

Instructions

1 Prepare your cart using materials of your choice. The crumple zone must be attached to the cart and must not drag along the ground impeding the motion.

2 Set up the crash test experiment as follows:
- Place a solid object on the lab bench to act as a wall for the impact.
- Place a ramp against this object to guide the cart into the wall. Prop the ramp up at one end using books to give a constant accelerating force for all tests.
- Set up a motion sensor at the opposite end of the ramp to record the velocity-time graph of the motion of the cart.

3 Once ready to test, place the cart on the track. Let the cart run into the wall while the motion sensor records the velocity-time of the cart.

4 Find the gradient of the velocity-time graph for the cart when it comes to a stop to work out the deceleration of the cart.

5 Measure the mass of the cart and then use Newton's second law to work out the average force acting on the cart during the collision.

6 Repeat with different designs.

Report

Produce a report of your findings from the above experiment. In your report you should include a description of the design and materials that you found to be most effective at reducing the average force acting on the cart during the collision.

Newton's third law

While the first two Newton's laws make perfect sense, the third law is a little more obscure. However, the third law is vital to space travel; it is the principle by which a rocket works. Newton's third law can also be used to explain how an aeroplane's wings produce lift and a jet engine produces thrust.

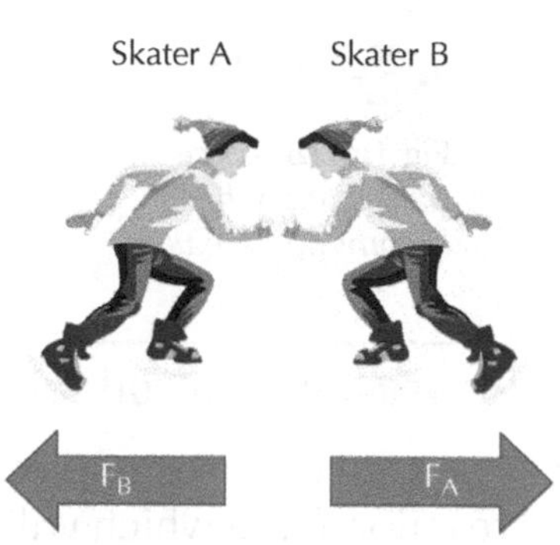

Fig 1.4.35

Introducing Newton's third law

Newton's third law is often misquoted as: 'every action has an equal and opposite reaction'. While this comes close, it misses out on the key component – the idea of a force. Newton's third law should be stated as:

'For every action force, there is an equal and opposite reaction force.'

We know from Newton's second law that if there is an unbalanced force this will result in an object accelerating. Consider two skaters on an ice rink. If skater A applies an unbalanced force, F_A, to skater B it will make skater B accelerate.

⚠ Think about it

When you are swimming, which way must you push the water in order to go forwards? Can you identify the Newton's pair for a swimmer, and draw the two forces on a diagram?

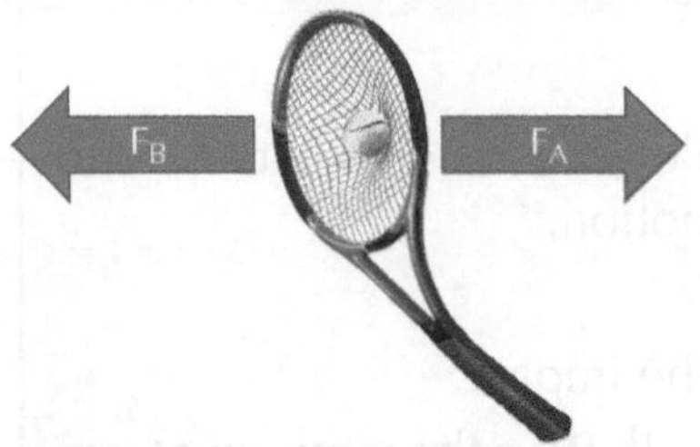

Fig 1.4.36: *Force pair acting between a racket and a ball*

Fig 1.4.37: *Force pair acting on a car tyre*

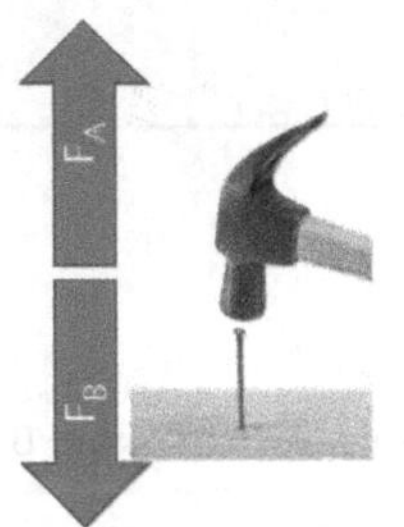

Fig 1.4.38: *Force pair acting when hammering a nail*

According to Newton's third law, there is an equal and opposite reaction force, F_B, applied to skater A making him accelerate in the opposite direction. The skaters push apart.

Newton's pairs

Considering the example of the skaters above, we can see that there are two forces acting. Forces come in pairs known as action reaction force pairs, or Newton's pairs. We can identify the Newton's pairs in a variety of different situations.

When you strike a tennis ball with a racket, the racket applies a force to the ball. The ball applies an equal and opposite force to the racket (which you can feel). This is a force pair – the force from the racket and the force from the ball. This causes deformation of the strings on the racket as shown.

The forces acting between a car tyre and the road are another example of a Newton pair. The tyre pushes backwards on the road, so the road pushes forward with an equal and opposite force. This propels the car forwards.

Another example is hammering a nail into a wall. The hammer exerts a force on the nail which pushes it into the wall. By Newton's third law, the nail exerts an equal force in the opposite direction on the hammer. This is why the surface of the hammer head is frequently damaged and dented by nails – the reaction force from the nail causes the damage! Similarly, if you push a drawing pin into the wall, you feel the force of the pin on your thumb. If you push the pin by the sharp end, it is the reaction force that will cause it to sink into your finger!

Exercise 1.4.7 Introducing Newton's third law

1 For each of the examples below draw a simple sketch indicating the action force and the reaction force which will be in the direction the object moves.

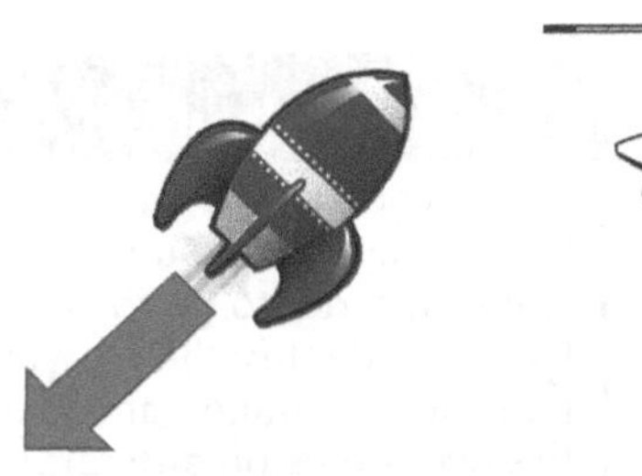
Fig 1.4.39

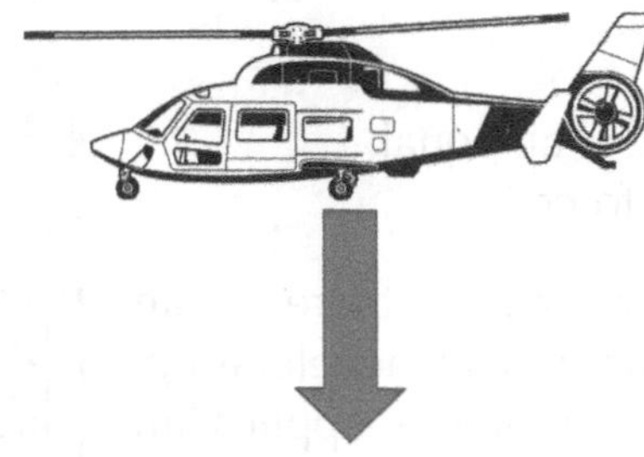
Fig 1.4.40

Fig 1.4.41

(continued)

2 Explain in your own words what causes a balloon to rise upwards when you let the air out of it.

3 A boat is sitting next to a pier. A man runs from one end of the boat towards the pier. As he does this, the boat moves away from the pier. Explain why this happens.

4 A skateboarder jumps off her skateboard. Explain the movement of the skateboard when she does this using Newton's third law.

5 A jet engine burns aviation fuel and ejects hot gases out of the back of the engine. Explain, using Newton's third law, how this acts to propel the aeroplane forwards. Draw a diagram showing the Newton pair.

Fig 1.4.42: *Skateboard*

SPOTLIGHT ON INDUSTRY: THE JET ENGINE

Newton's third law of motion is applied to the jet engine. A jet engine forces air backwards out of the exhaust. This action force has an equal and opposite reaction force which pushes the plane forwards.

The airflow through a jet engine is shown here. Blue arrows show the path of cold air; red arrows show the path of hot air leaving the combustion chamber. Cold air is sucked into the engine by the large fan at the front. This air is compressed as it moves towards the combustion chambers. Fuel is added and burned in the combustion chambers. The energy provided by burning the fuel is used to turn the compressors and fan to suck in more air. Hot gases are ejected out of the exhaust of the jet engine. Air is forced backwards through the engine, generating an equal and opposite forwards force to propel the plane.

Fig 1.4.43: *A jet engine*

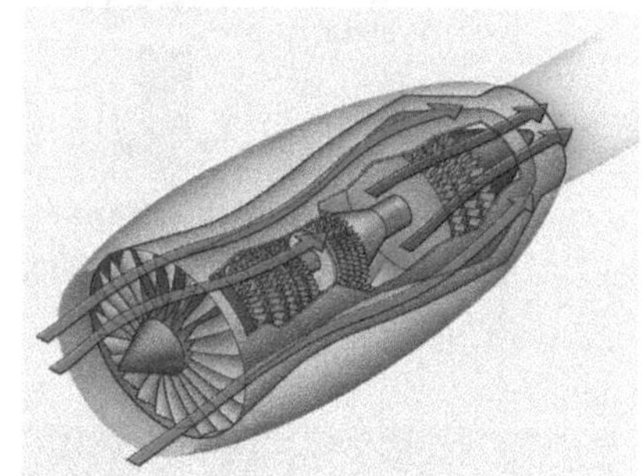

Fig 1.4.44: *Airflow through a jet engine*

Rockets

Just like the jet engine, a rocket engine forces air downwards in order to push the rocket upwards. The greater the force of the air downwards, the greater the unbalanced force acting on the rocket to force it upwards. The thrust produced by a rocket or jet engine is often measured in pounds (mass) – this relates to the mass that the rocket engine is able to lift. The rocket will always be propelled in the opposite direction to the gas flow.

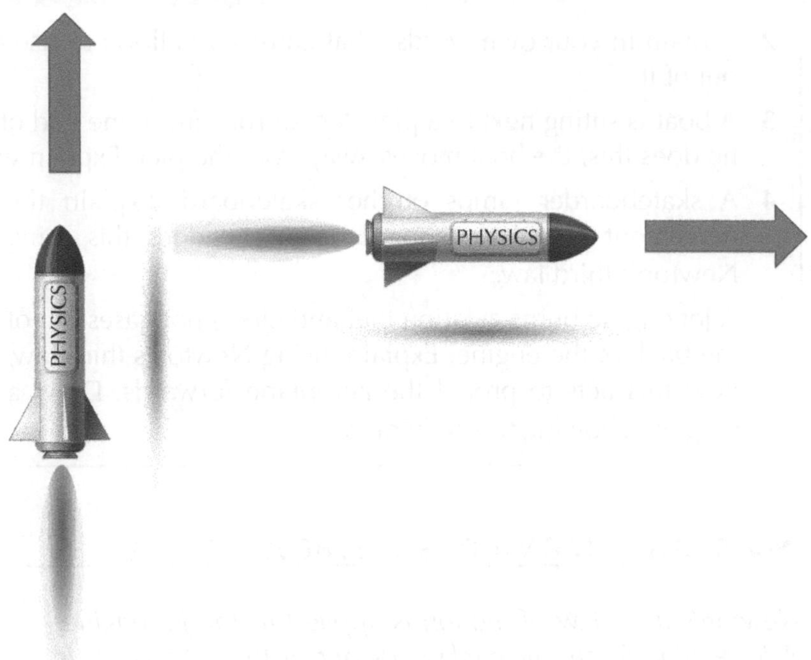

Fig 1.4.45: *Gas flow and direction of a rocket*

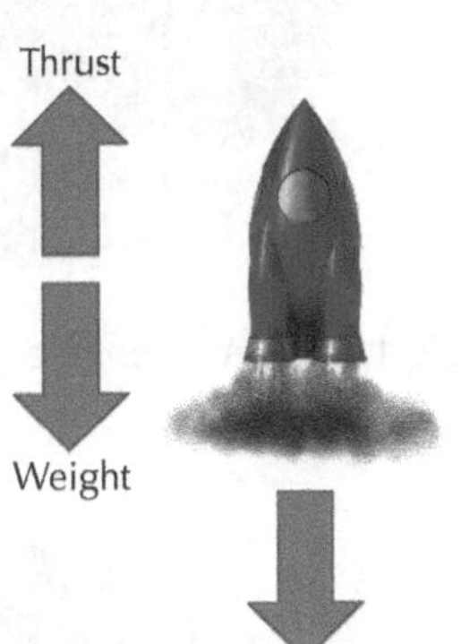

Fig 1.4.46: *Forces acting on a rocket*

How much thrust?

The thrust required to propel a rocket into space comes from the rocket's engines. These engines produce an unbalanced force which results in an acceleration, as demonstrated by the first worked example below.

In practice there are two main forces acting – weight downwards (due to gravity) and thrust upwards (reaction force to the gases being ejected from the rocket). A free-body diagram can be used to help calculate the forces required to propel a rocket into space.

Key point

When a rocket is lifting off from the surface of a planet, the thrust of the engines must overcome the weight of the rocket. The unbalanced force is always given by:

unbalanced force = thrust – weight

Technique: Rocket acceleration

1. Draw a free-body diagram of the problem showing all forces acting (weight downwards; lift upwards).
2. Calculate the weight of the rocket using:
$$W = mg$$
3. Find the unbalanced force acting on the object using:
$$F_{un} = \text{Thrust} - \text{Weight}$$
4. Use Newton's second law to find the acceleration of the object. Write down what we know from the calculations above:

   Unbalanced force, F_{un} = _______ N

(continued)

Mass, m = _______ kg

Acceleration = _______ ms^{-2}

5. Write down the equation that links unbalanced force, mass and acceleration:

$$F_{un} = ma$$

6. Substitute into the equation what we know.
7. Solve for the unknown.

Worked example

Consider a rocket of mass 800 kg. What is the unbalanced force required in space to propel the rocket forwards with an acceleration of 8 ms^{-2}?

Draw a free-body diagram of the forces acting. In space the only force contributing to the motion of the rocket is the reaction force (thrust) due to the hot gases being ejected from the engines.

The required force can then be worked out using Newton's second law:

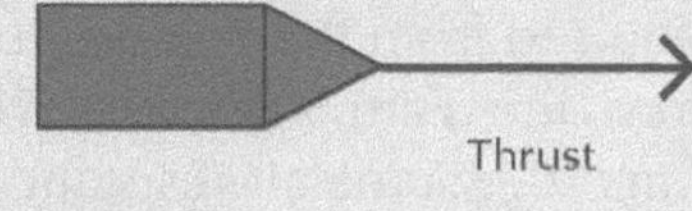

Fig 1.4.47

$$F_{un} = ma$$
$$F_{un} = 800 \times 8$$
$$F_{un} = 6400\,N$$

Worked example

A rocket of mass 1200 kg takes off from the surface of the moon where the gravitational field strength is 1·62 N/kg. Its engines produce a thrust of 4000 N. Calculate the acceleration of the rocket assuming air resistance is negligible.

First of all, draw a free-body diagram of the forces acting as shown on the left

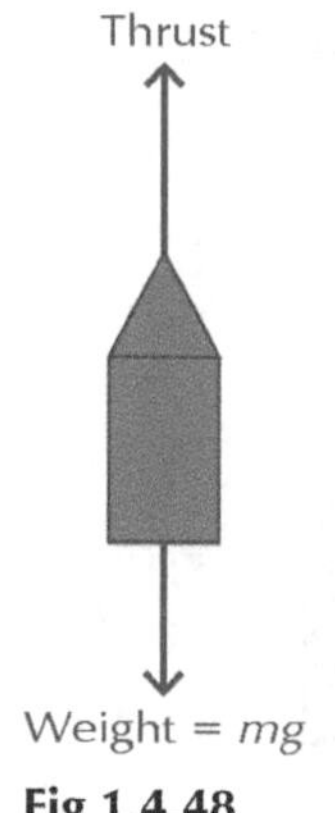

Fig 1.4.48

The weight of the rocket on the moon is found using:

$$W = mg$$

where g is the gravitational field strength on the moon (1·62 N/kg), so we have:

$$W = mg$$
$$W = 1200 \times 1{\cdot}62$$
$$W = 1944\,N$$

(*continued*)

The unbalanced force acting is then given by:

$$F_{un} = Thrust - Weight$$
$$F_{un} = 4000 - 1944$$
$$F_{un} = 2056\,N$$

The acceleration of the rocket is then calculated using Newton's second law:

$$F_{un} = ma$$
$$2056 = 1200a$$
$$a = \frac{2056}{1200}$$
$$a = 1{\cdot}71\,ms^{-2}$$

Think about it

Consider the different gravitational fields on the Earth and the moon. Can you explain why large rockets are used to get to the moon, but smaller ones are used to get back from the moon?

Worked example

Consider a rocket of mass 1200 kg. How much force is required from the engines to accelerate the rocket upwards from the Earth's surface with a constant acceleration of 4 ms^{-2}? Assume air resistance is negligible.

First of all, draw a free-body diagram of the forces that will be acting (Fig 1.4.49)

We calculate the weight of the rocket:

$$W = mg$$
$$W = 1200 \times 9{\cdot}8$$
$$W = 11760\,N$$

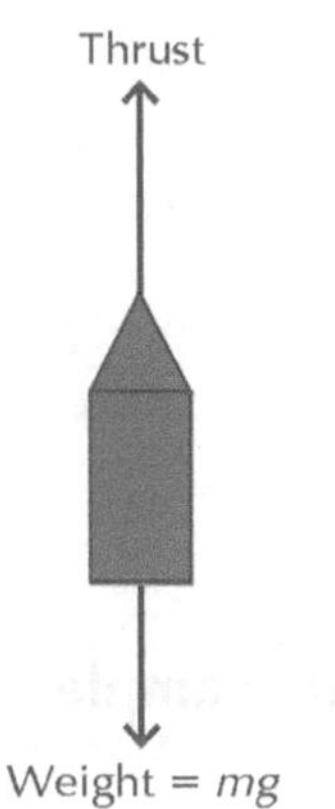

Fig 1.4.49

According to Newton's second law, if there is an acceleration of 4 ms^{-2} there must be an unbalanced force which is given by:

$$F_{un} = ma$$
$$F_{un} = 1200 \times 4$$
$$F_{un} = 4800\,N$$

Adding this to the free-body diagram, we can see that the thrust must be greater than the weight by the amount of the unbalanced force (4800 N) (Fig 1.4.50)

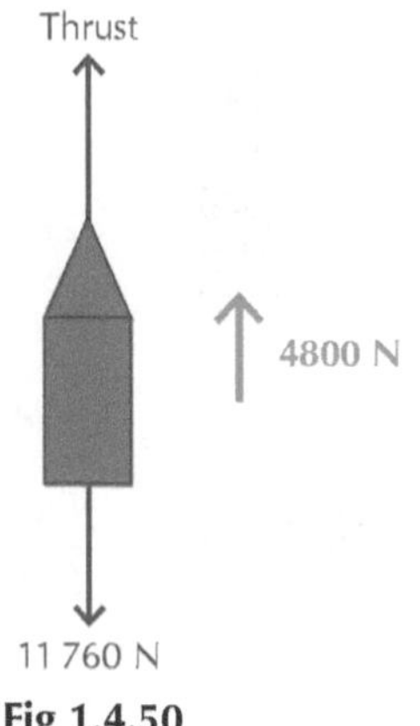

Fig 1.4.50

We can work out the thrust of the rocket by adding the 4800 N to the weight that the engines must overcome:

$$Thrust - Weight = 4800$$
$$Thrust - 11760 = 4800$$
$$Thrust = 16560\,N$$

Exercise 1.4.8 Rockets

All questions in this exercise assume air resistance is negligible.

1 A rocket of mass 1500 kg is propelled in space by engines with a thrust of 25 000 N. Calculate the acceleration of the rocket.

2 A rocket en route to Mars has a mass of 2500 kg. It needs to be accelerated in space with a minimum acceleration of 10 ms^{-2}. What is the minimum force the engines must provide?

3 Use the stages outlined in the previous technique box to fill in the missing values in the table below. You may assume that the rocket is being launched from Earth.

Mass (kg)	Weight (N)	Thrust (N)	Unbalanced force (N)	Acceleration (ms^{-2})
150	1500	2500		
210		4900		
175				2·5
240				4·1
1970				3·9

Table 1.4.4

4 The launch mass of Apollo 11 was 2 923 387 kg. It took off from Earth with an acceleration of 14 ms^{-2}. Calculate the force provided by the rocket engines.

5 A rocket of mass 12 500 kg is launched vertically upwards. Its engines produce a constant force of 600 kN.

a) Draw a diagram showing all the forces acting on the rocket.

b) Calculate the initial acceleration of the rocket.

c) As the rocket rises, its acceleration is found to increase. Give three reasons for this observation.

d) Calculate the acceleration of the same rocket from the surface of the moon where the gravitational field strength is 1·6 N/kg.

e) Explain, in terms of Newton's laws of motion, why a rocket can travel from Earth to the moon and for most of the journey not burn up any fuel.

6 A rocket of mass 6700 kg is launched from Earth with an initial acceleration of 4·7 ms^{-2}.

a) Calculate the unbalanced force required to produce the acceleration.

b) What is the weight of the rocket on Earth?

c) Calculate the thrust produced by the engines of the rocket.

d) What force is required to accelerate the rocket with the same acceleration off the surface of the moon where the gravitational field strength is 1·6 N/kg?

7 A water rocket is used to demonstrate the principle of Newton's third law. Water is forced out of the bottom of a bottle under high pressure and the bottle is propelled into the air.

(continued)

Fig 1.4.51: *Rocket launch*

a) Explain, using Newton's laws of motion, how a water rocket is propelled into the air. Include in your description a diagram which shows all of the forces acting on the rocket.

b) If the mass of the water rocket is 1·45 kg, calculate the acceleration of the rocket if it ejects water with a force of 40 N assuming air resistance is negligible.

8 You are an engineer on the next space mission programme. The mass of the rocket without engines is 4000 kg. You need to choose an engine from the following selection that will accelerate the rocket away from the surface of the Earth with an acceleration of at least 20 ms^{-2}. You must justify your choice by calculation.

Engine	Mass (kg)	Thrust (N)
Apollo Power XLX	2000	1000000
NASA Super	1500	990000
Breadalbane Power	1800	1189000
Sputnik ZXL	2500	1293500
Jupiter 11	1700	1083000

Experiment: Balloon rockets

This experiment demonstrates the principle of Newton's third law to propel a small rocket into the air using air ejected from a balloon.

You will need:

- Balloons
- Straw
- String
- Sticky tape
- Paper clips

Challenge

Construct a rocket that will lift as many paper clips (or other small masses) as possible from the ground to the ceiling. Cups or plastic bags can be used as holders for the paperclips. You must fit a straw to the rocket – a piece of string should be fed through the straw to guide the rocket upwards towards the ceiling as shown in the picture above.

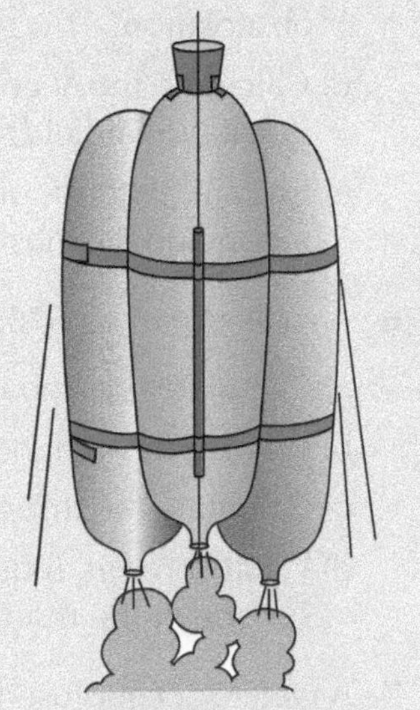

Fig 1.4.52

STEP BACK IN TIME: THE HERO ENGINE

Hero of Alexandria demonstrated a 'rocket engine' more than 2000 years ago. Hero's engine was a spinning copper sphere that was propelled by the thrust produced by a jet of steam. The engine was an early demonstration of Newton's third law. The steam was ejected out of L-shaped holes producing an action force. There was an equal and opposite reaction force that caused the sphere to rotate. Heat was applied externally in order to get engine to work.

Modern day rocket and jet engines are self-contained. That is, they produce the action force internally, e.g. in the combustion chamber of a jet engine.

Experiment: The Hero engine

Hero's engine can be demonstrated easily using a drinking can and some string. This experiment illustrates the basic principles of Hero's engine.

You will need:

- Two drinking cans
- Some string
- Water
- A nail or other sharp, pointed object

Fig 1.4.53

Instructions

1 Use a nail to punch holes in the bottom of the can as follows:

   a) Gently push the nail in the position where the hole is to be punched. Do not crumple the can!

   b) Bend the nail to the side once the hole has been made – this will angle the hole so the water shoots off in a certain direction.

   c) Rotate the can and repeat – but take care to bend the nail in the same direction each time!

2 Tie a piece of string to the ring pull at the top of the can.

3 Immerse the can in water and allow it to fill to the top.

4 Lift the can out of the water and watch the rotational motion.

5 Count the number of turns the can makes in ten seconds.

(continued)

6 Repeat the experiment for different numbers of holes on the can and record your results in a table similar to the one shown.

Number of holes	Size of holes	Rotations in 10 s
2	Small	
2	Large	
4	Small	
4	Large	

Table 1.4.5

7 Investigate other factors which may affect the number of rotations made by the Hero engine in ten seconds. Examples can include the amount the nail is bent when making the holes.

8 Write a short report that describes:

a) How you make a simple Hero engine.

b) How your Hero engine works (relate this to Newton's third law).

c) The effect of hole size on the performance of the engine.

d) The effect of number of holes on the performance of the engine.

Learning checklist

In this section you will have learned:

- How to apply Newton's laws and balanced forces to explain constant velocity (or speed), making reference to frictional forces.
- How to apply Newton's laws and unbalanced forces to explain and/or determine acceleration for situations where more than one force is acting.
- How to use an appropriate relationship to solve problems involving unbalanced force, mass and acceleration for situations where one or more forces are acting in one dimension or at right angles.
- How to use an appropriate relationship to solve problems involving weight, mass and gravitational field strength.
- How to explain the motion resulting from a 'reaction' force in terms of Newton's third law.
- How to explain free-fall and terminal velocity in terms of Newton's laws.

5 Energy

In this chapter you will learn about:

- The principle of conservation of energy.
- Energy transfers – energy efficiency and the generation of electricity.
- Kinetic and gravitational potential energy.
- Mechanical power

The principle of energy conservation

The principle of conservation of energy states that in an isolated system (a system without external influence) the total amount of energy is always the same. Energy can be transformed from one form to another, but the total amount of energy remains constant. Energy cannot be created or destroyed.

An example of this would be a light bulb. The light bulb converts electrical energy (from a battery or from the mains supply) into light energy and heat energy. The type of energy has changed. However, the total amount of electrical energy before is equal to the total amount of heat and light energy after.

In physics, we talk about stores of energy. Energy can be stored in one form or another – for example, energy can be stored as electrical energy.

Key point

In an isolated system the total amount of energy remains the same. Energy cannot be created or destroyed.

Fig 1.5.1: *A light bulb converts electrical energy to heat and light energy*

Energy transfers

Energy cannot be created or destroyed. However, energy can be transferred between different stores. It may be hard to believe, but we are currently recycling the energy that existed over 14 billion years ago at the time of the Big Bang!

Types of energy

There are many different types (or stores) of energy, and we can transfer between any two or more of these stores. We have so far considered the example of the light bulb transferring energy from electrical to heat and light. Energy is measured in joules (J).

Fig 1.5.2: *A car converts stored chemical energy to kinetic energy by burning fuel*

Fig 1.5.3: *The box is lifted off the ground. Gravity acts to accelerate the box down to the centre of the Earth*

Fig 1.5.4: Light energy from the sun travels through a vacuum to Earth

Word bank

- **electromagnetic spectrum** the spectrum of light waves from low energy to high energy, low frequency to high frequency and long wavelengths to short wavelengths. The electromagnetic spectrum is considered in detail in Chapter 9.

Fig 1.5.5: *At around 770 miles per hour a jet fighter will break the sound barrier and travel faster than the speed of sound in air*

Kinetic energy

Any object which is moving has kinetic energy. Think of Kinect for Xbox! The faster the object is moving, the greater its kinetic energy. An example of an object that has kinetic energy is a moving car.

Gravitational potential energy

Any object which has been lifted off the ground has gravitational potential energy. It has the potential to fall! An example of an object with gravitational potential energy is a box which has been lifted into the air.

Light energy

We normally associate light energy with a form of radiation that we can see. However, light energy includes light from the whole of the **electromagnetic spectrum**. Examples of this include light travelling between the Earth and the sun, X-rays taken at a hospital and the microwaves used by mobile phones. Light travels in small 'packets' called photons. In Greek, the term 'photo' means light.

Heat energy (thermal energy)

Heat energy is a measure of the thermal energy that an object has. Atoms in objects vibrate and the faster they vibrate the more energy they have. In a hot cup of coffee there are a huge number of particles vibrating. The higher the temperature of the coffee, the more the particles are vibrating.

Heat is different from temperature. Heat is a type of energy measured in joules (J), while temperature is a measurement of how hot or cold an object is, measured in degrees Celsius (°C) or degrees Kelvin.

Sound energy

Sound waves can only travel where there are particles. In a vacuum (like space), sound cannot travel. By vibrating, particles pass on energy. Sound travels faster through solids and liquids than it does through air because the particles are closer together in a solid or liquid.

Electrical energy

Electrical energy comes from the movement of charged particles which carry energy to appliances such as televisions, phones, radios and computer game consoles. Electrons (negatively charged particles) can travel through some materials. The rate of flow of charge is called current. Electrons move easily in conductors. A battery is a chemical store that converts chemical energy to electrical energy.

Fig 1.5.6: *Charged particles carry energy*

Nuclear energy

Nuclear energy is a very powerful form of energy. Nuclear energy is given out when an atom is involved in a nuclear reaction. Nuclear reactors, nuclear bombs and the cores of stars all have nuclear reactions occurring and all give out nuclear energy.

The sun emits a large amount of energy towards the Earth. This energy is produced by nuclear reactions at the sun's core.

Make the link

Nuclear energy is covered in more detail in Chapter 11.

Energy transfer examples

As discussed above, energy can be transferred between different stores. Some examples of these energy transfers are shown below.

	ENERGY IN		ENERGY OUT
Light bulb:	electrical energy	→	heat and light energy
Microphone:	sound energy	→	electrical energy
Wind turbine:	kinetic energy	→	electrical energy
Wind-up toy:	kinetic energy	→	mechanical stored energy AND kinetic energy

Fig 1.5.7: *A wind-up toy changes kinetic energy to mechanical stored energy back to kinetic energy*

The above are examples of everyday energy transfers. A more complex example of an energy transfer is a golf swing. It consists of many transfers taking place in a short time. As the club swings up, the body converts kinetic energy to gravitational potential energy. As the club is released and swings down towards the ball, the potential energy is transferred to kinetic energy. Kinetic energy transfers to sound energy as the club strikes the ball and to gravitational potential energy as the ball soars into the air. The ball also gains kinetic energy when it is struck.

Objects which transfer energy from one store to another are known as energy changers. The light bulb, for example, is an energy changer.

Fig 1.5.8: *Think about the energy transfers in a golf swing*

Think about it

Consider what you have done between waking up this morning and arriving at school. How many energy changers have you used? What were the energy transfers that took place?

Fig 1.5.9: *Toy car*

Fig 1.5.10: *The Xbox requires 110 V of electricity*

Fig 1.5.11: *Headphones emit sound energy*

Exercise 1.5.1 Energy transfers

1 For each energy changer below, write down the main energy transfer taking place:

a) LED (light emitting diode)

b) Loudspeaker

c) Mobile phone screen

d) Water running down a mountain

2 For each of the examples below, write down as many energy transfers as you can. Remember there may be more than one energy transfer for each!

a) Golf swing

b) Striking a match

c) Burning coal

d) Television

e) iPad

3 A toy car is released from the top of a slope. The pupil calculates that by the bottom of the slope the car should reach a speed of 2.0 ms^{-1}. However, when the pupil measures the speed of the car, he finds that the car only reached 1·8 ms^{-1}.

a) How much speed has been lost?

b) What percentage of speed has been lost?

c) What forms of energy has the 'lost' energy been transferred to?

4 The Xbox requires 110 V electricity to operate. In what form is energy lost from the electrical transformer that reduces mains voltage from 230 V to 110 V?

5 Headphones use energy to transfer electrical information to sound. What type of conversions take place in the headphones?

Useful and wasted energy: energy efficiency

So far we have seen that energy can be transformed from one form to another (from one store to another store). Energy changers are used to **transform** the energy. However, many energy changers transform energy from one form to many different forms. Some of these may be what we actually want (useful) while others may not (wasted).

Consider the light bulb. It transfers electrical energy to light energy and heat energy. The energy transformation that we want is electrical energy to light energy, so the light energy produced is useful energy. We do not want heat energy from a light bulb, so the heat energy produced here is wasted energy. In fact, for an old fashioned filament light bulb, the majority of the electrical energy is converted into heat energy. In other words, a filament light bulb creates a lot of wasted energy. Sometimes we talk of this energy being lost. However, as energy is not created or destroyed, energy cannot be lost. Rather, it is transferred to a form which is not useful. This is highlighted in the following diagram.

Hint

Energy can also be stored chemically. Chemical stored energy examples include the head of a match or inside a battery.

Word bank

- **electrical transformer**

a device that can increase or decrease voltage, current and power. Transformers are used after a power station to increase the voltage from around 20 000 to around 400 000 V for transmission along power lines. Transformers are also used to reduce the voltage in domestic devices such as games consoles, electric shavers and laptop chargers.

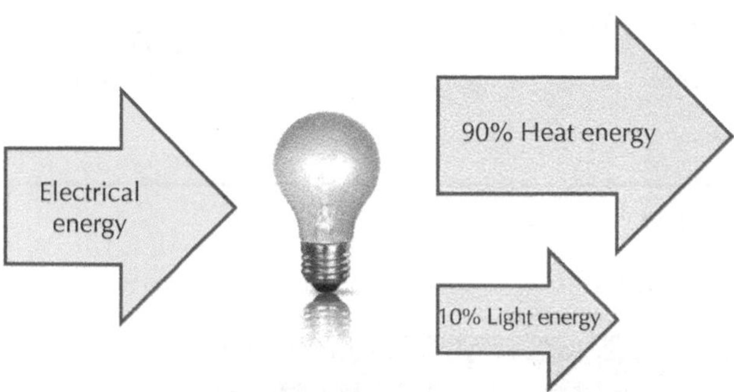

Fig 1.5.12: *Energy conversion in a filament light bulb*

The efficiency of an energy converter is a measure of how much of the converted energy is useful. For example, the filament light bulb above converts just 10% of the input electrical energy to light energy. This means the light bulb is 10% efficient. 90% of the electrical energy is wasted as heat energy.

Key point

The efficiency of an energy transformer is a measure of how much useful energy is output for a given input energy:

$$Efficiency = \left(\frac{Useful\ energy\ out}{Energy\ in}\right) \times 100\%$$

Technique: Energy efficiency

Write down what you know from the question, leaving blank what you are trying to find:

Energy input = ____ J

Useful energy output = ____ J

Efficiency = ____ %

Write down the equation linking the quantities above as you see it on the formula sheet:

$$Efficiency = \left(\frac{Useful\ energy\ out}{Energy\ in}\right) \times 100\%$$

Substitute into the equation what you know.

Solve for the unknown.

Worked example

An electric heater converts 4000 J of electrical energy every seconds. 3000 J of energy is converted into heat, and 1000 J of energy is converted into light. Calculate the efficiency of the heater.

Energy input = 4000 J

Useful energy output = 3000 J

Efficiency = ?

$$Efficiency = \left(\frac{Useful\ energy\ output}{Energy\ in}\right) \times 100\%$$

$$Efficiency = \left(\frac{3000}{4000}\right) \times 100\%$$

$$Efficiency = 0{\cdot}75 \times 100\%$$

$$Efficiency = 75\%$$

Fig 1.5.13: *An electric heater*

Exercise 1.5.2 Energy transformation efficiency

1 Complete the table below to find the missing values.

Energy input (J)	Useful energy output (J)	Percentage efficiency (%)
2000	1500	(a)
14 000	7500	(b)
2500	(c)	60
(d)	2500	50
(e)	4000	65

Table 1.5.1

2 Students are investigating the efficiency of an electric kettle. They find that every second, the kettle uses 2500 J of electrical energy. 1750 J of this energy is converted to heat energy, while the rest is converted to other wasted energy forms. Calculate the energy efficiency of the kettle.

3 An electric heater has an efficiency of 75%. It is connected to the mains, and it is found to convert 4000 J of electrical energy every second. Find the amount of useful heat energy that is output from the heater in one second.

4 Students are investigating energy efficient light bulbs.

One manufacturer claims that their light bulb (Bulb A), which uses 12 J of energy every second, produces the same light output as a traditional light bulb (Bulb B) that uses 100 J of energy every second.

Fig 1.5.14: *Gas discharge light bulb*

(*continued*)

Students investigating this claim find that the efficiency of the light bulbs are as follows:

Bulb A: 75%

Bulb B: 15%

a) Every second, the energy input to each bulb is 200 J. Find the amount of light energy output for each bulb.

b) Calculate the light output from Bulb B when it receives 100 J of input electrical energy.

c) Calculate the light output from Bulb A when it receives 12 J of input electrical energy.

d) Comment on the accuracy of the manufacturer's claim about their energy efficient light bulb.

e) For the same electrical input energy, the students find that one light bulb gets a lot hotter than the other one. Which bulb will produce the most heat energy? Explain your answer.

5 If a model electric motor is 60% efficient at converting electrical energy to kinetic energy, how much useful energy is produced from 100 joules of electrical energy?

Generation of electricity

The generation of electricity is a major world industry. As the human race becomes more carbon conscious, we seek new ways to generate electricity that emit as few pollutants as possible. New energy sources that are low carbon emitters are called clean energies.

Fig 1.5.15: *A coal fired power station with cooling towers releasing steam*

But first let's consider the method by which most of the electricity we use is generated – via a turbine and a generator in a power station.

A turbine is a machine with blades that rotate. It 'collects' energy of different forms, for example, from wind or steam, and transforms it to kinetic energy. The blades are specially designed to ensure that the turbine collects as much energy as possible. The image shows a steam turbine from a power station, consisting of many sets of blades. Steam passing through these blades causes the turbine to rotate.

Fig 1.5.16: *Steam turbine from a power station*

The generator transforms the kinetic energy from the rotating turbine into electrical energy. As the turbine rotates it forces the generator to rotate. The centre of the generator is a solid metal (iron) core. Around the outside of this core are many thousands of turns of wire. As the core rotates it causes a

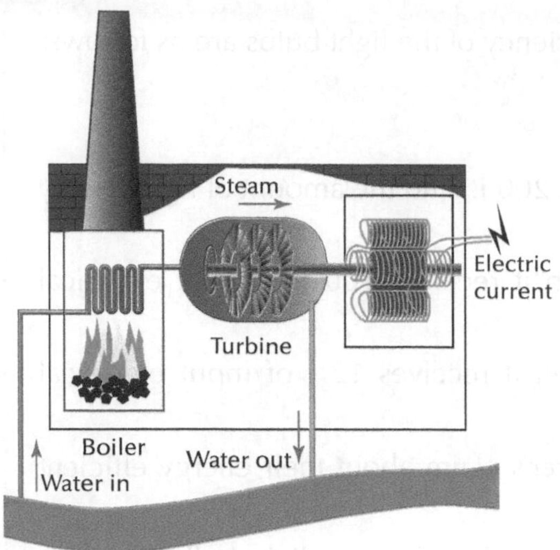

Fig 1.5.17: *Thermal power station*

current to be produced in the coils of wire – kinetic energy has been transformed into electrical energy!

Thermal power stations

In thermal power stations the driving force of the turbine is steam that is created by heating water to high temperatures. There are a number of different ways to heat up water in power stations. The two main methods are burning fossil fuels (coal, oil and gas) or using nuclear power. Both methods share the same basic principle though – heat up water to produce steam to drive the turbine!

The diagram shows a typical thermal power station. Water is heated in the boiler. This produces steam which turns the turbine (kinetic energy). This turns the generator, converting the kinetic energy into electrical energy.

Key point

The energy transformations in a thermal power station are:

Boiler: chemical → heat

Turbine: heat → kinetic

Generator: kinetic → electrical

Not every transformation is 100% efficient – there will be some wasted energy at each stage.

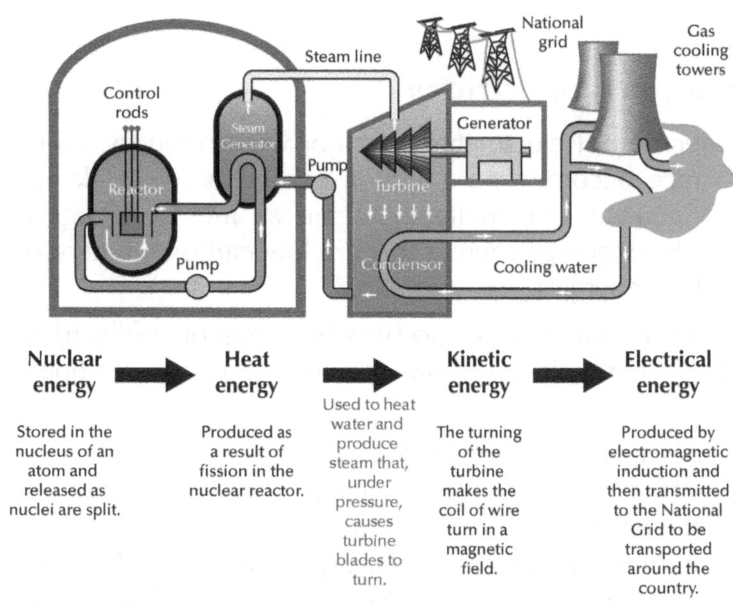

Fig 1.5.18: *Nuclear power planet energy conversion and production process*

Make the link

You will learn about the energy released from nuclear reactions in Chapter 11: Nuclear radiation.

Hydroelectric power stations

Hydroelectric power stations produce the greatest electricity output in the world. They use stored gravitational potential energy to drive their turbines. As of 2013, the largest hydroelectric power station in the world is the Three Gorges Power Station on China's Yangtze River, shown below.

Fig 1.5.19: *Three Gorges Power Station sits on the Yangtze River*

Key point

The following energy transformations take place in a hydroelectric power station:
Gravitational potential energy (dam, mountain loch) → kinetic energy (turbine) → electrical energy (generator)

Hydroelectric power requires water to run downhill. The source of this water is usually a dam which has trapped the water at a height. As the water travels through the pipe it meets the turbine and the kinetic energy of the flowing water causes the turbine to spin as shown in the diagram. The turbine turns the core of the generator to create electricity so, as in thermal power stations, hydroelectric power stations convert kinetic energy to electrical energy.

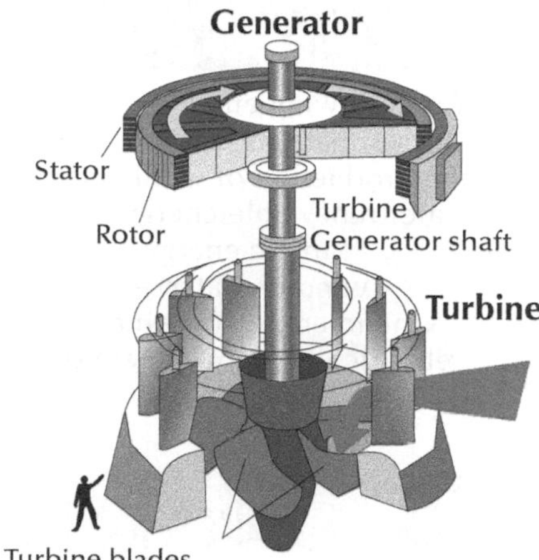

Fig 1.5.20: *Hydroelectric power station generator and turbine*

Hydroelectric power suits countries with certain geographical features. The water needs to fall from a height to transfer its stored gravitational potential energy to kinetic energy. For this reason, mountain lochs make Scotland an especially suitable place for the production of hydroelectric energy.

Wind and water power

Fig 1.5.21: *Wind turbines are often located on hill tops where it is usually most windy*

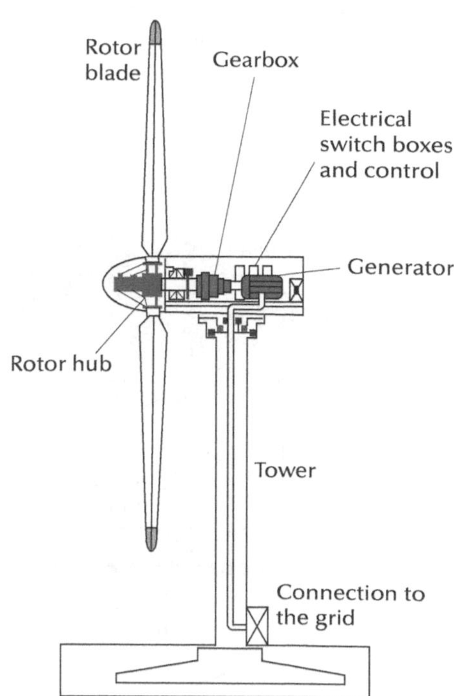

Fig 1.5.22: *The structure of a modern wind turbine*

Renewable energy is increasingly becoming part of life in the 21st century. Because of the dwindling resources of fossil fuels

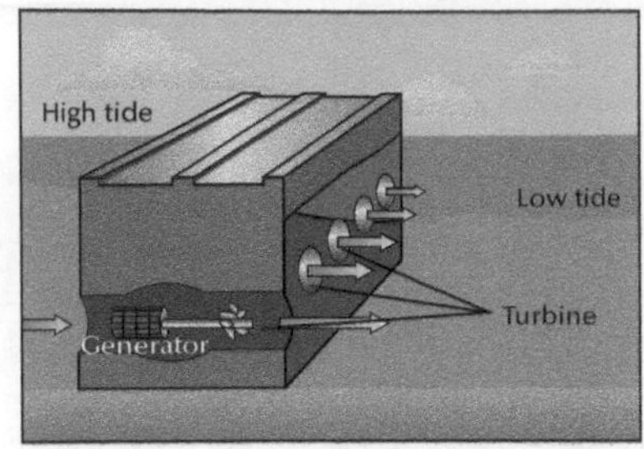

Fig 1.5.23: *An illustration of a tidal electrical energy generation system*

we must find new ways to produce our electricity. Scotland is positioned to take advantage of both wind and wave energy. Wind turbines are visible in many parts of the country, both on land and at sea. Advanced wave and tidal technologies are improving efficiency to allow large-scale implementation across the UK. Wind power can achieve 35% efficiency while tidal wave power can achieve up to 90% efficiency.

These technologies use the same combination of turbine and generator: the turbine is rotated either by wind or by the flow of water from tidal changes. As tides are caused by the moon's orbit, they are well established and predictable. This results in a reliable source of electricity.

Key point

The following energy transformations take place in the production of wind and water renewable energy:

Kinetic energy (wind or water) → kinetic energy (turbine) → electrical energy (generator)

Solar power

All methods of generating electricity that we have considered so far have involved kinetic energy – the iron core of the generator rotating to produce electricity. The main exception to this method is solar power. No kinetic energy is involved in this electricity generation process – the transformation is light energy to electrical energy.

Key point

The following energy transformations take place in the production of solar power:

Light energy (the sun) → electrical energy (solar cell)

Solar energy is becoming more popular in the UK. By having solar panels on the roof of your home it is possible not only to create electricity but also to sell any surplus to the electricity companies!

Fig 1.5.24: *Solar panels*

There are two types of solar panels:

- Photovoltaic (PV) – produce electricity
- Thermal – use energy from the sun to heat up water or liquid inside the panel

Current
Sunlight
Sunlight (photons)
Electron flow
'hole' flow

Fig 1.5.25: *Diagram of the inside of a PV solar panel*

PV solar panels

PV solar panels capture visible and ultraviolet light from the sun. A PV solar panel is typically made up of many silicon cells. These cells create a flow of electrons when they are hit by light rays from the sun. The rays have so much energy that they remove electrons from an atom (a process called **ionisation**). These electrons flow, causing an electrical current.

However, photovoltaic solar panels are not very efficient. They generally operate at around 15%. As their popularity increases, work is being done to improve their efficiency for the future.

Make the link

Ionisation is picked up again in Chapter 3: Current, voltage and Ohm's law.

Make the link

PV solar panels are looked at in more detail in the Higher Physics course.

Thermal solar panels

Thermal solar panels are normally filled with water which is heated by infrared radiation from the sun. The main energy transformation in these solar panels is light energy to heat energy. Thermal solar panels are slightly more efficient than PV panels, operating at around 20% efficiency.

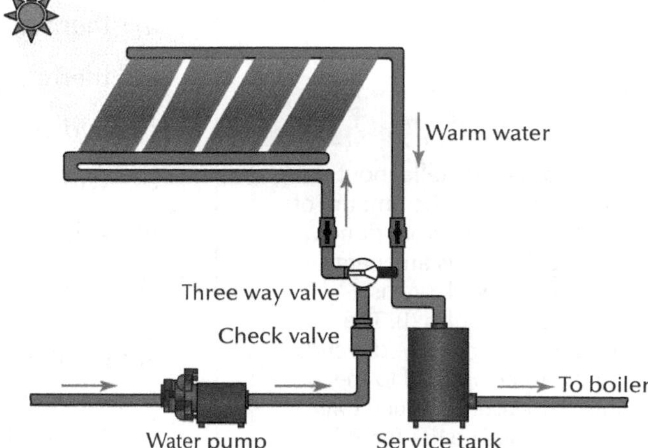

Fig 1.5.26: *How a thermal solar panel works*

Efficiency of energy generation

All the methods described above have strong points to support their development but each also has its challenges. The efficiency of the electricity generation is important because the more efficient the method, the less energy is wasted. The graph below shows the approximate efficiency of each form of energy generation.

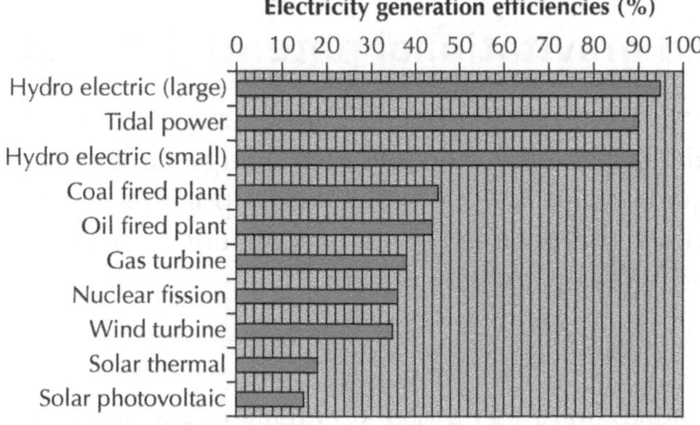

Fig 1.5.27

While hydroelectric is clearly the most efficient form of energy production, there are limitations on where a hydroelectric plant can be constructed.

Physicists and engineers work with governments to find the most efficient, cost effective and most environmentally friendly way to produce energy.

When designing a power system for a town, city or country there are many parameters to consider. One of the most important is the cost of implementation as compared to the value the plant could generate.

Exercise 1.5.3 Generation of electricity

1 Describe the main energy transformation for each of the methods of electricity generation below:

   a) thermal power station (coal fired)

   b) thermal power station (nuclear fuel)

   c) hydroelectric power station

   d) wind farm

   e) solar panels (PV)

2 You are working for a consultancy firm advising the government on electricity generation for your school. A recent decline in fossil fuel resources has necessitated a move to renewable energy. By considering the location of your school and the typical year-round weather, produce a short report of 80 to 100 words that details which method of electricity generation you would choose for your school. Justify your choice and describe the main energy transformations that take place.

Key point

When physicists talk about using energy, they call the amount of energy used the 'work done'. Technically this is an energy transfer and work (W) is measured in joules (J). The amount of work done depends on: the force applied to the object, measured in newtons (N), the distance the force is applied for, measured in metres (m), and the angle between the force and the displacement. Displacement is considered in detail in Chapter 12.

Kinetic and gravitational potential energy

As with all forms of energy, it is possible for energy to transform between kinetic and gravitational potential energy. An example is a falling object: the object has gravitational potential energy when held above the ground which is converted to kinetic energy as the object falls to the ground.

Work

If you want to make an object move, you have to supply it with kinetic energy. Think about pushing a shopping trolley. In order to make the trolley move, you have to apply a **force** to it. It takes energy from you to make the trolley move. In physics, we describe this process as doing work (W) on the object.

Fig 1.5.28: *Kinetic energy in a supermarket*

The greater the force you apply to the trolley, the faster it will move. In other words, the greater the applied force, the greater the work done. Similarly, if you apply the force for a larger distance, the trolley will gain more kinetic energy and therefore move faster. Work done depends on both the force applied (F) and the distance the force is applied for (d). We summarise this in the following formula:

$$W = Fd$$

It should be noted that $W = Fd$ only applies when a constant force is applied in the same direction as the **displacement** of the object.

Technique: Calculations involving work done

Write down what you know from the question, leaving blank what you are trying to find:

Work done = ____ J

Force = ____ N

Distance = ____ m

Write down the equation linking the quantities above as you see it on the formula sheet:

$$W = Fd$$

Substitute into the equation what you know.

Solve for the unknown.

Word bank

- **force**

an action, either internal or external, that can cause a change. This might be a change to shape, direction, structure or something else. Force is measured in newtons.

Word bank

- **displacement**

the change in position from the start point to the end point as the crow files; the route taken between these points is irrelevant

Word bank

- **newton (N)**

the unit of measurement of how strong or weak the force acting on an object is. 1 newton is equal to the force of 1 kilogram metre per second per second, i.e. $1\text{N} = 1\text{kgms}^{-2}$

Worked example

Calculate the work done by the applied force if a horizontal force of 18 N is applied to push a box a distance of 3 m across the floor.

Write down what we know and what we are looking for:

Work = _____ J

Force = 18 N

Distance = 3 m

We are using the equation that links work, force and distance:

$$W = Fd$$

Substitute what we know:

$$W = 18 \times 3$$

Solve for work done:

$$W = 54J$$

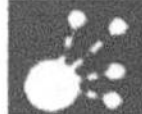

Make the link

Vectors and displacement are considered in more detail in Chapter 12.

Exercise 1.5.4 Work

In each of the following questions you should assume a horizontal force is applied in the same direction as the displacement of the object.

1 Calculate the work done when a force of 65 N is used to move an object a distance of 8 m.

2 A pool cue exerts a force of 90 N on the cue ball for a distance of 0·008 m. Calculate the work done.

Fig 1.5.29

3 The work done moving an object a distance of 7 m is 49 J. Calculate the average force exerted on the object.

4 A force of 16 N is applied to an object. The work done on the object is 48 J. Calculate the distance over which the force is applied.

5 A golf club exerts a force of 170 N on a golf ball. The work done on the ball is 19 kJ. Calculate the distance over which the force is applied.

6 A force of 20 mN is applied to an object for a distance of 3 cm. Calculate the work done.

7 If the work done moving an object a distance of 40 cm is 370 kJ, calculate the average force that must have been supplied.

8 A car's engine exerts an average force of 19 kN. Calculate the work done over a journey of distance 3 km.

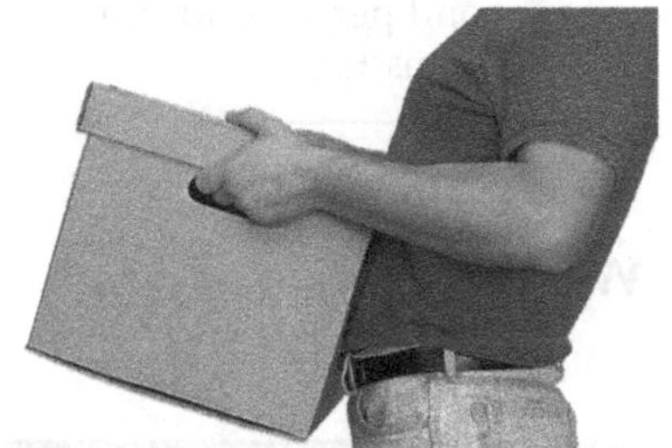

Fig 1.5.30: *Lifting an object involves doing work against gravity*

Potential energy

When an object is lifted from the ground up to a height (h) it gains gravitational potential energy. This is often shortened to just potential energy and is given the symbol E_p. When lifting an object you have to do work against gravity. This means you are supplying energy to the object. If you drop the box, it will fall back to the ground. So, once off the ground, the box has the potential to fall.

The equation for potential energy comes from the definition for work done:

$$W = Fd$$

Consider lifting a football of mass (m) to a height (h). When lifting the ball you are working against the force of gravity which means you must apply a force. When lifting the football at constant speed, you must apply a force:

$$F = mg$$

where g = gravitational field strength, measured in newtons per kilogram.

You apply this force for a height (h) so the work done in lifting the ball is:

$$W = Fd$$

$$W = (mg)(h)$$

$$W = mgh$$

Make the link

Unit 3: *Dynamics and space* discusses the force of gravity in more detail.

This allows us to work out the equation for gravitational potential energy:

$$E_p = mgh$$

Technique: Calculations involving gravitational potential energy

Write down what you know from the question, leaving blank what you are trying to find:

Gravitational potential energy = ____ J

Mass = ____ kg

Height = ____ m

Gravitational field strength = _____ N/kg ($g = 9{\cdot}8\ ms^{-2}$ on Earth)

Write down the equation linking the quantities above as you see it on the formula sheet:

$$E_P = mgh$$

Substitute into the equation what you know.

Solve for the unknown.

Key point

Gravitational potential energy depends on the mass (kg) of the object, the height (m) to which the object is lifted and the gravitational field strength (N/kg) of the planet.

Key point

As gravitational field strength is measured in N/Kg and 1 N/kg = 1 ms^{-2} physicists typically refer to the gravitational field strength (g) on Earth as:

$$g = 9{\cdot}8 ms^{-2}$$

Worked example

A weightlifter lifts a total mass of 50 kg to a height of 2 m. Calculate the potential energy gained by the weight.

$$m = 50\ kg$$
$$h = 2\ m$$
$$g = 9{\cdot}8\ ms^{-2}$$
$$E_P = mgh$$
$$E_P = 50 \times 9{\cdot}8 \times 2$$
$$E_P = 980J$$

Fig 1.5.31: *Weightlifter lifting 50 kg*

Fig 1.5.32: *Escalator*

Fig 1.5.33: *Car ramp lift*

Exercise 1.5.5 Gravitational potential energy

1 A crane on a ship lifts an object of mass 2000 kg to a height of 35 m.

a) Calculate the work done lifting the mass.

b) What is the potential energy of the mass?

2 A man of mass 80 kg uses an escalator to go up to the top floor of a shopping centre. The height gained is 20 m. Calculate the potential energy of the man on the top floor of the shopping centre.

3 Consider an object of mass (m) being lifted up to a height (h).

a) What is the work done lifting the object?

b) Hence derive the equation for gravitational potential energy.

4 The potential energy of a ball is 23 J at a height of 60 m. Calculate the mass of the ball.

5 In a garage, a ramp lift holds the car in the air so that a mechanic can work underneath it.

a) How much energy must be supplied to the lift to allow it to lift a car of mass 1200 kg to a height of 1·5 m?

b) Assuming that the lift is only 75% efficient at converting electrical energy to gravitational potential energy, find the electrical energy that must be supplied to the lift motor to raise the car in part (a) above.

Word bank

- **speed**

the rate of change of distance with time measured in metres per second, i.e. 1 ms^{-1} = 1m ÷ 1s

Kinetic energy

Any moving object will have kinetic energy. The amount of kinetic energy an object has depends on its mass (kg) and its speed (ms^{-1}). The faster the object moves, the greater this kinetic energy. Also, the greater the mass of the object, the greater the kinetic energy.

The kinetic energy of an object can be found from the following equation:

$$E_K = \frac{1}{2}mv^2$$

where m = mass (kg) and v = speed (ms^{-1})

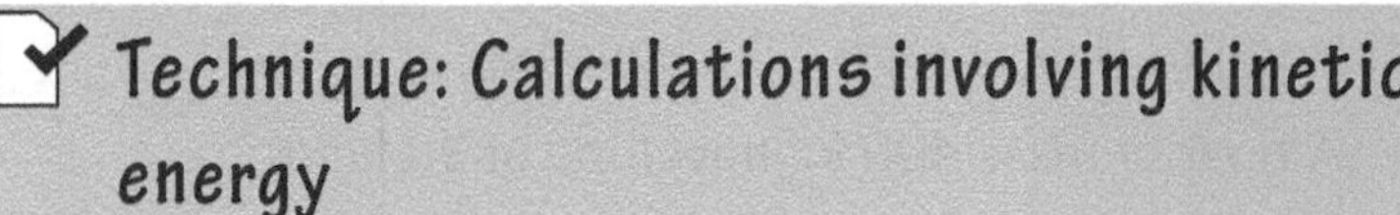

Technique: Calculations involving kinetic energy

Write down what you know from the question, leaving blank what you are trying to find:

Kinetic energy = ____ J

Mass = ____ kg

Speed = ____ ms^{-1}

Write down the equation linking the quantities above as you see it on the formula sheet:

$$E_K = \frac{1}{2}mv^2$$

Substitute into the equation what you know.

Solve for the unknown.

Fig 1.5.34: *Bowling ball with kinetic energy*

Worked example

Calculate the kinetic energy of a bowling ball of mass 3 kg when it is moving with a speed of 10 ms^{-1}.

Write down what we know and what we are looking for:

Kinetic energy = _____ J

Mass = 3 kg

Speed = 10 ms^{-1}

Use the equation that links kinetic energy, mass and speed:

$$E_K = \frac{1}{2}mv^2$$

Substitute what we know:

$$E_K = \frac{1}{2} \times 3 \times 10^2$$

Solve for kinetic energy:

$$E_K = 150 \text{ J}$$

Exercise 1.5.6 Kinetic energy

1 Calculate the kinetic energy of the following objects (taking care to ensure that mass is measured in kilograms and speed in metres per second!):

a) A car of mass 1100 kg travelling at 20 ms^{-1}.

b) A lorry of mass 38 000 kg travelling at 15 ms^{-1}.

c) A crate of mass 10 kg travelling at 8 ms^{-1}.

(continued)

Fig 1.5.35: *The moving tennis ball has kinetic energy*

d) A tennis ball of mass 50 g travelling at 50 ms^{-1}.

e) A boat of mass 400 kg travelling at 30 km/h.

f) An electron of mass $9{\cdot}11 \times 10^{-31}$ kg travelling at 15 000 ms^{-1}.

2 During a tennis serve, the tennis ball of mass 150 g is hit over the net. It is calculated that the kinetic energy of the ball as it leaves the racket is 215 J. Calculate the speed of the serve.

3 The kinetic energy of a moving object is measured to be 12 J. The speed of the object is 2 ms^{-1}. Calculate the mass of the object.

Fig 1.5.36: *Car brake disc and caliper*

Energy transfer between potential and kinetic

We have seen that energy can be transformed from one form to another. Kinetic energy, for example, can be transformed into heat energy in the brakes of a car when they slow the vehicle down.

It is important to remember that energy cannot be created or destroyed. The principle of energy conservation can be used to solve a variety of problems. Let's look at the conversion between potential energy and kinetic energy.

GO! Experiment: The bouncing ball

This experiment investigates the energy transfer between potential and kinetic for a ball bouncing off the ground and considers the efficiency of the energy transfer.

You will need:

- A ball (for example a tennis ball)
- A metre stick

Fig 1.5.37: *A ball being released towards the ground*

Instructions

1. Using a metre stick to measure the distance, hold the ball one metre above the ground.
2. Drop the ball so that it bounces off the ground. Do not throw the ball upwards or downwards, just let the ball drop freely.
3. Use the metre stick to measure the maximum height that the ball bounces back up to.

(*continued*)

4. Write a short report about the experiment, including the following points:
   - How did the height of the rebound compare to the height the ball was dropped from?
   - Explain why there is a difference between the initial drop height and the rebound height.
   - Calculate the efficiency of the bounce by comparing the potential energy of the ball before it was dropped and after it rebounded.

Falling objects

As we have seen, when an object is lifted to a height (h) it has potential energy given by:

$$E_P = mgh$$

and when the object falls back to the ground, this potential energy is converted to kinetic energy:

$$E_K = \frac{1}{2}mv^2$$

Some energy is transferred to heat energy due to friction acting on the object. However, if we ignore friction, the kinetic energy on the ground must be the same as the potential energy at the height from which the object was dropped. We can therefore set the two above equations equal to each other because energy is conserved:

$$\frac{1}{2}mv^2 = mgh$$

$$v = \sqrt{2gh}$$

You do not need to remember this equation. However, understanding it is important because it shows that the mass of an object does not affect its final speed. In the absence of air resistance and friction, a brick and a feather would hit the ground with the same speed at the same time. This experiment was conducted on the moon as part of the Apollo 15 mission. On the moon there is almost no atmosphere to cause air resistance. It was found that a hammer and an eagle feather did indeed hit the ground at the same time, verifying the energy conservation equation above!

Think about it

The hammer and feather experiment was conducted on the moon. Would the results of this experiment have been the same on Earth? Explain the reason for your answer.

The pendulum

A pendulum has a mass which swings backwards and forwards between two high points (1 and 3) and a low point (2).

At points 1 and 3, the pendulum has maximum potential energy which is given by:

$$E_P = mgh$$

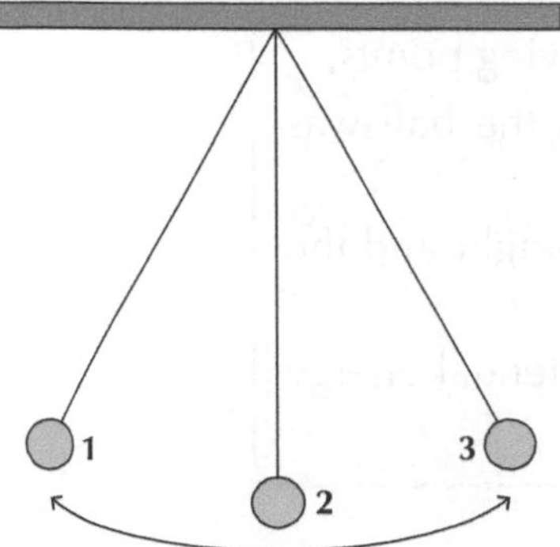

Fig 1.5.38: *Pendulum*

Here, h is the height of the mass above the lowest point of the swing. The kinetic energy at these points is zero because the speed of the pendulum is zero.

At point 2 the pendulum has maximum kinetic energy, i.e. it is moving at its fastest. At this point, the potential energy is zero because it has all been converted into kinetic energy. We can therefore find the speed of the pendulum at point 2 using the same equation as we did for falling objects:

$$v = \sqrt{2gh}$$

Exercise 1.5.7 Transfers between kinetic and potential energy

1 When a pendulum swings back and forth, it continually transfers energy between kinetic and gravitational potential. Assume the pendulum in this example has a mass of 0·75 kg.

a) At which point in the swing is the kinetic energy at its maximum?

b) At which point(s) in the swing is the potential energy at its maximum?

c) Calculate the speed of the pendulum at the bottom of its swing if it starts from a height of 0·3 m.

d) Explain why the maximum height reached by the pendulum decreases with each swing.

2 A tennis ball of mass 150 g is dropped from a height of 2·5 m. At what speed does the ball hit the ground?

3 A 500 kg car is parked at the top of a hill at a height of 40 m when its hand brake fails.

a) Ignoring frictional effects, what would be the speed of the car at the bottom of the hill?

b) The actual speed of the car at the bottom of the hill is lower. Why is the speed lower than calculated in part (a)?

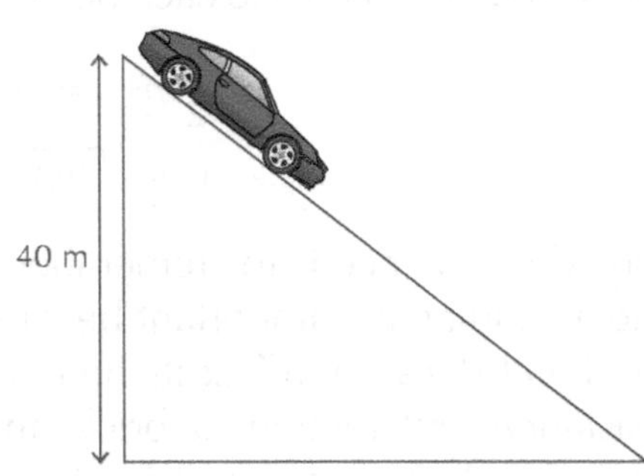

Fig 1.5.39: *Car at a height of 40 m*

4 A 0·2 kg ball is launched from the ground with a speed of 18 ms^{-1}. Calculate the maximum height that it will reach.

5 A ball of mass 0·6 kg is launched up a ramp from point A to point B as shown in the diagram below. The ball is launched with a speed of 14 ms^{-1}. It comes to rest at point B, at the top of the slope. You can assume that friction is negligible.

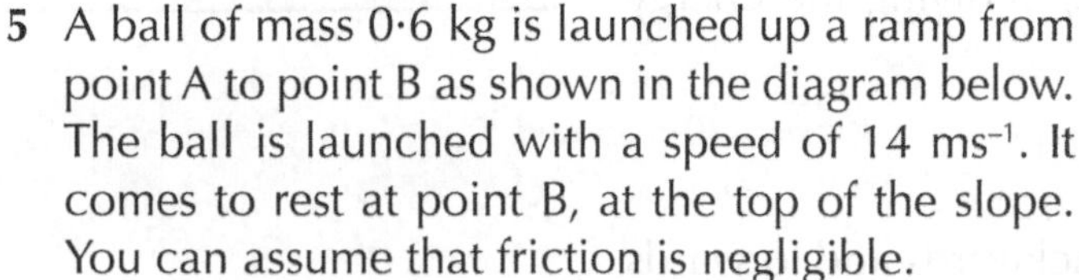

a) Calculate the kinetic energy of the ball the instant it is launched.

b) What is the potential energy gained by the ball when it reaches the top of the ramp?

c) Find the height of the ramp.

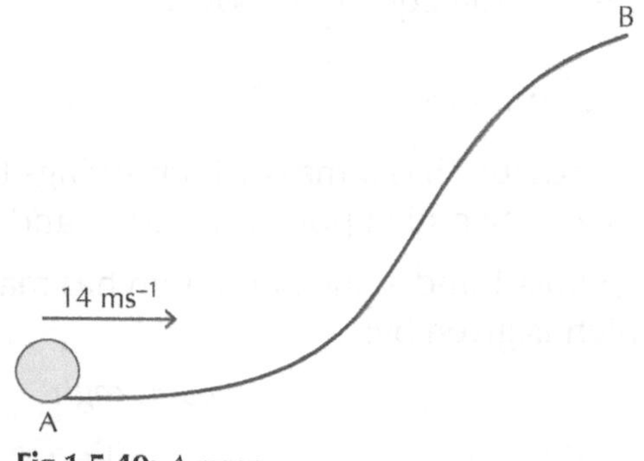

Fig 1.5.40: *A ramp*

Mechanical power

We looked earlier at kinetic energy and potential energy. Power is the rate at which energy is transformed into other forms. We can link many different forms of energy to each other. For example, mechanical energy transfers involving potential energy, kinetic energy and work done.

The work done moving an object a distance, d, with a constant force, F, is given by:

$$W = Fd$$

The gravitational potential energy from lifting an object of mass, m, to a height, h, is given by:

$$E_P = mgh$$

The kinetic energy of an object of mass, m, moving with a velocity, v, is given by:

$$E_K = \frac{1}{2}mv^2$$

Power is the rate of change of energy. The power for an energy change, E, in a time, t, is given by:

$$P = \frac{E}{t}$$

Word bank

- **dissipate**

to disperse, to use up, to transfer to; the power dissipated in the iron in question 5 is the power used by the iron

Experiment: Power of an electric motor

This experiment investigates the power of an electric motor.

You will need:

- A pulley
- An electric motor
- String
- A hanging mass
- A metre stick
- A stop clock
- A DC power supply
- Wires

Instructions

1. Connect the motor to the DC power supply.
2. Hook the hanging mass onto the electric motor using the string.
3. Switch on the motor and time how long it takes the motor to lift the mass through a height of 1 metre. Measure the distance using a metre stick.
4. Calculate the potential energy using the equation:

$$E_P = mgh$$

5. Find the power of the electric motor using the equation:

$$P = \frac{E}{t}$$

Experiment: Measuring your power

This experiment will allow you to estimate your own power as you climb a flight of stairs.

You will need:

- A set of bathroom scales
- A metre stick
- A stop clock
- A flight of stairs

Fig 1.5.41

Instructions

1 Measure the height of the flight of stairs (or alternatively measure the height of one stair and multiply by the number of stairs).

2 Measure your mass, in kilograms, using a set of bathroom scales.

3 Starting at the bottom of the stairs, make your way to the top of the stairs.* Record how long it takes you. You may need a second person to operate the stop clock.

4 Calculate the potential energy gained using the equation:

$$E_P = mgh$$

where m is your mass, g is the gravitational field strength on Earth and h is the total height of the flight of stairs.

5 Calculate your power using the equation,

$$P = \frac{E}{t}$$

where E is the potential energy gained and t is the time (in seconds) taken to climb the stairs.

*Take care when climbing a set of stairs – the faster you go the greater your power, however, you should be careful when running up a flight of stairs!

Technique: Calculating mechanical power

1. Calculate energy transferred using one of three equations:

   The equation for potential energy,

   $$E_P = mgh$$

   or *the equation for kinetic energy,*

   $$E_K = \frac{1}{2}mv^2$$

   or *the equation for work done,*

   $$W = Fd$$

2. Use the equation for power to work out the power or the length of time taken,

   $$P = \frac{E}{t}$$

Worked example

A crane is used to lift containers off cargo ships. The container has a mass of 8000 kg. The crane can lift the container up to a height of 40 m in 2 minutes.

a) Calculate the potential energy gained by the container.

b) What is the power of the crane?

a) The potential energy gained by the container is given by the equation,

$$E_P = mgh$$

Substitute what we know from the question,

$$E_P = 8000 \times 9{\cdot}8 \times 40$$

Solve for the potential energy,

$$E_P = 3\,136\,000 J$$

b) The power of the crane will depend on how long it takes the crane to lift the container. We know the energy gained by the container from part (a), so we can use the following equation to find the power:

$$P = \frac{E}{t}$$

It is important to remember to convert the time into seconds:

$$t = 2 \times 60 = 120\ s$$

Substitute into the power equation,

$$P = \frac{3\,136\,000}{120}$$

Solve for the power of the crane,

$$P = 26\,133 W$$

Fig 1.5.42: *Crane carrying container*

Worked example

Fig 1.5.43: *Time taken depends on energy gained and power*

A girl has a mass of 45 kg. Her average power when running up a flight of stairs is 800 W. Calculate the length of time it takes for the girl to climb a flight of stairs with a height of 4 m.

The time it takes will depend on energy gained and power.

To find the energy gained by the girl running up the flight of stairs, we use the equation for potential energy,

$$E_P = mgh$$

Substitute what we know,

$$E_P = 45 \times 9{\cdot}8 \times 4$$

Solve for the potential energy,

$$E_P = 1764 J$$

Now we can work out the length of time taken using the equation for power, energy and time:

$$P = \frac{E}{t}$$

Substitute what we know from the question and what we have worked out above,

$$800 = \frac{1764}{t}$$

Solve for time,

$$800t = 1764$$

$$t = \frac{1764}{800}$$

$$t = 2{\cdot}21 s$$

Exercise 1.5.8 Mechanical power

1 Calculate the power in the following examples:

a) A car engine producing 30 000 J of energy in 10 s.

b) A man producing 500 J of energy in 20 s.

c) A horse producing 1200 J of energy in 1 minute.

d) An electric motor producing 5000 J of energy in 2 hours.

2 In a garage, a ramp lift hoists a car into the air to allow the mechanic to work on it. The car has a mass of 1700 kg and the lift hoists it to a height of 2 m.

a) Calculate the work done lifting the car to a height of 2 m.

b) If the lift takes 30 s to lift the car, calculate the power of the lift.

Fig 1.5.44

(continued)

3 A man has a mass of 90 kg. An escalator carries him to a height of 40 m in a time of 50 s.

a) What is the weight of the man?

b) Calculate the potential energy of the man at a height of 40 m.

c) Calculate the power of the escalator.

4 An electric motor can lift a crate of mass 5 kg to a height of 2 m in a time of 90 s. Calculate the power of the electric motor.

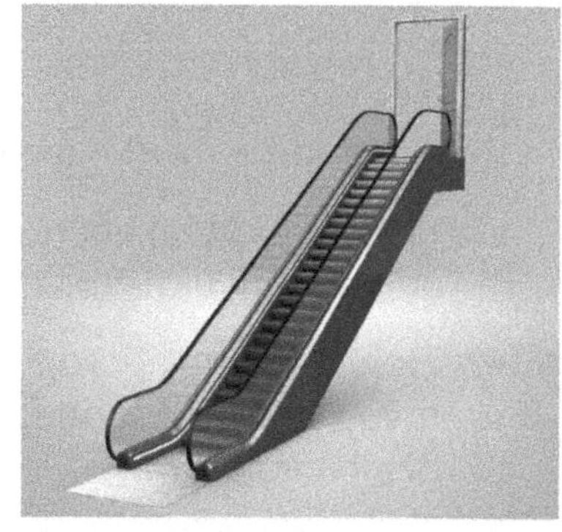

Fig 1.5.45: *Escalator*

5 A woman of mass 70 kg walks up a flight of stairs to a height of 25 m in a time of 1 minute. Calculate her power.

6 An electric hoist has a power of 1000 W. It is used to lift a mass of 200 kg on to a loading bay at a height of 6 m.

a) Calculate the weight of the mass.

b) Calculate the potential energy of the mass at a height of 6 m.

c) How long will the hoist take to lift the mass?

7 Calculate the time taken for a lift to carry passengers of combined mass 250 kg to a height of 300 m if it has a power of 2000 W.

8 An electric motor with a power of 150 W is used to lift a small mass of 100 g. What height will it lift the mass to in a time of 30 s?

9 A weightlifter has a maximum power of 200 W. In a competition, the mass must be lifted in a time of 20 s to a height of 3 m. What is the maximum mass that the weightlifter can lift?

Fig 1.5.46: *Weightlifter*

Learning checklist

In this section you will have learned:

- How to explain energy conservation, energy conversion and energy transfer.
- How to use an appropriate relationship to solve problems involving work done, unbalanced force and distance/displacement.
- How to define gravitational potential energy.
- How to use an appropriate relationship to solve problems involving gravitational potential energy, mass, gravitational field strength and height.
- How to define kinetic energy.
- How to use an appropriate relationship to solve problems involving kinetic energy, mass and speed.
- How to use appropriate relationships to solve problems involving conservation of energy.
- How to use an appropriate relationship to solve problems involving mechanical power, energy and time.

6 Projectile motion

In this chapter you will learn about:

- Projectile motion – an explanation of projectile motion, including range and time of flight.
- Horizontally launched projectiles, including calculations of projectile motion.
- Satellites – types of satellite, how they work and an explanation of satellite orbits.
- The benefits of satellite technology.

Projectile motion

When we throw a ball into the air it will rise to a maximum height and then fall back down again. If you throw the ball forward as well as upwards, it will follow a curved path like the one shown below:

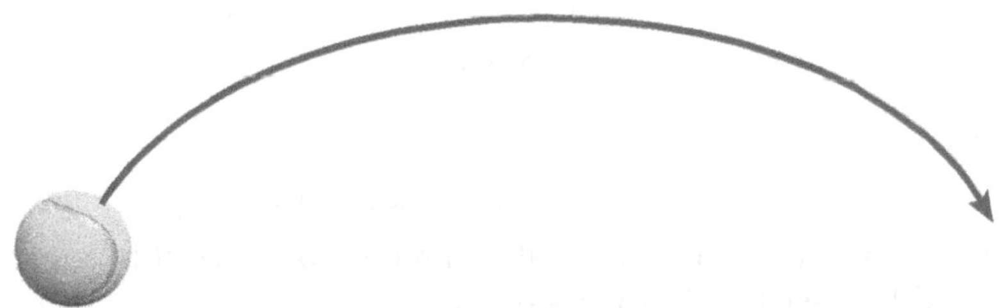

Fig 1.6.1: *Curved path of a projectile*

Projectile motion is a combination of horizontal and vertical motion. These motions work independently from each other and can be analysed separately. When combined, the result is a curved path called a **trajectory**.

Word bank

- **trajectory**

the flight path of the projectile

Vertical motion (recap)

When a ball is thrown vertically upwards, it will reach a maximum height and then fall back down to Earth. At all times, the ball experiences a downward force due to gravity. This means the ball is constantly accelerating downwards. Taking upwards to be positive, we can say that the ball has a constant acceleration of:

$$a = -9{\cdot}8\,\mathrm{m\,s^{-2}}$$

If you throw the ball with larger velocity then it will reach a greater maximum height. This is because the ball takes longer to come to a stop so it covers a greater vertical distance. We can use a velocity-time graph to calculate the maximum height that an object will reach when it is thrown vertically.

Consider throwing a ball with an initial upwards velocity of 30 ms^{-1}. This velocity will be positive because it is directed upwards. The ball will decelerate due to gravity with a constant deceleration of 9·8 ms^{-2}. (This is a negative acceleration.) A velocity-time graph representing this motion is shown below.

Fig 1.6.2: *Vertical motion of a ball*

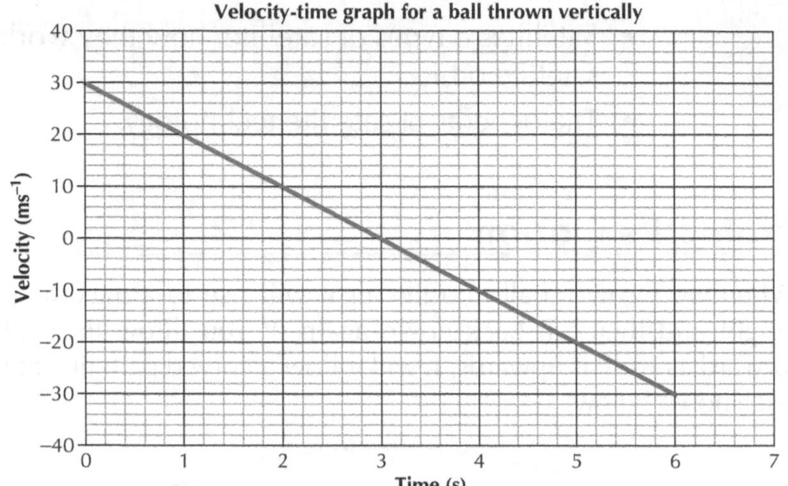

Fig 1.6.3

This graph shows that the velocity of the ball decreases with a constant deceleration to zero after three seconds. At this point, the ball has reached its maximum height.

After this, the velocity of the ball becomes negative – this means that the ball is now moving downwards – the ball is returning to the thrower's hand.

Our equations of motion for acceleration covered in Chapter 12, and the velocity-time graph, can be used to find the time of flight of the ball and the maximum height that the ball will reach.

Technique: Finding the time of flight

1. Write down what you know, including that the final velocity is zero at the maximum height and the acceleration is $-9.8\ ms^{-2}$ due to gravity, leaving blank what you are trying to find:

   Initial speed, $u =$ ____ ms^{-1}

   Final speed, $v = 0\ ms^{-1}$

   Time, $t =$ ____ s

   Acceleration, $a = -9.8\ ms^{-2}$
2. Write down the equation linking the quantities above as you see it on the formula sheet:
$$a = \frac{v-u}{t}$$
3. Substitute into the equation what you know.
4. Solve for the time of flight.
5. For the total time of flight, you need to double the time taken to reach the maximum height.

Technique: Finding the maximum height reached

1. Plot a graph of the velocity of the ball against time:

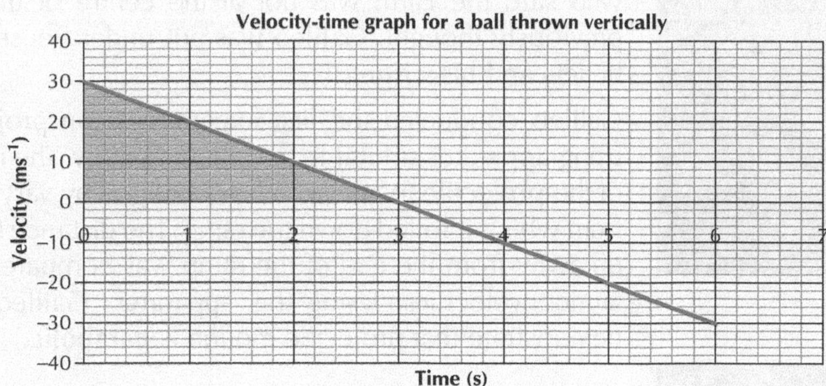

Fig 1.6.4

2. Find the area under the graph from the point of launch to the time when the ball's velocity is zero using the formula for the area of a triangle:
$$A = \frac{1}{2}bh$$

Exercise 1.6.1 Vertical motion

1 A ball is thrown vertically upwards with an initial velocity of 25 ms^{-1}. It returns to the thrower's hand at the same height as its take off.

a) Calculate the time taken for the ball to reach its maximum height.

b) Calculate the total time of flight for the ball.

c) Plot a velocity-time graph of the ball showing the velocity of the ball throughout its flight from the moment its leaves the thrower's hand until the moment it returns.

d) Find the maximum height reached by the ball.

Fig 1.6.5: *Ball*

2 A ball is thrown vertically upwards. It takes 1·4 s to reach its maximum height.

a) What was the initial velocity the ball was thrown with?

b) Find the maximum height reached by the ball.

Fig 1.6.6: *Galileo*

Horizontally launched projectiles

Combining vertical motion with horizontal motion gives us a projectile. The initial experiments carried out on projectile motion were by Galileo.

Galileo's initial experiments

Galileo (1564–1642) was an Italian physicist and mathematician who said the Earth was not at the centre of the universe as previously thought. Galileo was put under house arrest for his beliefs and his writings!

Galileo conducted the first experiments on projectile motion using apparatus similar to that shown below. The **initial velocity** of the projectile (e.g. a ball) was changed by varying the height from which it rolled down the ramp. The distance the ball ended up away from the end of the ramp was compared for different launch velocities. Using this apparatus, Galileo was able to demonstrate that projectile motion is **parabolic.**

• **initial velocity**

the velocity that the object starts with

• **parabolic**

curved path, from the parabolic equations (those involving an x^2 term) in mathematics

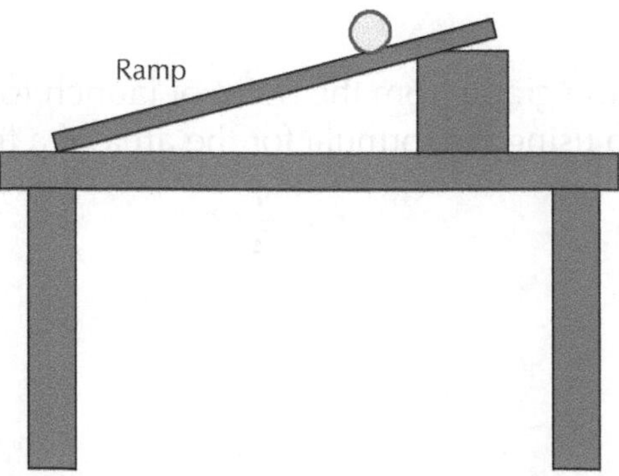

Fig 1.6.7: *Galileo's apparatus (simplified!) to show projectile motion is parabolic*

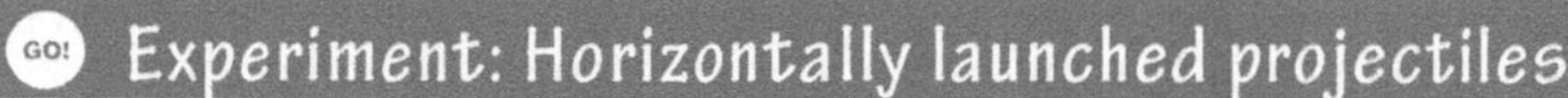

You are going to carry out Galileo's experiment to investigate two things:

- How the maximum horizontal distance of the projectile depends on the launch velocity.
- How the time of flight of the projectile depends on the launch velocity.

You will need:

- A ramp
- A ball
- A metre stick
- A tub of sand
- A timer

Instructions – range of projectile

1 Set up the experiment as shown in Galileo's apparatus above, using a tub of sand in the path of the ball as a marker for where the ball lands. Ensure a small flat region on the table after the ramp to make sure the ball is launched horizontally.

2 Mark a starting position on the ramp to ensure the ball is dropped from the same place every time.

3 Prop the ramp up to a height using a wedge or books and measure the height of the ramp above the table.

4 Let the ball roll down the ramp and land in the tub of sand.

5 Use a ruler or metre stick to measure the distance the ball has landed away from the edge of the table – this is the range of the projectile.

6 Record your results in a table similar to the one shown below:

Ramp height (m)	Range (m)

Table 1.6.1

7 Use the results above to write a short report about the effects of the launch velocity on the range of the projectile – remember that the greater the ramp height, the greater the launch velocity.

The launch velocity at the bottom of the ramp can be approximated using the following formula:

$$v = \sqrt{\frac{2ghs}{L}}$$

(continued)

where *s* is the starting distance up the ramp, *h* is the height of the ramp and *L* is the total length of the ramp, as shown in the diagrams below:

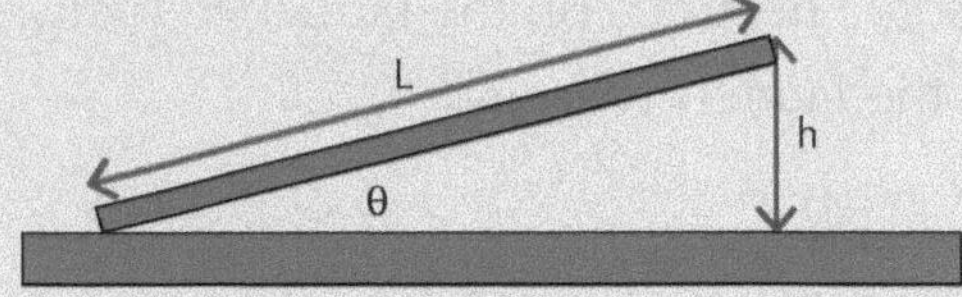

Fig 1.6.8

Fig 1.6.9

Use the launch velocity equation above to work out the velocity of the ball for each of the ramp heights. This can be used to plot a graph of launch velocity against range.

Instructions – time of flight

1 Set up and carry out the same experiment as detailed above, only this time measure the time of flight, i.e. the time taken for the ball to fall to the ground from the moment it leaves the edge of the table.

2 Record your results in a table similar to the one shown below:

Ramp height (m)	Time of flight (s)

Table 1.6.2

3 Use the results above to write a short report about the effects of the horizontal launch velocity on the time of flight of the projectile – remember that the greater the ramp height, the greater the launch velocity.

In these experiments we have considered a projectile which has been launched with an initial horizontal velocity. It follows a curved path to the ground as shown in the diagram below.

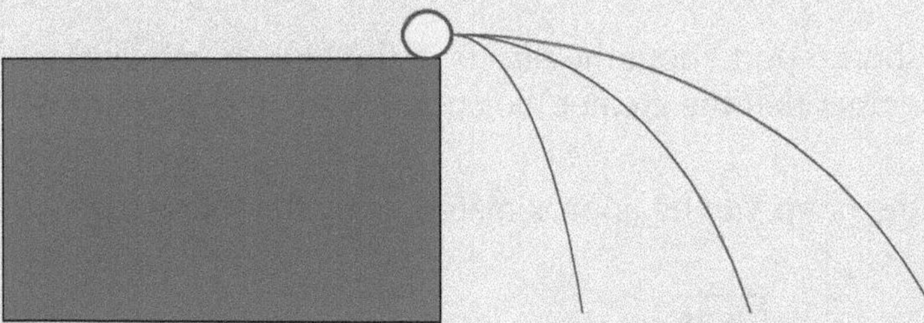

Fig 1.6.10

The greater the launch velocity, the greater the range of the projectile. The time of flight of the projectile is the same for all horizontal launch velocities.

Analysing the motion of a projectile

When analysing the motion of a projectile, it is important to split it into horizontal and vertical components. These can be analysed separately.

Ignoring friction, the horizontal motion of a projectile is constant. The vertical motion of a projectile is subject to acceleration under gravity – so its velocity will continue to increase during the flight. Combining these velocities will lead to the curved path which is shown below:

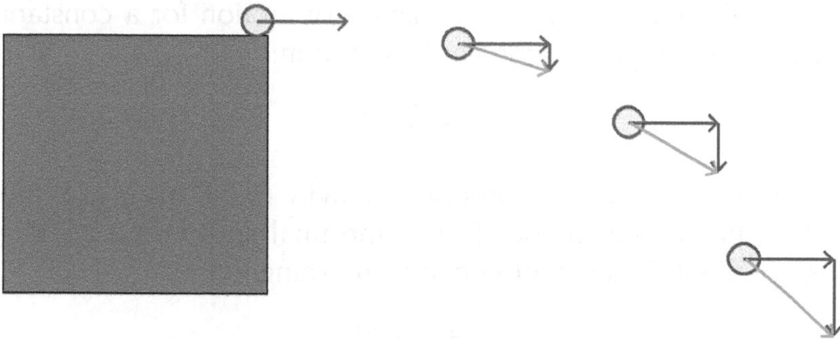

Fig 1.6.11

The resultant velocity (shown in green) is found by adding the vertical and horizontal velocities as vectors.

Horizontal motion

As the horizontal motion is constant, the equation linking velocity, distance and time applies:

$$d = vt$$

Here, t is the time of flight and d is the range of the projectile. As the horizontal velocity is increased, the range of the projectile is also increased. This result was also shown in Galileo's experiments above.

A graph of the horizontal motion of a projectile is shown below. The area under the graph will give the horizontal distance travelled.

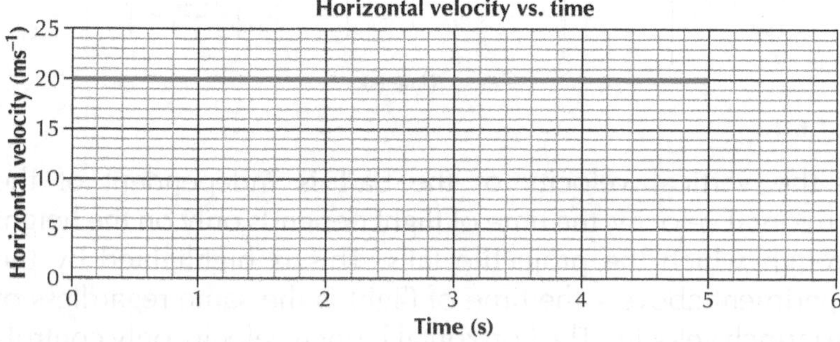

Fig 1.6.12

Vertical motion

The vertical motion of a projectile is subject to acceleration due to gravity. In this section, we will assume that acceleration due to gravity is positive. As discussed in the vectors section, we can do this by simply assuming that the downwards direction is positive. You can also assume acceleration due to gravity to be negative if the upwards direction is positive, as we have done in the section above with the thrown ball.

Just like the motion of a ball thrown vertically upwards, the vertical motion of a projectile will increase or decrease due to gravity. For this reason, the equation of motion for a constant acceleration applies to the vertical motion:

$$a = \frac{v - u}{t}$$

where a is the acceleration due to gravity (= 9·8 ms^{-2}), u is the initial speed (launch speed), v is the final speed, and t is the time of flight. This equation can be rearranged to:

$$v = u + at$$

A graph of the vertical motion of the projectile is shown below. It assumes the projectile is launched with zero vertical velocity initially (the same as in the experiments above). The area under the graph gives the height through which the projectile has fallen – notice that this is positive which shows we have taken the downwards direction to be positive.

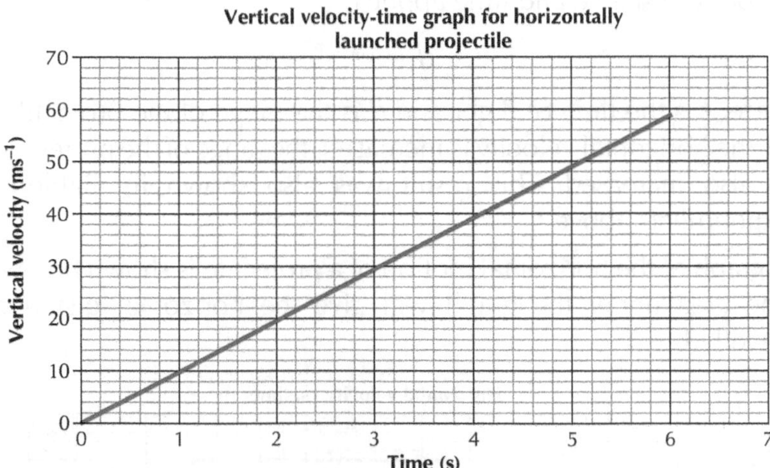

Fig 1.6.13

As the vertical velocity of the ball is independent of the horizontal velocity, the time of flight depends only on the height through which the projectile falls. This is highlighted by the experiment above – the time of flight is the same regardless of the launch velocity! The horizontal launch velocity only controls the range of the projectile.

Worked example

A cannonball is fired from the top of a cliff at an enemy ship. The cliff has a height, *h* metres, and the ship is a distance, *R*, away as shown in the diagram below:

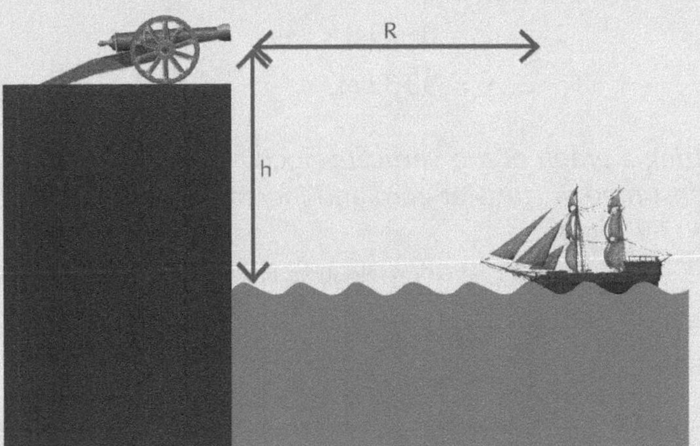

Fig 1.6.14

The cannonball strikes the ship a time of 3·4 s after it was fired. The initial velocity of the cannonball was 90 ms^{-1} horizontally.

a) Calculate the distance, *R*, the ship is away from the cliff.

b) Find the vertical velocity of the cannonball as it lands on the ship.

c) By plotting a graph of the vertical velocity against time, find the height, *h*, of the cliff.

a) *The distance, R, is a horizontal distance so we use the equation for horizontal motion (which has a constant velocity):*

$$d = vt$$
$$R = vt$$

Substitute what we know and solve for the distance, R:

$$R = 90 \times 3{\cdot}4$$
$$R = 306\ m$$

b) *The vertical velocity can be found by using the equation for vertical motion (which has a constant acceleration due to gravity):*

$$v = u + at$$

The initial vertical velocity, $u = 0\ ms^{-1}$.

(*continued*)

The acceleration is due to gravity so, assuming downwards is positive, $a = 9{\cdot}8\ ms^{-2}$

Solving for the final vertical velocity, v:

$$v = 0 + 9{\cdot}8 \times 3{\cdot}4$$

$$v = 33{\cdot}32\ ms^{-1}$$

c) *Plotting a graph of the vertical velocity of the cannonball against time, it starts at zero and increases to 33·32 ms⁻¹ after 3·4 seconds:*

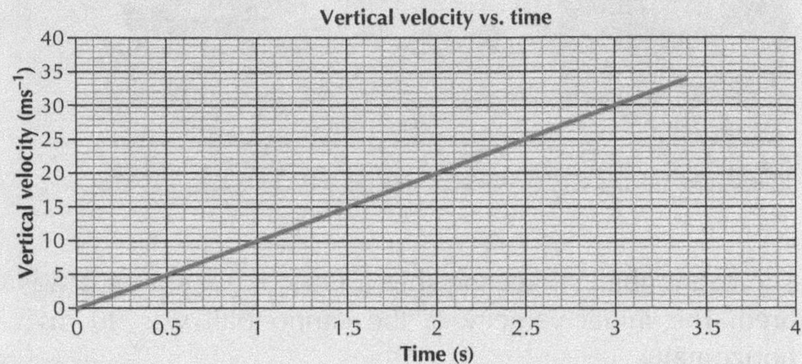

Fig 1.6.15

The height of the cliff can then be found by finding the area under the graph of the vertical motion:

$$h = Area = \frac{1}{2}bh$$

$$h = \frac{1}{2} \times 3{\cdot}4 \times 33{\cdot}32$$

$$h = 56{\cdot}64\ m$$

Exercise 1.6.2 Projectiles

1 A ball rolls off of a table at 7 ms^{-1} and lands on the ground 0·3 s later.

a) What is the initial vertical velocity of the ball?

b) How far does the ball travel horizontally? (i.e. what is the range of the projectile?)

c) What is the final vertical velocity of the ball?

d) Plot a graph of the horizontal velocity of the ball against time.

e) Plot a graph of the vertical velocity of the ball against time.

f) Use the graph in part (e) to find the vertical distance through which the ball fell.

g) The ball now rolls off the table at 14 ms^{-1}. Describe the effect this change has on:

(i) the time of flight

(ii) the range of the projectile

(continued)

2 An admiral wants to know the velocity at which he should fire a cannonball in order for it to reach an enemy ship. The cannon is situated on a cliff top. The time taken for a cannonball to drop into the water is measured to be 1·8 s. If the ship is 400 m away, with what velocity must the cannonball be launched to strike the ship? You may ignore air resistance.

3 An archer fires an arrow at a target which is 27 m away. He aims at the target so that the arrow leaves the bow level with the target, moving horizontally. The arrow leaves the bow with a velocity of 40 ms^{-1}.

a) Explain why the arrow does not hit the centre of the target.

b) How long does it take the arrow to reach the target?

c) How far below the target does the arrow hit?

Fig 1.6.16: *Archer firing an arrow*

4 A golfer strikes a golf ball such that it has an initial horizontal velocity of 15 ms^{-1} and an initial vertical velocity of 20 ms^{-1}.

a) What is the vertical velocity of the golf ball when it reaches its maximum height?

b) How long does the ball take to reach its maximum height?

c) By plotting a graph of the vertical velocity against time, find the maximum height reached by the ball.

d) What horizontal distance has the ball travelled in reaching its maximum height?

5 When taking a penalty kick, a rugby player kicks the ball with an initial horizontal velocity of 12 ms^{-1} and an initial vertical velocity of 8 ms^{-1}. Does the ball's maximum height allow it to clear the goal posts which are 4 m high?

Fig 1.6.17: *Rugby player kicking the ball*

6 A plane is flying horizontally with a velocity of 250 ms^{-1} when it drops a parcel to the ground. The parcel hits the ground 8 seconds later.

a) What is the initial horizontal velocity of the parcel?

b) What is the initial vertical velocity of the parcel?

c) Calculate the vertical velocity of the parcel when it hits the ground.

d) At what altitude was the plane flying?

e) What is the horizontal range of the parcel?

f) At what position above the parcel will the plane be when the parcel lands, assuming it does not change its course or velocity?

Satellites

Satellites affect our day to day life without us even noticing. Satellites make worldwide communication possible. We use global positioning satellites to help us find our way with a satellite navigation system. The same global positioning satellites guide aircraft around the globe and are now so precise they can be used to help land a plane automatically!

Satellites come in many forms, some are natural and some are man-made. A satellite is essentially a projectile in orbit around a star or planet. Pretty much the same rules apply to satellites in orbit to those in free-fall near the surface of the Earth. That is, they both involve a constant acceleration towards the star or planet as well as a constant horizontal velocity. As mentioned previously, these two motions acting simultaneously results in a satellite falling towards the celestial body (planet or star) it is orbiting, but at the same time the celestial body curves away from the satellite at the same rate as it falls towards it thus ensuring that the satellite remains in orbit. There is a very important balance between the constant horizontal speed and the constant vertical acceleration, without which the satellite would either fly off in a straight line (if the horizontal speed is too large) or fall directly down towards the celestial body (if the gravitational pull is too large). By getting the balance between these two factors just right, satellites can remain in orbit around a celestial body.

Fig 1.6.18: *The moon is a natural satellite of the Earth*

Natural satellites

When we think of satellites, we often think of the man-made ones which are launched to carry out specific jobs. However, there has been a satellite in orbit around Earth for billions of years – the moon! The moon is said to be a natural satellite of the Earth. In turn, the Earth is also a natural satellite of the sun.

Using technologies developed for exploring the universe, astronomers have been able to track the development of the solar system. The moon was formed during the early stages of the development of the solar system. When the Earth was a young planet, another planet collided with it. The collision blasted the Earth's crust into orbit around the Earth and destroyed the other planet. The crust in orbit around the Earth coalesced (joined together) and formed the moon which remains in orbit around the Earth to this day.

Fig 1.6.19: *Newton's thought experiment*

A history of satellites

The concept of putting an object into orbit around the Earth was considered by Newton in his cannonball experiment. He visualised a cannon firing from a high mountain at a speed fast enough that the cannonball doesn't fall back to Earth but slow enough that it does not escape. Instead, it does a full circle round the Earth as shown in the diagram.

It was 1957 when the first man-made satellite was launched. Sputnik 1 was designed and built by the Russians, and launched on 4th October. Sputnik 2 was launched just one month later, and carried a passenger – a dog named Laika. The success of the Russian satellites prompted the space race with the USA.

Fig 1.6.20: *Sputnik 1*

Group exercise: A history of satellites

Since first launched in 1957, satellites have become an important part of all of our lives – yet we take their presence for granted.

Research the history of satellites and present your results accordingly.

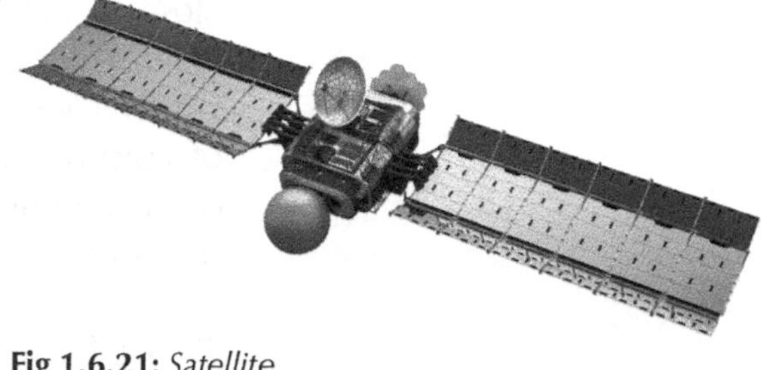

Fig 1.6.21: *Satellite*

Uses of satellites include:

Weather forecasting

Meteorological satellites are used primarily to monitor weather patterns around the world which allows predictions of weather fronts advancing long before they actually occur. These satellites are often polar satellites which can monitor changes in the atmosphere more easily than non-polar satellites. In addition to phenomena such as tsunamis, hurricanes, tornados, polar ice melting, wildfires, drought monitoring, sea level changes, sand and dust storms these satellites can also be used to monitor volcanic eruptions producing plums of dust which is very important for aircraft which would normally fly through these regions.

Global positioning

Global positioning satellites or GPS can be used to accurately pinpoint the position of individuals or mobile transportation such as planes or ships. It can be used to provide emergency services with exact locations of an individual in need of assistance. It can be used to track mobile phones, luggage, laptops etc...It can even provide directions for individuals using mobile phones or in-car sat-navs. There are hundreds of applications which can use GPS technology.

Scientific discovery

Many scientific discoveries have been made using satellite technology. These include environmental factors such as the melting of the polar ice caps resulting in rising sea levels as well as ocean pollution. The detection of depleted regions of the ozone layer in the atmosphere has also led to concerns regarding chlorofluorocarbon (CFC) release into the atmosphere since this has resulted in an increase in certain types of cancers particularly in the southern hemisphere. The Hubble telescope and the ISS are also used to monitor the movement of meteorites passing close to the Earth as well as discovering exoplanets and new regions of star formation. They can also monitor solar activity which is very important since a solar storm could result in a loss of communications around the world as well as damage to other satellites. Much more understanding of the universe is one direct result of the use of telescopes such as 'Hubble'.

Orbit height and geostationary satellites

Nowadays there are many satellites orbiting the Earth. Some of them orbit close to the Earth's surface, while others orbit further away. The time it takes for a satellite to orbit the Earth, known as the period, depends on the height of the satellite above the surface of the Earth. The greater the height, the longer the period of the orbit. This is partly because the distance covered by the orbit is greater, and partly because the speed of the satellite is slower for higher orbits.

Fig 1.6.22a: Sir Arthur C Clarke

Geostationary satellite

A geostationary satellite is one which takes 24 hours to orbit the Earth. In other words, it is travelling around the Earth at the same speed as the Earth is rotating. This means that a geostationary satellite will stay above the same point on the Earth's surface. This allows the satellite to be easily used for communication as it will always be in the same place relative to Earth.

In order for geostationary satellites to maintain a 24 hour synchronous orbit of the Earth, Sir Arthur C Clarke worked out in 1945 that they must be placed at a height of 36,000 km above the Earth's equator. By using a series of these satellites 24 hour around-the-world communication is possible.

Fig 1.6.22b: *Sports commentator*

Satellite communication

Satellites are used today in communication applications. You may have watched a broadcast from a sporting event far away on another continent. There is often a delay between the commentator asking a question and an answer coming. This is because the transmission is via satellite and it takes time for the satellite signal to travel from the commentator via the satellite to the sports person at the event.

The first transatlantic transmission involving a satellite occurred in 1962 using a satellite called Telstar. As this satellite was in a low orbit, it travelled round the Earth quickly. It was not a geostationary satellite and thus was only in a suitable place for satellite transmission for a short time. Nevertheless, a short satellite transmission from America was successfully received in both Britain and France using the Telstar satellite.

Fig 1.6.23: *Satellite dish*

Curved reflector

A satellite dish has a very familiar shape (shown opposite). We see them attached to the walls and roofs of many houses and probably don't give them a second thought! However, the shape of a satellite dish is very special and gives it some very useful properties!

Experiment: The curved reflector

This experiment demonstrates the effects of a curved reflector on incoming rays of light.
You will need:

- A ray box and power supply
- A three slit mask for the ray box
- A curved reflector

Instructions

1. Fit the three slit mask to the ray box.
2. Connect the ray box to the power supply and switch on. Ensure the ray box is producing three parallel rays of light. If not, a collimating lens may be required.
3. Direct the three parallel rays of light at the curved reflector.
4. Draw a diagram of the effects of the curved reflector on the rays of light.
5. Describe the effects of the curved reflector on the three rays of light.

The diagram on the left shows the effect of a curved reflector on parallel rays of light coming from a distant source. It could also be radio waves or microwaves hitting the reflector as the physics is the same!

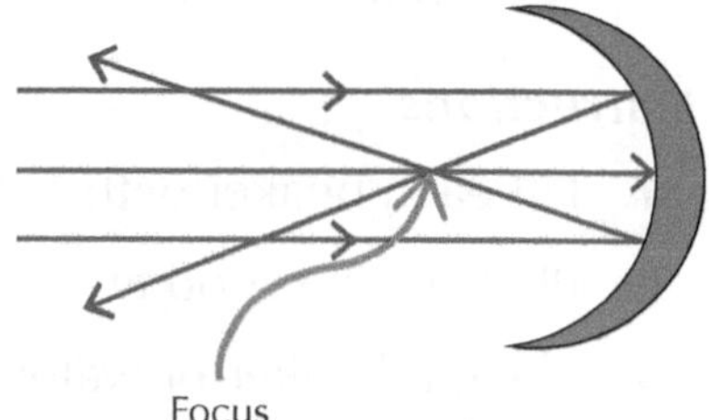

Fig 1.6.24

The curved reflector reflects all of the rays of light hitting it back through the same focus point. Weak signals can be collected over a large area and focused to a single point to give a much stronger signal. The effect is similar to collecting rain water – if

Fig 1.6.25: *Curved reflector with a large surface area*

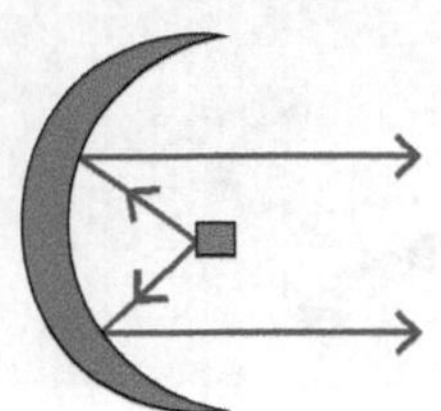

Fig 1.6.26: *Curved reflector*

Fig 1.6.27: *Car headlight*

you collect over a large area, you will collect more water. For a curved reflector, if you have a large surface area you will collect a large amount of signals and then reflect them all to the focus point to give you a stronger resulting signal. This is why the radio telescopes used for SETI (Search for Extra Terrestrial Intelligence) have traditionally got very large dishes – they are searching for a weak and distant signal so they need to collect over a wide area.

A curved reflector can also be used in a transmission system. Light rays are reversible, so placing a transmitter at the focus of a curved reflector will result in the signal being transmitted as a beam in a straight line as shown here.

This effect is used in practice in a car's headlight. A curved reflector at the back of the headlight reflects the light from the bulb out into a beam – this ensures the light is sent out and directed towards the road in front of the car where it is needed.

Research exercise: Curved reflectors for SETI

Large curved reflectors can collect signals over a large area. This allows weak signals to be focused to one point, so very weak signals can be detected.

Use the internet to research the curved reflectors used for SETI. What are the typical dish sizes?

GO! Experiment: Solar heating

This experiment demonstrates the use of a curved reflector for heating water using light.

You will need:

- Two high power light bulbs
- A power supply to run power to the bulbs
- Two tripods
- Two glass beakers
- A thermometer
- A curved dish reflector

Instructions

1 Fill each beaker with 100 ml of cold water.

2 Place each tripod an equal distance from the light source.

3 Place a beaker of water on each tripod, ensuring the distance between the beaker and the light source is equal for each beaker.

(*continued*)

4 Place a curved reflector behind one of the beakers.
5 Measure the starting temperature of the water in each beaker using a thermometer.
6 Switch on the bulbs and leave the experiment to run for 10 minutes.
7 Carefully measure the end temperature of the water in each beaker using the thermometer. In which beaker was the water the hottest? Can you explain this result?
8 Can you make the water boil using the curved reflector? Take care – the water will be extremely hot!

A solar heater uses a curved reflector to focus the sun's rays to a single point. The curved reflector can gather the energy from the sun over a wide area and focus it to a single point where the object to be heated is placed. This results in a large amount of energy being focused on a relatively small spot. This intense energy can be used to heat objects. Some solar furnaces can be used to melt metals!

Long distance communication

Satellites in orbit around the Earth are used to assist long range communication across the globe. One of the main issues with long distance communication is that there is no direct line of sight between cities which are far apart. This is because the Earth is round – the curvature of the Earth gets in the way! A satellite in orbit is in direct line of sight of both the transmitter and the receiver so can be used to relay the signal between two distant places as shown in the diagram.

It is important to note that the satellite in orbit does not reflect the signal. Instead, it receives the signal and then re-transmits it to the receiver on Earth.

The time taken for the signal to reach the receiver depends on the distance the signal travels and the speed of the signal:

$$d = vt$$

This is the same equation as we would use for the average or instantaneous speed of objects considered in Chapter 12. All radio signals travel at the speed of light, $v = 3 \times 10^8$ ms^{-1}.

Fig 1.6.28: *Using satellites to send signals around the globe*

Key point

A satellite in orbit receives the signal sent up from Earth. It then re-transmits the signal back down to the receiving station on Earth. Satellites do not reflect the signal!

Key point

All radio signals travel at the speed of light:

$v = 3 \times 10^8 \text{ ms}^{-1}$

Technique: Calculations involving signal transmission

1. Write down what you know from the question, leaving blank what you are trying to find:
   Distance = ____ m (remember to convert km to m!)
   Time = ____ s (or hours)
   Signal speed = $3 \times 10^8 \text{ ms}^{-1}$
2. Write down the equation linking the quantities above as you see it on the formula sheet:
   $$d = vt$$
3. Substitute into the equation what you know.
4. Solve for the unknown.

Worked example

A radio signal is transmitted across the globe using a satellite as shown. The total distance travelled by the signal is 40 000 km. Calculate the time taken for the signal to reach the receiver. Assume the signal is re-transmitted instantaneously by the satellite.

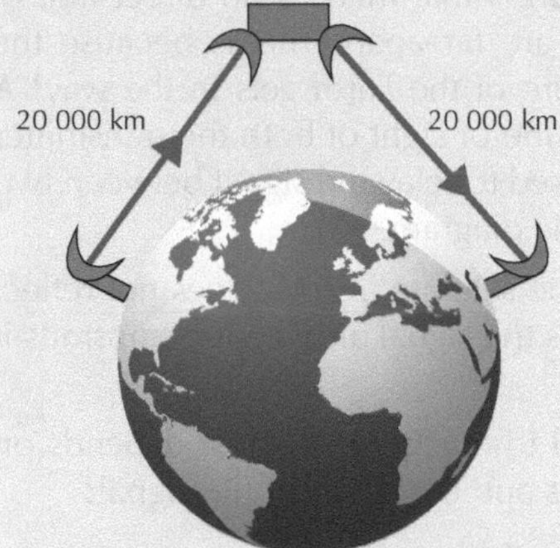

Fig 1.6.29: *The transmission of radio signals by satellite*

Write down what we know and what we are looking for:

Distance, d = 40 000 km = 40 000 000 m

Speed $v = 3 \times 10^8 \text{ ms}^{-1}$

Time, t = ? s

We are using the equation which links speed, distance and time:

$$d = vt$$

(continued)

Substitute what we know:

$$4 \times 10^7 = 3 \times 10^8 t$$

Solve for the unknown:

$$t = \frac{4 \times 10^7}{3 \times 10^8}$$

$$t = 0{\cdot}13\ s$$

Exercise 1.6.3 Satellite communication

1 Calculate the time taken for a radio signal to travel:

a) 50 000 m

b) 450 km

c) 29 000 km

d) 5×10^6 km

2 A satellite signal is sent from a sporting event to another country across the globe. It travels a total distance of 72 000 km as shown in the diagram. How long does it take for the signal to travel from the transmitter to the receiver?

3 A GPS satellite is directly above the surface of the Earth. It orbits at a height of 2400 km. How long does it take for a signal to travel from the satellite to a GPS receiver on Earth?

4 A signal takes 40 ms to travel from the transmitter to receiver via a satellite. Calculate the total distance travelled by the signal?

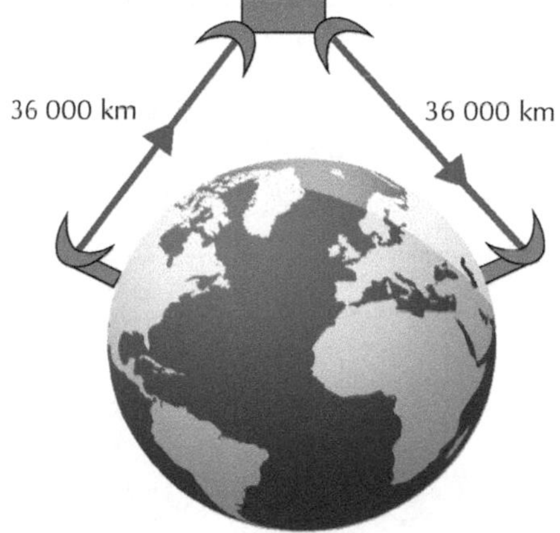

Fig 1.6.30

5 During a TV broadcast of a news event across the globe, a signal is transmitted from the USA to China. It takes 0·9 s to travel up to the satellite and back down to the receiver. The distances between the satellite and the receiver and satellite and the transmitter are equal. Calculate them.

Learning checklist

In this section you will have learned:

- How to explain projectile motion in terms of constant vertical acceleration and constant horizontal velocity.
- How to use appropriate relationships to solve problems involving projectile motion from a horizontal launch, including the use of motion graphs.
- How to explain satellite orbits in terms of projectile motion, horizontal velocity and weight.
- The benefits of satellites in weather forecasting, GPS, communications, scientific discovery and space exploration.
- The facts surrounding the position and use of geostationary satellites.

7 **Space exploration**

- Current understanding of the universe
- Space terminology
- Optical telescopes
- The challenges of space travel
- The risks associated with manned space exploration

8 **Cosmology**

- Terminology
- Big Bang theory
- Age of the universe
- Using the e-m spectrum to analyse astronomical objects
- Line spectra
- Identifying star composition

AREA 2

Space exploration and cosmology

7 Space exploration

In this chapter you will learn about:

- Space exploration, including rockets.
- Optical telescopes – how the process of refraction is applied to produce an instrument that can be used to view stars billions of miles away.
- How to use 'light' do determine the composition of astronomical objects.
- Space exploration – the use of technologies arising from space exploration.
- The challenges of space travel.
- The risks associated with manned space exploration.

Word bank

- **nebula(e)**
floating dust cloud(s) in space where stars are born

The universe

The universe is everything around us. Planets, moons, stars, galaxies and **nebulae** all make up what we call the universe. Study of the universe helps our understanding of how it is changing and how new galaxies, stars and solar systems come into existence. Study has also allowed us to estimate how long the universe itself has been in existence.

Fig 2.7.1: *Planet Earth*

Word bank

- **natural satellite**
an object in orbit around a planet or star which is not man-made, e.g. the moon is a natural satellite of the Earth

Planets and moons

A planet is an object that orbits a star. Earth is an example of a planet. Planets form after the birth of a star. Stars form within a nebula. Any left over material in the nebula settles into a rotating disc about the star. Within this rotating disc small bodies form, collide, and eventually coalesce to form a planet.

Planets in the solar system can be grouped into terrestrial (rocky) planets of which Earth and Mars are examples, and gas giants, such as Jupiter and Saturn.

A moon is an object which orbits a planet. Moons are **natural satellites** to a planet.

Planets and moons do not emit light. They can only be seen in space because they reflect the light from the star they are orbiting.

Exoplanets

An exoplanet or extrasolar planet is a planet which is not a part of our own solar system. Over 700 such planets have been discovered so far, the first one being discovered in the early 1990s. Almost all of the exoplanets discovered so far are the same size as Saturn and Jupiter, though this does not mean this is the case for every exoplanet.

Fig 2.7.2: *An exoplanet*

There are several methods that can be used to detect an exoplanet. One of these is the transit method where astronomers detect very small changes in the brightness of stars. When a planet passes in front of its star, it causes the light level from the star to drop slightly. The changes in brightness can be observed over the course of years to prove that a planet is orbiting a star. This research has only been made possible by recent advancements in technology which allow astronomers to measure these very small variations in brightness.

Research exercise: Exoplanets

Since the early 1990s, astronomers have been discovering exoplanets around stars.

Use the internet to research exoplanets that have been discovered to date. What technological advances are assisting the discovery of exoplanets?

Dwarf planets

Fig 2.7.3: *Pluto*

A dwarf planet is a celestial body resembling a small planet but lacking certain technical criteria that are required for it to be classed as such. It can be found in orbit around a star, having enough mass to produce a gravity which is able to sustain a nearly round shape (some even have their own moons). There are five officially recognised dwarf planets in our solar system, Pluto (which prior to 2006 was considered the ninth planet orbiting our Sun), Ceres, Makemake, Eris and Haumea.

Asteroids

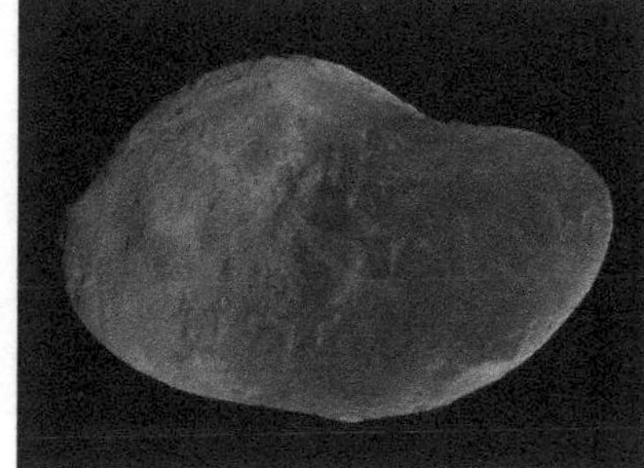
Fig 2.7.4: *An asteroid*

Asteroids are rocky materials which float around in space orbiting a star varying in size from a few metres to hundreds of miles across. They are often grouped according to their composition. Many examples of these can be found between the orbits of Mars and Jupiter (in an area called the 'asteroid belt', though some do have more eccentric orbits. It is believed that these are the left over remnants from the formation of the inner solar system. There are three main types – C-types (chondrites) which are made of clay and silicates, S-type (stony) which are made mostly of silicates and nickel-iron, and finally M types (metallic) which are made from metallic nickel-iron (the same composition as the core of the inner planets). Speculation also resides in the fact that since the composition of these asteroids is very similar to that of the rocky planets in the inner solar system that many of them may have come about as the result of a collision between a rocky planet further out than Mars is today and Jupiter very early in the formation of the solar system. Some interesting facts about asteroids include:

Asteroid impacts with the Earth are a major concern since one day it could lead to the destruction of life on Earth. Asteroid impacts were very numerous in the early days of the formation of the solar system but are much less common today (mainly thanks to Jupiter swallowing so many of them when they approach too close). It is thought that an asteroid impact about 65 million years ago led to the extinction of the dinosaurs. Some asteroids are comets which have lost their icy coating leaving behind a rocky surface. Some asteroids are so large that they even have their own moons and as a consequence are often referred to as minor planets or planetoids.

Fig 2.7.5: *The sun*

Stars

A star is a large ball of gas which is held together by gravity. Fueled by nuclear fusion, the energy released by stars generates large amounts of heat and light. At the core of a star, hydrogen is fused together to form helium. The central temperature of the star governs the rate of fusion in a star. Heavier elements can also be formed through the process of fusion.

Our nearest star is the sun which is responsible for all life on Earth. The sun is a relatively small star. Its small size makes life on Earth possible – the radiation from much larger stars would make life impossible.

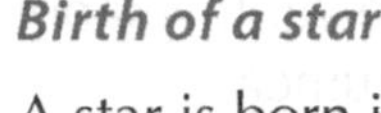

Birth of a star

A star is born in a nebula – a giant cloud of dust that is floating in space. Nebulae can form very interesting shapes. Perhaps the best known is the Horsehead Nebula. The nebula contains the matter that is required to form a star. As a nebula collapses a star can form once enough mass has come together. Eventually the protostar will have sufficient mass that the temperature and pressure will be high enough for fusion to begin.

Fig 2.7.6: *Horsehead Nebula*

Only by viewing infrared radiation coming from inside the nebula has the birth of stars been witnessed and allowed us to understand this process.

Death of a star

Throughout a star's life, it fuses hydrogen at its core into helium. When a star runs low on hydrogen, the fusion process slows down. When this happens, the nuclear fusion no longer pushes the star outwards with the same force and gravity begins to take over. The star collapses inwards, increasing the temperature at the core of the star. This results in a new pulse of energy from nuclear fusion which causes the star to grow outwards into a red giant. When our own sun becomes a red giant, it will engulf the inner planets of the solar system, including the Earth. It will be the end of the world as we know it!

A red giant star is unstable and undergoes pulsations. The outer layers are blown off as the star pulsates. What remains at the centre will be pure carbon. This is known as a white dwarf, a cosmic diamond that is the end result of the death of a star.

Stars which are much larger than our sun will usually die in a different way. The fusion at the centre of large stars can form heavier and heavier elements. It can keep fusing until it forms iron. Iron cannot be fused into another element, and it will result in fusion stopping instantly at the centre of the star. The resulting quick collapse of the star will cause a supernova explosion which ejects material into space which will go on to form more stars and planets.

Fig 2.7.7: *Our solar system*

Solar system

A solar system is a system of planets orbiting a star. The Earth is part of a solar system orbiting the sun. Our solar system contains seven other planets. These planets are, in order of their distance from the sun:

- Mercury (closest)
- Venus
- Earth
- Mars
- Jupiter
- Saturn
- Uranus
- Neptune (furthest)

Many stars have their own solar systems. Research into the development of other solar systems has led to discoveries about how our own solar system developed.

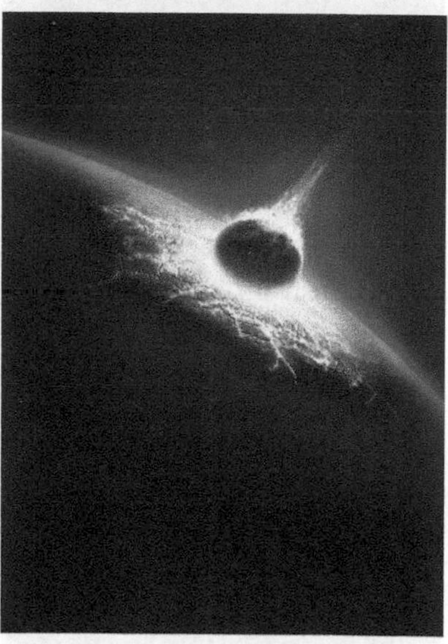

Fig 2.7.8: *Asteroid colliding with a planet*

In the early solar system, there were many more planets than there are now. They collided into each other, and the collisions destroyed some planets and made others larger. These collisions were also responsible for producing moons.

Galaxy

A galaxy is a collection of stars orbiting a central point. Recent research into the motion of stars close to the centre of galaxies has allowed us to conclude that at the centre of a galaxy is a super massive black hole. There are many galaxies in the universe of varying shapes and sizes. The Milky Way is the galaxy within which our own solar system exists.

Fig 2.7.9: *A galaxy is a collection of stars orbiting a central point*

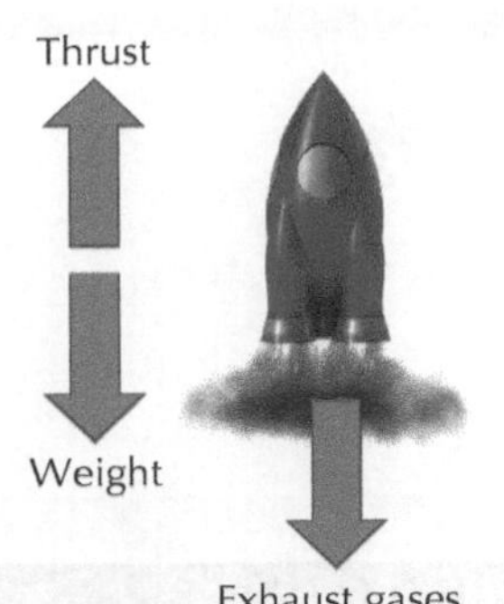

Fig 2.7.10: *Forces acting on a rocket*

Fig 2.7.11: *Rockets used to power missiles*

Fig 2.7.12: *Sputnik1*

Fig 2.7.13: *Rocket on launch pad*

Fig 2.7.14: *Moon landing*

Space exploration

Looking at the night sky on a clear night, it is possible to see many stars and planets. Occasionally we may see meteors, also called shooting stars. Studies of astronomy over the years have revealed that we are just a tiny part of a very large universe and humans have long pursued investigating this universe through space travel.

Rockets

We have already investigated the physics of rockets as an example of the applications of Newton's laws. By forcing exhaust gases downwards there is an equal and opposite force upwards called thrust. When this thrust is greater than the weight of the rocket (as shown on the diagram opposite), the unbalanced force upwards causes an upwards acceleration according to Newton's second law,

$$F_{UN} = ma$$

In space, the rocket can continue moving without its engine. According to Newton's first law the rocket will continue moving at a constant speed unless acted on by an unbalanced force. In the vacuum of space, there are no resistive forces to slow the rocket down.

STEP BACK IN TIME: HISTORY OF ROCKETS

The story of modern rockets has its foundations in Newton's laws of motion, published in his book Philosophiae Naturalis Principia Mathematica. *Rockets were not initially used for going into space. Instead, they were used for warfare, competing with cannons. To this day, rockets are still used to power missiles used in combat.*

The space age began in earnest after the second world war, using rockets that were designed as weapons for the war – the German Vergeltungswaffe 2 *(V2). The V2 was the foundation of ballistic missile development that led to a manned space programme. Russia had a space programme, and engaged with the USA in a 'space race'. The Russians took an early lead when they launched Sputnik 1 in 1957.*

In 1961, Yuri Gagarin became the first human to enter space. The rocket, Vostok 1, *propelled him into space to a maximum altitude of 315 km. His flight lasted nearly two hours and on re-entry, Gagarin ejected from the capsule at a height of 6100 metres and parachuted safely to Earth. Later in 1961, American astronaut Alan Shepard Jr made a suborbital flight in the space capsule* Freedom 7 *which was attached to a Redstone rocket for propulsion. It wasn't until 1962 that an American went into orbit courtesy of the* Atlantis.

In 1969, American astronaut Neil Armstrong made history by being the first man to walk on the moon as part of the Apollo 11 *mission. This was the first of six moon landings between 1969 and 1972.*

Word bank

- **ascent stage**

when the rocket is gaining altitude

Word bank

- **descent stage**

when the rocket is coming back down to a planet's surface

Research exercise: Rockets today

Use the internet to research the use of rockets today in modern space travel, and their current applications in warfare. Present your findings in a suitable format.

The challenge of space travel

So far, mankind has only travelled as far as the Moon but it is hoped that as technology advances, one day mankind will venture to further outreaches such as Mars and beyond. There are many benefits of such explorations since this may eventually lead to colonisation of other worlds as well as extraction of valuable mineral resources which could be used to benefit mankind.

There are however many challenges to space travel such as travelling large distances within a suitable timeframe. These will require very high velocity rockets. One suggestion is that the rockets use 'ion drive' which produces a small but sustainable force over extended periods of time. Another proposal involves using the 'catapult effect' produced by fast moving asteroids or even planetary bodies. This technique was used very successfully by the two Voyager spacecraft both launched in 1977 and has allowed Voyager 1 to actually leave our solar system. It is hoped that Voyager 2 will eventually follow suit.

Another challenge currently facing astronauts when manoeuvring in space is how to control the direction in which the spacecraft travels. This is particularly challenging when a spacecraft docks with the ISS (International Space Station). This requires manoeuvring of the spacecraft in the zero friction environment of space. This is necessary to allow the docking of the spacecraft with an ISS docking port. The procedure is carried out manually by firing small gas canisters located around the outside of the spacecraft to provide the thrust required to change direction and using a joystick to manoeuvre the ship into place. Technically the ship must be turned around and reversed into the docking port.

When travelling long distances in space, one way to conserve energy is to use the 'catapult' or 'slingshot' effect from a fast moving asteroid, moon or planet. This involves the use of the gravity and motion of the asteroid/moon or planet to pull a spacecraft into a new path without using additional valuable energy. One of the most recorded uses of this means of propulsion involved Voyager 2 which toured vast regions of our solar system. It relied particularly on the gravitational assist of Jupiter. The spacecraft's arrival was carefully timed so that it would pass behind Jupiter in its orbit around the Sun. As the spacecraft came into Jupiter's gravitational field, it fell towards Jupiter, increasing its speed to a maximum at its closest approach. Since all masses in the universe attract each other, Jupiter sped up the spacecraft substantially, and the spacecraft slowed down Jupiter in its orbit by a tiny amount. At this point, Voyager 2 had been sped up enough by Jupiter's gravity to get a speed greater than Jupiter's escape velocity. As it left, it slowed down again, but it never slowed all the way to the speed it was before getting to Jupiter. It consequently gained some kinetic energy from the planet which allowed it to change direction and travel further. As it left region around Jupiter, Voyager 2 travelled further out of the solar system passing other gas giants such as Saturn and Uranus. The same slingshot effect just described was also used with both of these planets.

Gravity assists can be also used to decelerate a spacecraft by flying in front of a planet or moon. When the unmanned Galileo spacecraft was launched in October 1989 it first used Venus and then the Earth to provide the initial slingshots before eventually arriving at Jupiter in December 1995. When it arrived it passed in front of one of Jupiter's moons, Io, which caused it to decelerate and enter into orbit around Jupiter. This then allowed Galileo to study more carefully the moons of Jupiter as well as the planet itself.

One major concern regarding long-term space travel is the provision of sufficient energy to sustain life support systems on board the spacecraft. One proposal involves using solar cells which have an area that can vary as the spacecraft ventures further from the Sun. These would be used to capture sunlight which is then converted into other forms of energy, some of which will be used to sustain life support systems on board the spacecraft. Another proposition (which still requires energy) is cryogenic technology where astronauts are put into a hyper sleep which slows down their respiration rate allowing them to travel greater distances without the need to carry a lot of provisions. This however is still very far off in the future...

The risks of manned space exploration

There are several risks that are associated with manned space travel:

A considerable amount of fuel is required in order to provide the necessary energy for a rocket to escape the Earth and achieve enough altitude to reach space. For example, each of the two Solid Rocket Boosters on the Space Shuttle carried more than **one million pounds** of solid propellant. Other types of rockets use a liquid propellant. At lift-off, a spacecraft can carry 835,958 gallons of the principle liquid propellants: hydrogen, oxygen, hydrazine, monomethylhydrazine, and nitrogen tetroxide, the total weight being 1,607,185 pounds. If a fault should develop during lift-off there is sufficient propellant to incinerate the rocket and all on board!

On Earth mankind is protected from high levels of radiation exposure from the Sun and beyond by the Earth's own magnetic field. This magnetic field can deflect sub-atomic particles which would otherwise result in high levels of radiation exposure. Once in space however astronauts do not have the protection of the Earth's magnetic field and are therefore subject to much higher levels of radiation than back on Earth. The radiation can take many forms, but is generally made up of sub-atomic particles, x-rays and gamma rays from the Sun and beyond. The sub-atomic particles will over time tear through DNA molecules splitting them up or altering the replication process required to repair and replace damaged or obsolete cells. Damaged DNA can result in cancers and other diseases which could prove fatal for any crew on board a spacecraft. Whether it is low level radiation over long periods of time (chronic radiation exposure) or high levels of radiation over short periods of time (acute radiation exposure), the effects can be devastating.

In space the pressure outside of a spacecraft is so low that many consider it as non-existent. It has a pressure of about 1.3×10^{-11} Pa. By comparison, on Earth (at sea level) air pressure averages approximately 1.01×10^5 Pa, which is about 8×10^{15} times greater than that in space. This huge pressure difference can cause catastrophic effects if for example the hull of a satellite or a spacesuit was to be punctured by debris such as space junk or small meteorites orbiting the Earth at extremely high speeds. Currently it is estimated that there are over 19,000 pieces of space junk out there, some which are very small and others as big as a house. If the skin of a spacesuit was punctured the pressure difference would cause the suit to depressurise losing air rapidly. Chances are though, that the item puncturing the suit would probably cause considerably more damage to the astronaut travelling faster than a bullet. Depending upon the size of the hole or tear an astronaut would attempt to reach an air-lock as quickly as possible. If the hole was particularly large then the astronaut would instantly internally liquefy as the result of the bodily fluids boiling and a rapid freezing of the body would occur. Fortunately to-date this has never occurred...

When a spacecraft returns to Earth it must enter the Earth's atmosphere at a very shallow angle of inclination between about 5° and 8°. If it comes in too steeply it would burn up on re-entry and if it entered too shallow then it would bounce off the Earth's atmosphere and be propelled back into space. When a spacecraft enters at the correct angle it encounters air resistance which causes friction resulting in the spacecraft slowing down and also (particularly in the nose cone) a rapid build-up of heat energy resulting in an increase in temperature of the outside of the spacecraft. To protect the spacecraft and its occupants special heat-resistant coatings (TPS) are used to dissipate the heat to the surroundings. The composite of materials used can withstand temperatures in excess of 1,260 °C (see the chapter relating to 'specific heat capacity' and 'latent heat' later on in the 'Properties of Matter' topic).

Obtaining information about astronomical objects

Optical telescopes

For hundreds of years, the human race has been fascinated with space. The question of whether or not we are alone in the universe causes much debate. Studying signals from space allows us to search for life in distant worlds as well as helping us to understand the foundations of our universe and how the solar system came into existence.

The first optical telescopes to be recorded were in the Netherlands in the early 17th century. Since then they have been developed and optimised to allow us to view objects many light years away. This has furthered our knowledge of the universe and allowed us to study the development of solar systems and the birth of stars.

There are three main types of optical telescope. The **refracting telescope** is a simple design consisting of two lenses – one which gathers the light from the object and focuses it and a second which magnifies the image so that it can be viewed. Such a design is also used for long focus camera lenses. The refracting telescope is considered in more detail below.

The **reflecting telescope** was invented in the 17th century as a replacement for the refracting telescope. It uses a mirror to gather light from a distant object and focus it onto an eyepiece

Word bank

- **optical telescope**
an instrument used to produce an image of a visible light-emitting object (e.g. a star) which is a very large distance away

Fig. 2.7.16: *Optical telescope*

that can be used to magnify and view the image. The majority of telescopes used in astronomy are sophisticated reflecting telescopes. Many use additional optics to improve the quality of the image.

Catadioptric telescopes are preferred by astronomers and scientists. They use specially shaped mirrors and lenses to improve the images and allow users to view objects that are dimmer and deeper into space. However, their complexity means they are used less often than reflecting telescopes.

The refracting telescope

We have seen how light **refracts** when it passes through a lens. A refracting telescope has two lenses:

- Objective lens – this gathers the light from distant objects and focuses it inside the telescope. The bigger the objective lens, the greater the amount of light gathered, allowing a deeper look into space.
- Eyepiece lens – this acts like a magnifying glass to enlarge the image of distant objects.

Consider the ray diagram for a simple refracting telescope.

The objective lens forms an image of the distant object inside the telescope. This image is very small, and it is also upside down. A second eyepiece lens magnifies the initial image so that it can be easily viewed.

The ideal separation of the two lenses is equal to the sum of the focal lengths of the lenses. This is because the first lens forms the image at a distance equal to one focal length away. A magnifying lens forms its largest image when held just under one focal length away from the object.

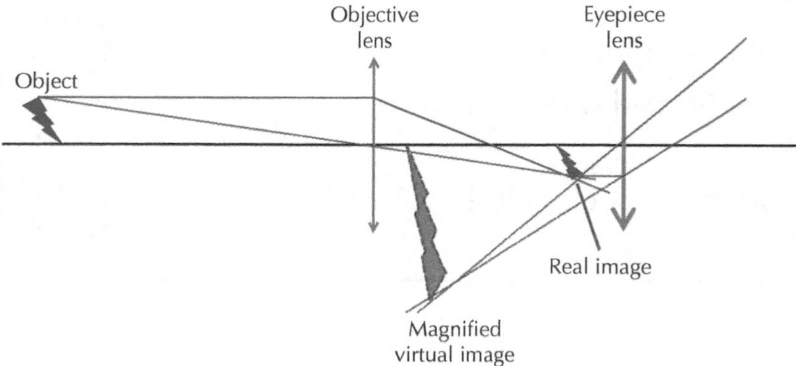

Fig 2.7.17: *Ray diagram for a refracting telescope*

Word bank

- **refract**
  light slowing down and in certain circumstances changing direction when travelling from one medium to another

Think about it

Consider the effects of the size of the objective lens on the image produced by the refracting telescope. What is the effect of making the diameter of this lens bigger?

Experiment: The refracting telescope

This experiment builds a refracting telescope.
You will need:

- Cardboard tubes (from kitchen roll)
- Convex lenses
- Sticky tape
- Ruler
- Screen

Instructions

1 Measure the focal length of the objective lens using a ruler and the screen – instructions for how to measure this focal length can be found in Chapter 19.
2 Measure the focal length of the eyepiece lens.
3 Calculate the total length of the telescope by adding together the focal lengths of the magnifying and eyepiece lenses.
4 Use sticky tape to securely fit the lenses to either end of the cardboard tube. The tubes can be fitted into each other.
5 Fit the cardboard tubes together and adjust the length until equal to the length calculated in part 3 above.
6 Use the telescope to look at a distant object. **Never use the telescope to look at the sun!**

Learning checklist

In this section you will have learned:

- A basic awareness of our current understanding of the universe.
- How to use the terminology associated with space exploration correctly and in context.
- An awareness of the challenges of space travel.
- An awareness of the risks associated with manned space travel.
- How to obtain information about astronomical objects using telescopes.

8 Cosmology

In this chapter you will learn about:

- The term 'light year' and the conversion between light years and metres.
- The age of the universe.
- The universe - what makes up the universe and theories as to the beginning of the universe.
- Spectra – emission spectra and absorption spectra and how these can be used to identify the elements that make up a star.

Distances in space: the light year

Distances in space are, quite literally, astronomical! Our nearest neighbour in space, the moon, is on average 384 400 km away. The sun is a massive 149 600 000 km away! Our solar system is only a very small part of the Milky Way galaxy. The Milky Way galaxy is just one of many observable galaxies in the universe. Clearly, the use of metres or kilometres to describe the distances in space is not going to be adequate!

For this reason, astronomers developed a new unit to measure distances in space – the **light year (Ly)**. One light year is equal to **the distance light will travel in one year**, measured in metres, *m*. The key to working with light years is to understand the light year is derived from the familiar equation for distance, speed and time:

$$d = vt$$

Light travels at a speed of 300 000 000 metres per second. There are 31 557 600 seconds in a year. Thus, one light year is equal to:

$$d = 300\,000\,000 \times 31\,557\,600$$
$$d = 9\,467\,280\,000\,000\,000$$
$$d = 9{\cdot}5 \times 10^{15} m$$

The light year can also be considered to be a measure of the length of time taken for light travelling from an object to reach us. For example, light from a star which is 2 Ly away has taken two years to reach us. When we are looking at the night sky, we are seeing light that was emitted from distant stars many years ago, in some cases thousands of years ago! We are looking into the past!

Word bank

- **light year (Ly)**
  a unit of measurement for very large distances in space
  $1\ Ly = 9{\cdot}5 \times 10^{15} m$

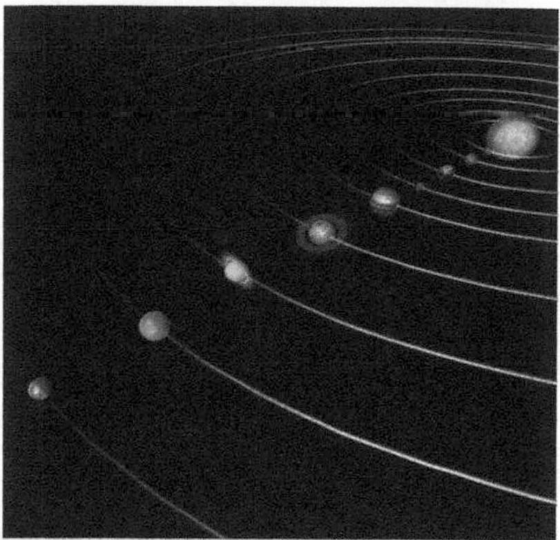

Fig 2.8.1: *Light years measure large distances in space*

Worked example

Proxima Centurai was discovered by Scottish astronomer Robert Innes in 1915. It is approximately $4{\cdot}03 \times 10^{16}$ m away from our star. Calculate this distance in light years.

There are two methods here. The first involves knowing that there are $9{\cdot}5 \times 10^{15}$ m in a light year. Thus converting metres to light years, we divide by $9{\cdot}5 \times 10^{15}$.

$$t = \frac{4{\cdot}03 \times 10^{16}}{9{\cdot}5 \times 10^{15}}$$

$$t = 4{\cdot}24\ Ly$$

We can also work out how long it takes light to travel the distance of $4{\cdot}03 \times 10^{16}$ m using the equation for distance, speed and time:

$$d = vt$$

Substituting what we know and solving for time in seconds:

$$4{\cdot}03 \times 10^{16} = 3 \times 10^{8} t$$

$$t = \frac{4{\cdot}03 \times 10^{16}}{3 \times 10^{8}}$$

$$t = 1{\cdot}34 \times 10^{8} s$$

Now convert the time in seconds to the time in years:

$$t = \frac{1{\cdot}34 \times 10^{8}}{60 \times 60 \times 24 \times 365{\cdot}25}$$

$$t = 4{\cdot}24 \text{ years}$$

Thus, the distance in light years is:

$$d = 4{\cdot}24\ Ly$$

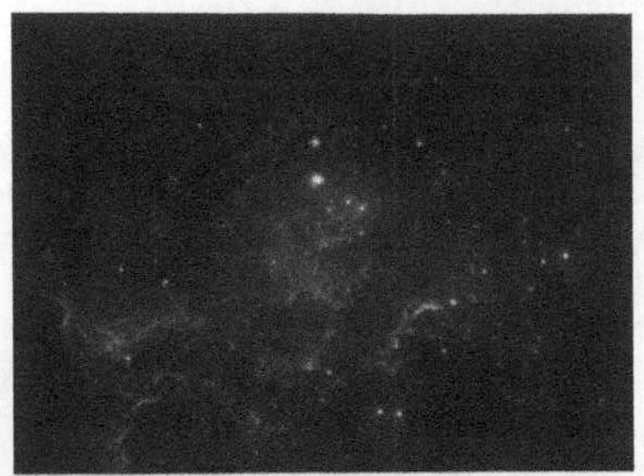

Fig 2.8.2: *Proxima Centurai*

Word bank

- **spiral galaxy**
  a galaxy with a shape that looks like a disc with a bulge at the centre; first described by Hubble in 1936, spiral galaxies have spiral arms that extend from the centre into the disc

The Andromeda galaxy is the nearest spiral galaxy to our own Milky Way. It is $2{\cdot}5 \times 10^{6}$ Ly from Earth. Calculate how far from Earth the Andromeda galaxy is in metres.

To convert light years into metres use the fact that there are $9{\cdot}5 \times 10^{15}$ m in a light year, and multiply the distance in light years by the conversion factor:

$$d = 2{\cdot}5 \times 10^{6} \times (9{\cdot}5 \times 10^{15})$$

$$d = 2{\cdot}4 \times 10^{22} m$$

Fig 2.8.3: *Andromeda galaxy*

Exercise 2.8.1 Distances in space

1 Convert the following distances from light years into metres.

a) 1 Ly

b) 4.5 Ly

c) 34 Ly

d) 0.056 Ly

2 Convert the following distances from metres into light years.

a) 4×10^{20} m

b) 7.8×10^{29} m

c) 3.1×10^{11} m

d) 6.53×10^{15} m

3 It takes light approximately 8 minutes to travel from the sun to Earth. How far away, in metres, is the Earth from the sun?

4 How long does it take light to travel from a distant star which is 6.8 Ly away?

5 How long does it take light to travel from a distant planet which is 5.9×10^{21} m away?

6 Betelgeuse is a reddish star found in the constellation Orion. It is the ninth brightest star in the night sky. It is approximately 643 Ly away from Earth. Calculate the distance to Betelgeuse in metres.

7 The star Pollux is a part of the Gemini constellation. It is a distance of 3.78×10^{17} m from Earth. Convert this distance into light years.

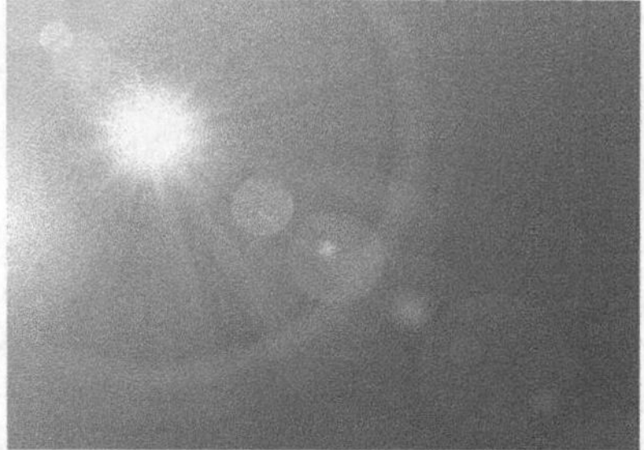

Fig 2.8.4: *The sun*

Fig 2.8.5: *Betelgeuse in the Orion constellation*

Fig 2.8.6: *The Gemini constellation*

The universe

The 'universe' is the term used to describe all of space and time (spacetime) and all of its contents, including all mass and energy. At the moment, and probably for the foreseeable future the size of the universe is unknown but we can make reference to the 'observable universe' which as the name suggests is that part of the universe that we can currently 'see' with telescopes like the Hubble telescope in outer space. Much speculation surrounds the beginning of the universe but currently the 'Big Bang Theory' looks like the most promising.

Fig 2.8.7: *The Universe*

Fig 2.8.8: *Artist's depiction of the Big Bang*

Origin and age of the universe

The Big Bang is thought to be the start point of the expansion of the universe, from a point infinitesimally small to the billions of stars and galaxies that astronomers observe today. It is important to note that the universe is not expanding within space; it is, in fact, the expansion of space itself.

One way of calculating the age of the universe is to study distant objects such as far away galaxies. The distance to a galaxy can be found by measuring its brightness and speed. The furthest away galaxy that has been measured is 13·2 billion light years away. This means that light has travelled for 13·2 billion years from the distant galaxy to reach us on Earth. Light must therefore have been in existence in the universe for 13·2 billion years, so the universe must be at least 13·2 billion years old.

Composition of the universe

Astronomers can determine the composition of the universe from the 'light' (electromagnetic radiation) that it emits using telescopes other than optical ones. The whole of the electromagnetic spectrum (see 'Waves' topic later on…) can be used to examine the universe. Radio, infrared, ultraviolet and x-ray telescopes can all be used to examine events such as the formation of new stars in the interstellar dust clouds as well as x-ray emissions from distant galaxies.

One of the most useful means of determining the composition of the galaxies, stars and planets from the light that they emit uses a technique known as, 'spectral analysis'…

Spectra

When light is shone through a prism a spectrum of different colours can be observed (see Chapter 19, page 326). White light is made up of different colours and when it passes through a prism, each colour is refracted by a slightly different amount. The result is that the white light is split into the different colours that make it up.

The effect can be seen in nature when a rainbow forms. White light from the sun is split into the separate colours that make it up when it passes through a rain drop. Looking at a rainbow, the white light from the sun looks to contain every colour. However, if you examine the spectrum closely, you will notice that some of the colours are missing! In the 19th century a professional lens maker, Joseph von Fraunhofer, discovered dark lines in the spectrum coming from the sun. These lines were to become known as Fraunhofer lines and are shown in the picture opposite.

Fig 2.8.9: *Fraunhofer lines*

White light containing all the colours is emitted from the centre of the sun. It passes through the gases in the cooler atmosphere of the sun and different gases absorb different colours. The missing colours are therefore a signature of the gases in the sun's atmosphere. This can be used by astronomers to estimate the life expectancy of the sun!

Word bank

- **spectroscope**

a device used to split light into the different colours that make it up, for example, splitting white light into all of the colours of the visible spectrum

Line spectra (emission spectra)

A line spectrum is an emission spectrum, as given out by an excited element, such as sodium. A **spectroscope** can be used to examine the spectra in detail. A spectroscope splits the spectrum into its separate colours just like a prism splits up white light.

Experiment: Line spectra

In this experiment you will use a spectroscope to examine the spectra emitted by different gas discharge lamps.

You will need:

- Gas discharge lamps
- Spectroscope

Instructions

1 Connect the discharge lamp to a high voltage power supply and switch on. Allow the lamp to warm up, during which time you will notice the colour changing.

2 Aim your spectroscope at the discharge light and observe the spectrum.

3 Note any colours that you see as part of the spectrum.

Some sources of light, for example, mercury and sodium vapour lamps, produce line spectra where there are specific colours spaced out by specific amounts. These lines are like an *optical signature* of the element producing them. Each element has its own distinct spectrum. In a gas, the presence of an element can be determined by looking at the emission spectrum of the gas.

Absorption spectra

If white light (all colours) passes through a cool gas, some of the colours of the spectrum will be absorbed. This results in black lines being visible in the spectrum – exactly the same as the Fraunhofer lines found in spectra from the sun.

Consider the example of sodium gas. If white light is shone through the sodium gas, black lines will form in the spectrum. These black lines correspond to the same positions of the lines in the emission spectrum from sodium. This is shown in the diagrams below.

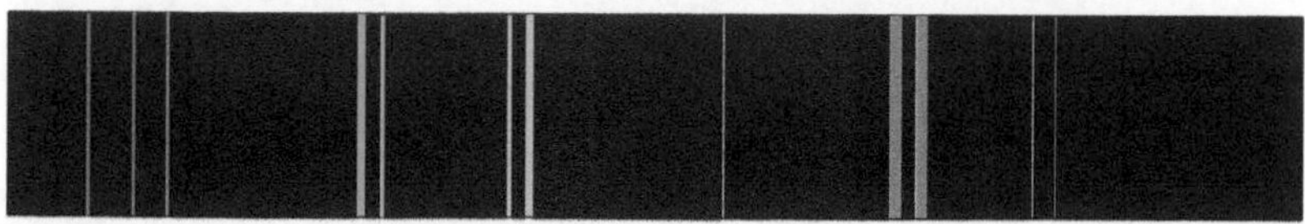

Fig 2.8.10A: *Sodium emission spectrum (line spectrum)*

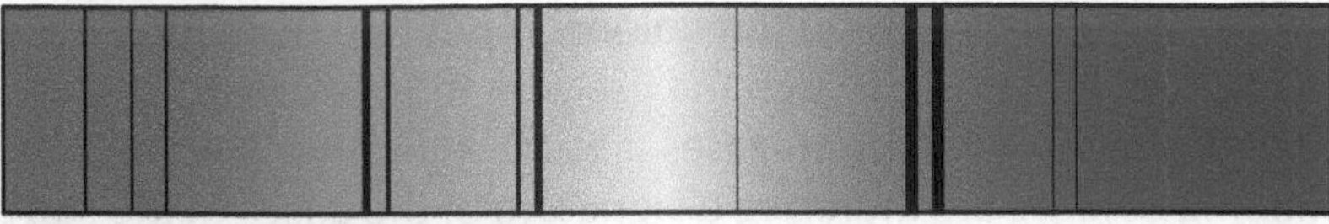

Fig 2.8.10B: *Sodium absorption spectrum*

The 'signature' of sodium has been left in the spectrum. This allows you to identify the presence of sodium. If you carefully examine the spectrum from the sun, or from any other incandescent (or luminous) object, you can identify elements present in its atmosphere.

Worked example

Helium

Cadmium

Hydrogen

Sodium

Using the line spectra shown above, identify the elements present in the spectrum below.

By placing the spectra next to each of the gas spectra above we can see which ones match and which ones do not. The ones that match are the gases present in the spectrum. These are helium and hydrogen.

Learning checklist

In this section you will have learned:

- How to use the term 'light year' and the conversion between light years and metres.
- How to describe the 'Big Bang' theory of the origin of the universe.
- How to estimate the age of the universe.
- How to use the electromagnetic spectrum to obtain information about astronomical objects.
- How to identify continuous and line spectra.
- How to use spectral data for known elements, to identify the elements present in stars.

9 **Electrical charge carriers**

- Definition of electrical current
- Alternating and direct current
- Instruments used to identify a.c. or d.c. sources

10 **Potential difference (voltage)**

- Force on charged particles in an electric field
- Definition of potential difference (voltage)

11 **Ohms law**

- V-I graphs and resistance
- Temperature and resistance
- Experiment to verify Ohm's law

12 **Practical electricity and electronic circuits**

- Measurement of current, potential difference (voltage) and resistance
- Electrical symbols
- Transistor switching circuits
- The rules for series and parallel circuits

13 **Electrical power**

- Definition of electrical power
- The effect of potential difference (voltage) and resistance on current in and power developed across components in a circuit
- Fuse ratings for electrical appliances

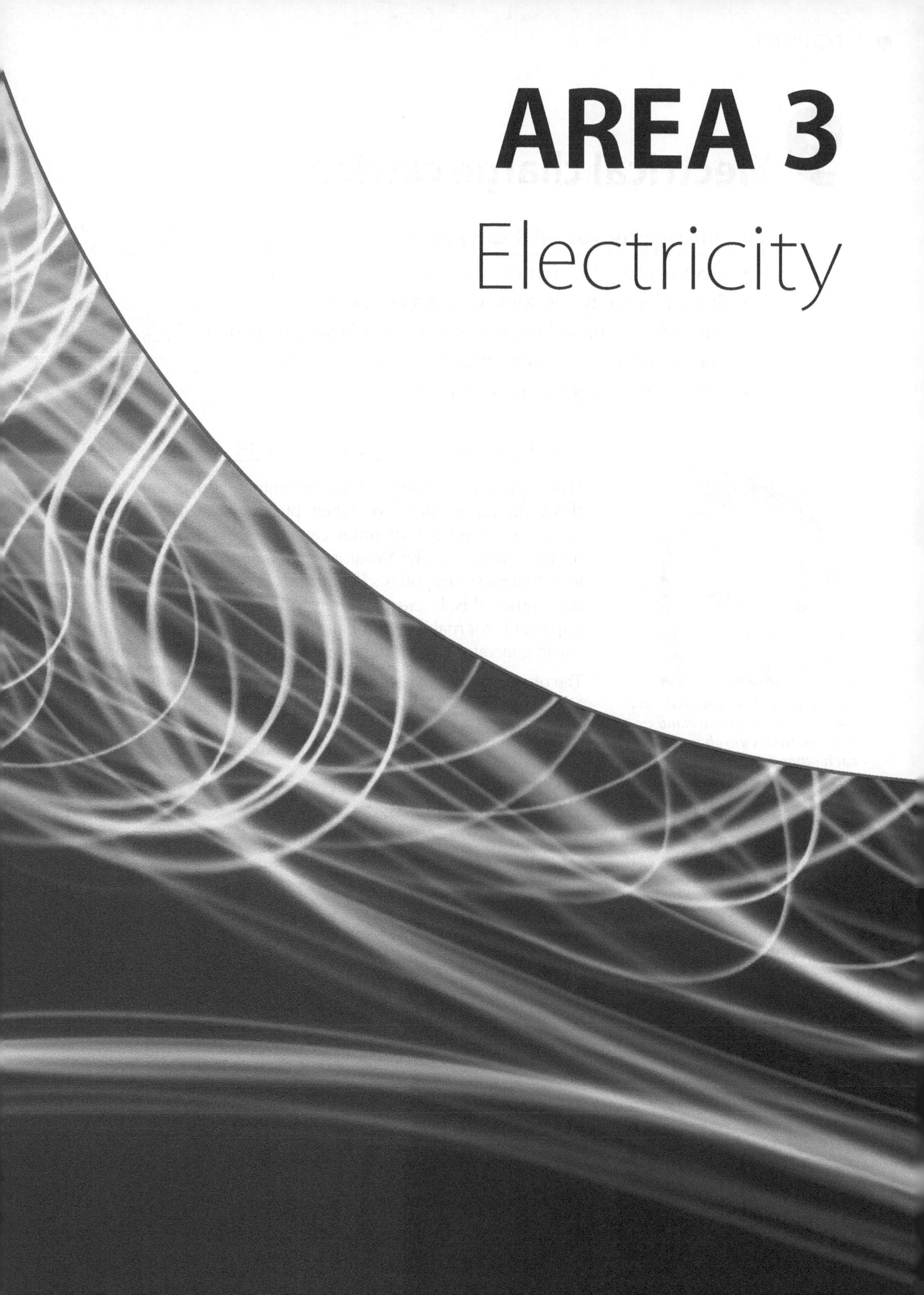

AREA 3

Electricity

9 Electrical charge carriers

In this chapter you will learn about:

- Electric charge in terms of positive and negative charge.
- Static electricity and some of its applications.
- Electric current as the rate of charge that flows per second.
- The relationship between charge, current and time.
- The difference between AC and DC.

Fig 3.9.1: *Rubbing your hair on a balloon can cause it to stand on end due to electric charges repelling each other*

Positive and negative charge

Have you ever wondered why sometimes you get an electric shock on an escalator or when pushing a trolley around a supermarket? What sometimes causes a spark when you shake hands with somebody? What is happening to cause an invisible force that makes us pull back our hands in pain? The phenomenon that causes this is electric charge. The same phenomenon is responsible for making your hair stand up when using the Van de Graaff generator and for lightning strikes during thunder storms.

The phenomenon of electric charge was first discovered by the Greeks around 600 BC. The experiment below replicates what the Greeks saw.

Experiment: Discovering electric charge

This experiment demonstrates the effects of electric charge, and how the force from electric charge can be used to pick up small objects.
You will need:

- A woollen cloth or duster
- A polythene rod
- Paper

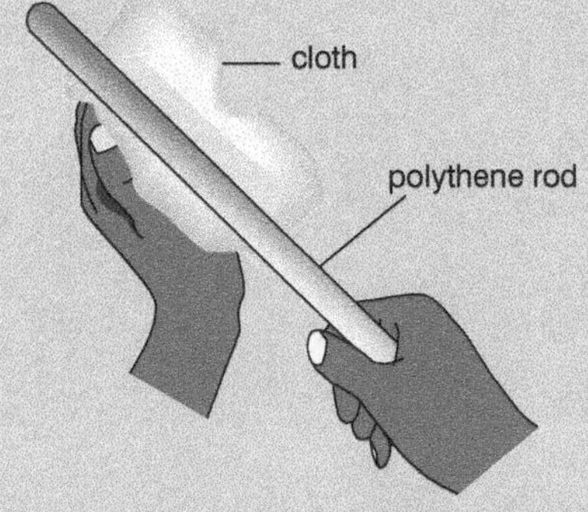

Fig 3.9.2: *Charging a polythene rod*

(continued)

Instructions

1. Rip up some paper to make small light objects and lay them out on a desk.
2. Rub the polythene rod with the woollen duster for around one minute.
3. Place the polythene rod above the pieces of paper – what happens?
4. Think about what happened – can you explain the effect that you are seeing?

The Ancient Greeks rubbed a piece of amber with wool and found it could pick up light objects just like the polythene rod did in your experiment! We say the amber or polythene rod is gaining an electrical charge when being rubbed by the wool.

The Greek word for amber is *elektron*. As the first experiments recorded were carried out with amber, we use this word to name the charge – we now talk of **electrons**! Rubbing the rod with a woollen cloth either adds or removes electrons, which makes the rod electrically charged. This force can then be used to pick up the pieces of paper.

Word bank

- **electron**

a sub-atomic particle (i.e. smaller than an atom) with a negative charge. Current flow in a circuit is a flow of electrons. The electron is a fundamental particle as it does not break down to smaller particles.

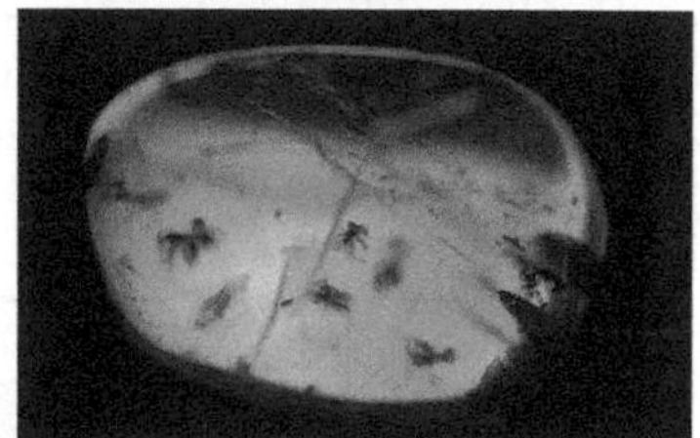

Fig 3.9.3: *A piece of amber*

Experiment: Positive and negative charge

This experiment investigates the difference between positive charge and negative charge and the effects of these charges on each other.
You will need:

- 2 polythene rods
- 2 Perspex rods
- Woollen duster
- String

Instructions

1. Tie a piece of string around each of the rods so that they can be dangled but are still free to move.
2. Charge one polythene rod by rubbing it with the woollen duster for one minute.
3. Repeat to charge one Perspex rod.
4. Bring the two rods close to each other. What happens when the rods come close to each other.
5. Now charge both Perspex rods by rubbing with the woollen duster. Bring these charged rods close to each other. What happens when these rods come close to each other?
6. Repeat the above experiment with the two polythene rods – what happens when they come close to each other?

Think about it

Find out who was the first person to measure the specific mass of an electron. Why do you think this was such an important experiment?

When the polythene rod is rubbed with the woollen cloth electrons from the cloth are added to the rod (below left). These extra electrons make the rod negatively charged because electrons have a negative charge and leave the cloth positively charged. When the Perspex rod is rubbed with the cloth electrons are removed from the rod (below right). This has the effect of making the rod positively charged and the cloth negatively charged, as the cloth gains electrons.

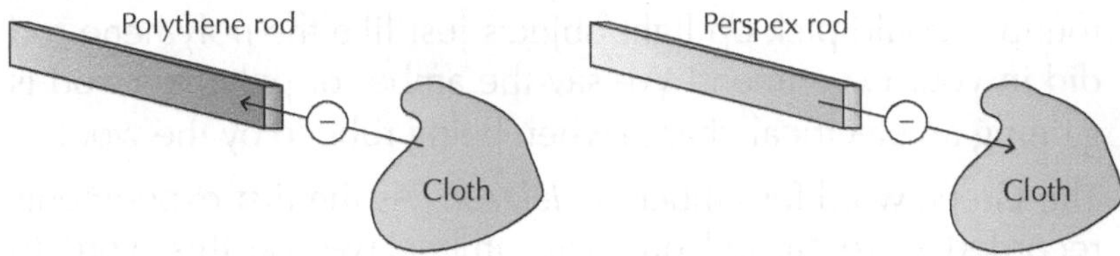

Fig 3.9.4: *By rubbing the plastic rods with cloth a static charge can be created*

Word bank

• proton

a sub-atomic particle (i.e. smaller than an atom) with an overall positive charge. The proton is not a fundamental particle – it is made up of smaller particles called quarks.

It was in the 1700s that Benjamin Franklin described the idea of two types of charge – positive and negative. The electron is negatively charged and the particle known as a **proton** is positively charged. A neutron has zero charge.

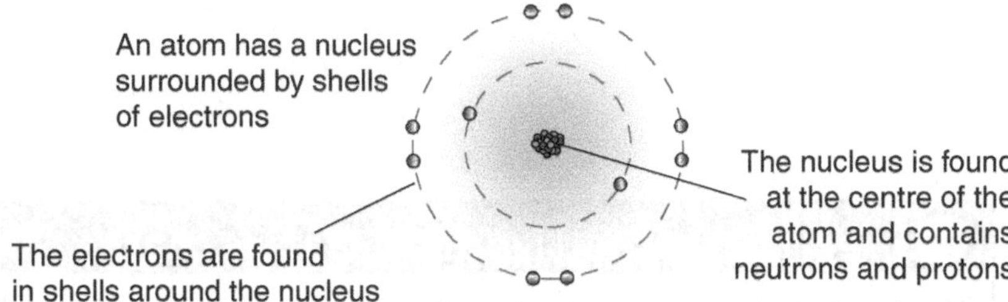

Fig 3.9.5: *The atomic model showing the arrangement of protons, neutrons and electrons*

Key point

There are two main charge carriers that we consider: electrons and protons.

- Electrons have a negative charge.
- Protons have a positive charge.

The magnitude (size) of the charge of a proton is the same as that of an electron. The only difference is that one is positive and the other is negative.

In electricity, opposites attract: electrons and protons are attracted to each other by a force because they have opposite charges. A positive charge repels another positive charge so protons repel protons. The same is true of negative charges, i.e. electrons repel other electrons. Therefore, in the above experiment the different rods were attracted to each other because they had opposite charges. The identical rods, with identical charges, repelled each other.

Key point

Opposite charges attract each other.

Like charges repel each other.

Exercise 3.9.1 Defining electric charge

1 State the two different charge carriers and say whether they are positively or negatively charged.

2 A Van de Graaff generator consists of a metal dome which is charged with positive charges as shown in the diagram.

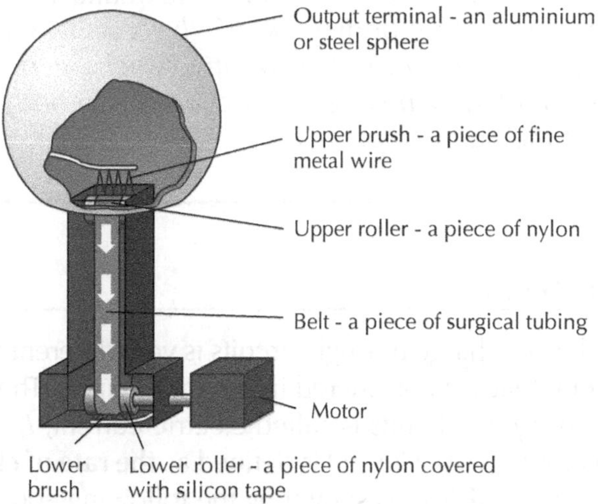

Fig 3.9.6: *Van de Graaff*

a) Explain why your hair stands on end when you touch the sphere of the generator when it is charged.

b) Explain why a negatively charged sphere can cause a spark when it is brought close enough to the positively charged metal dome. What weather phenomenon does this remind you of?

3 When you brush your hair, the hair strands often move apart from each other. When you remove your hair brush from your hair, the hair is often pulled away with the brush and stands on end. Explain these effects in terms of electric charge.

Applications of static electricity

In the above examples, we have considered static electricity. This is where an object can become electrically charged (for example a Perspex rod rubbed with a cloth). However, this charge is trapped on the object – it does not go anywhere, so it is said to be static. The charge remains on the object until it gets a path to flow to Earth, for example, through an electrical circuit or by discharging through the air or through a person.

Static electricity has many applications based on the principle of opposite charges attract and like charges repel; this force can be put to use in laser printers, paint spraying or dust removal.

Fig 3.9.7: *Brushing your hair with a plastic brush can cause it to stand on end. Why?*

Fig 3.9.8: *Painting a car's bodywork*

SPOTLIGHT ON INDUSTRY: CAR PAINTING

A car's bodywork is painted to prevent it from rusting. The painting process can be improved by using static electricity.

Paint from the spray gun is electrically charged before it leaves the gun. This is achieved by removing electrons (like the wool on the polythene rod above) to make the paint positively charged. The bodywork of the car is given a negative charge. Therefore, the positively charged paint droplets are attracted to the negatively charged car bodywork and this helps the painting process.

Electric current

The flow of electric charge through circuits is very different to the static electricity that we considered in the last chapter. The flow of electrical charge in circuits is called electric current, I, and is measured in amperes (A). Current is defined as **the rate of charge that flows per second**. This is shown by the following equation:

$$I = \frac{Q}{t}$$

where Q is the charge measured in coulombs (C) and t is the time measured in seconds (s). Current is measured in amperes (A) and it is important not to shorten this to 'amps'! This equation is usually rearranged to give:

$$Q = It$$

Key point

From the equation $Q = It$,
1 ampere = 1 coulomb per second
1 A = 1 C/s

Experiment: Electric current

This experiment investigates the effects of changing the electric current on the brightness (energy supplied) of a light bulb.

You will need:

- 12 V filament bulb
- Ammeter (multimeter)
- Variable voltage DC supply
- Wires

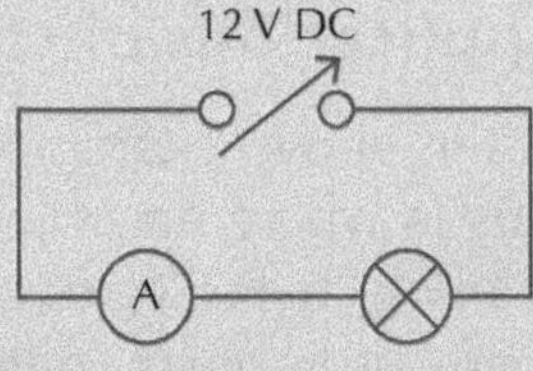

Fig 3.9.9

Instructions

1. Set up the circuit shown in the diagram above (**series circuit**), consisting of a bulb and an ammeter connected to a DC power supply.
2. Set the DC power supply to 2 volts and switch it on.
3. Using the ammeter, measure the current in the circuit in amperes.

(continued)

4 Note the brightness of the bulb.

5 Change the voltage to 6 volts and measure the new current in the circuit. Comment on the relative brightness of the bulb.

6 Compare futher readings of current to the brightness of the bulb – how does the brightness of the bulb depend on the current?

We can see that current is linked to the amount of energy carried in the circuit – the greater the current, the brighter the bulb. This is because when the current is larger, more charge passes through the circuit in a given time. A greater amount of charge leads to a greater energy flowing round the circuit.

Technique: Calculations involving charge, current and time

Write down what you know from the question, leaving blank what you are trying to find:

Charge, Q = ____ C

Current, I = ____ A

Time, t = ____ s (remember to convert time to seconds!)

Write down the equation linking the quantities above as you see it on the formula sheet:

$$Q = It$$

Substitute into the equation what you know.

Solve for the unknown.

Word bank

- **series circuit**

a single loop of wires and components. In a series circuit components are attached one after each other. Current is constant and has only one path to follow. By tracing your fingers around a series circuit you will have only one route and no junction choices.

Worked example

A capacitor is a device that can store charge. This device can be used as a timer for flashing lights. How much charge is stored on a capacitor if a current of 0·1 A flows to it for 30 s?

Write down what we know and what we are looking for:

Charge, Q = _____ C

Current, I = 0·1 A

Time, t = 30 s

We are using the equation that links charge, current and time:

$$Q = It$$

Substitute what we know:

$$Q = 0{\cdot}1 \times 30$$

Solve for charge:

$$Q = 3\ C$$

Worked example

A radio telescope measures the radiation incident on the pixels of a detector. The radiation creates a current by stripping electrons from the pixel material. How much current would be produced if a charge of $1{\cdot}6 \times 10^{-19}$ C is removed from the detector in 100 ns?

Fig 3.9.10: *A radio telescope with a parabolic dish*

Write down what we know and what we are looking for:

Charge, $Q = 1{\cdot}6 \times 10^{-19}$ C

Current, $I = ?$ A

Time, $t = 100 \times 10^{-9}$ s (note: nano, n = $\times 10^{-9}$)

We are using the equation that links charge, current and time:

$$Q = It$$

Substitute what we know:

$$1{\cdot}6 \times 10^{-19} = I \times 100 \times 10^{-9}$$

Solve for current:

$$I = \frac{1{\cdot}6 \times 10^{-19}}{100 \times 10^{-9}}$$

$$I = 1{\cdot}6 \times 10^{-12} A$$

Word bank

- **ionisation**

the process where an atom has a charge removed. The removal of an electron leaves the atom with an overall positive charge and the atom becomes an ion.

Exercise 3.9.2 Electric current

1 Find the missing variables in the table below.

Charge, Q	Current, I	Time, t
(a)	1 A	60 s
(b)	3 A	30 s
(c)	1×10^{-3} A	60 s
(d)	200 mA	2 s
960 C	3 A	(e)
1200 C	(f)	60 s
175 μC	(g)	500 ms
1 kC	5 A	(h)

Table 3.9.1

2 A small generator can produce 0·5 mA of current every minute. How long will the generator take to produce a charge of 2 C?

3 Electrostatic precipitators are devices that can use electrostatic charge to remove small particles from gases. Positively charged plates are held near to plates with a negative charge. The current across the plates is so high that the device can **ionise** the gas between the plates. This gives the gas a charge by stripping electrons from the atoms.

a) If the current is 100 A and the device has been operating for 5 hours, how much charge will it have removed from the air?

b) If each air particle has the same charge as an electron ($1{\cdot}6 \times 10^{-19}$ C), how many particles have been removed?

4 An LCD monitor draws (uses) a current of 1·5 A. How much charge flows to the monitor if it is left on for 8 hours?

5 An LED is a device that converts electrical energy to light energy. A charge of 0·001 C flows through the LED in 4 s. Calculate the current flowing through the LED.

Fig 3.9.11: *LCD monitor*

Alternating and direct current

Current is the flow of electric charge (the flow of electrons). It is a measure of the amount of charge that flows in a given time. There are two types of current that we must consider: **direct current** and **alternating current**.

The oscilloscope

An oscilloscope is a device that is used to look at electrical signals.

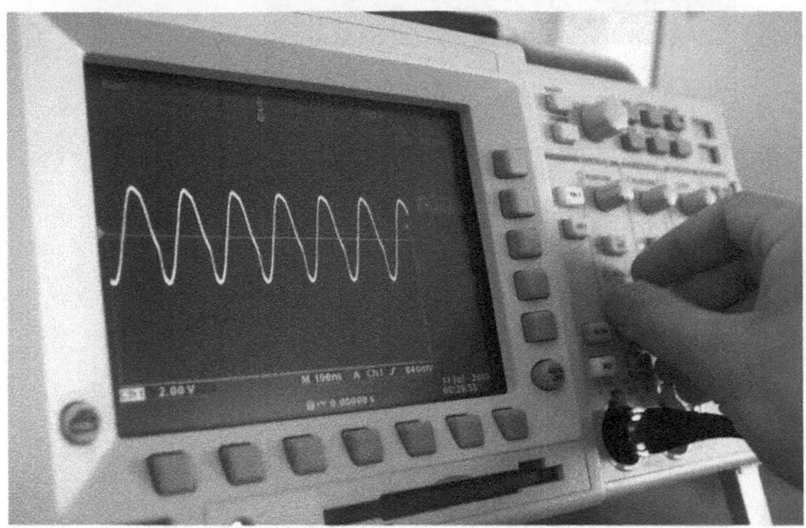

Fig 3.9.12: *An oscilloscope screen*

Direct current

Direct current (DC) flows in one direction around a circuit.

A battery is an example of a supply of direct current. Electrons travel from the negative terminal to the positive terminal. Sometimes when an ammeter is attached to a circuit you may get a negative reading. By switching the connectors around, the reading will become positive. This reading allows you to determine the direction of flow of the electrons and hence which side of the circuit is attached to the negative terminal of the battery.

Viewed using an oscilloscope, direct current appears as a straight line as shown in the diagram below.

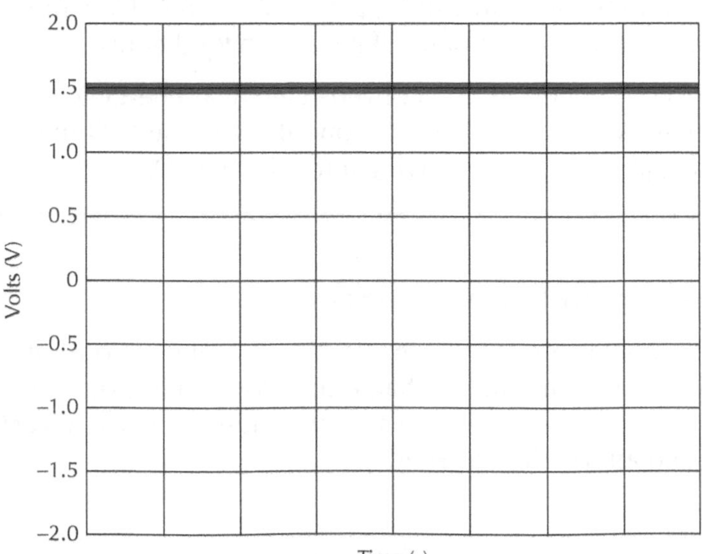

Fig 3.9.14: *Oscilloscope trace of a 1·5 V battery*

Word bank

- **direct current (DC)**

current which travels in one direction around a circuit. A battery is a source of direct current.

Fig 3.9.13: *A battery is a source of direct current*

Hint

The horizontal axis on an oscilloscope shows the timebase. This shows any changes in the wave over time and illustrates the frequency of the wave pattern.

Alternating current

Alternating current (AC) is where the current repeatedly changes direction. It will flow from negative to positive and then switch to positive to negative. The electricity that we get from the mains is alternating current. The reason for this is explained in the *step back in time* feature below.

Fig 3.9.15: *Mains electricity is AC. It changes direction 50 times per second, i.e. it operates at 50 Hz*

Viewed on an oscilloscope, alternating current forms a **sinusoidal wave** which highlights the changing direction of the electron flow. This is shown in the diagram below.

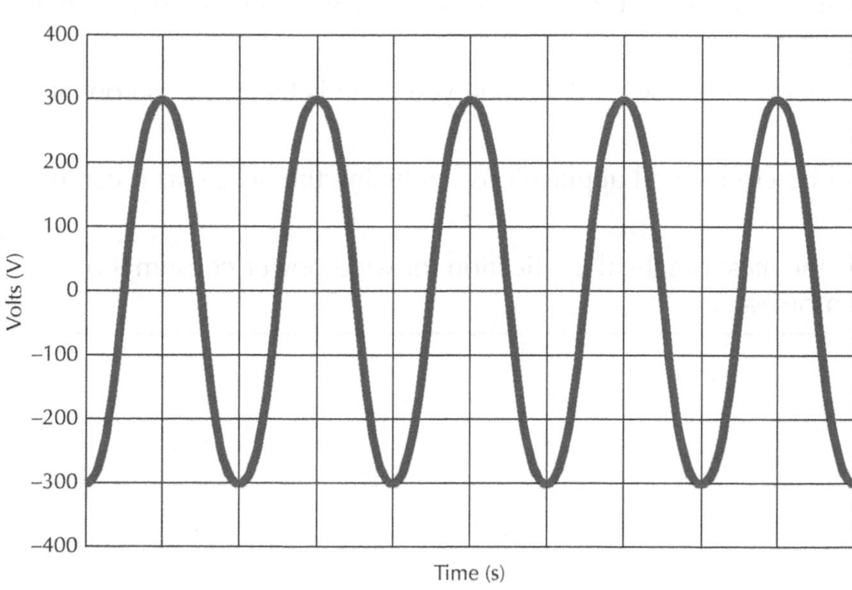

Fig 3.9.16: *Alternating current constantly changes the direction of supply – hence the direction of the current changes*

Word bank

- **alternating current (AC)**

current which changes direction after a certain amount of time. The frequency of UK mains current is 50 Hz. This means that the current changes direction 50 times per second flowing negative to positive, then positive to negative.

Word bank

- **sinusoidal wave**

a repetitive wave motion that describes the oscillation of an object.

Make the link

Waves are considered in detail in Chapter 8.

Fig 3.9.17: *Nikola Tesla*

STEP BACK IN TIME: THE WAR OF THE CURRENTS

AC was not always the first choice. In the 1880s, American scientist Thomas Edison owned and distributed electricity using DC current. However, this required large currents that could be dangerous. Another scientist, Nikola Tesla, had devised a more efficient transportation system using AC current. This enabled long distance distribution with a smaller current. Funded by George Westinghouse, Tesla was successful with his system of production and distribution of electricity, which is why we have AC in our homes today.

Exercise 3.9.3 Alternating and direct current

1 An oscilloscope is used to examine AC and DC signals.

  a) Sketch the signal that would be observed on the oscilloscope when it is connected to a DC power supply. Explain your sketch.

  b) Sketch the signal that would be observed on the oscilloscope when it is connected to an AC power supply. Explain your sketch.

  c) Give one example of a DC power supply.

  d) Give one example of an AC power supply.

2 Is the current from the mains supply in the UK AC or DC? Explain why current is distributed over long distances in this form.

3 Consider the following statements. Choose which one you think is the most appropriate and explain your choice.

  i) Direct current should be used for all applications, including the home, as it can be easier to produce.

  ii) Alternating current is the most practical application for wide power consumption as it can be distributed more safely.

Learning checklist

In this section you will have learned:

- The definition of electrical current in terms of the electric charge transferred per unit time.
- How to use an appropriate relationship to solve problems involving charge, current and time.
- The difference between alternating current (AC) and direct current (DC)
- How to identify a source as AC or DC based upon an oscilloscope trace or an image from data logging software.

10 Potential difference (voltage)

In this chapter you will learn about:

- The effect of an electric field on a charge.
- Potential difference (voltage) as a measure of the energy given to the charge carrier in a circuit.

Fig 3.10.1: *Alessandro Volta*

Key point

We can now consider a concise definition of the volt. This definition might form part of your assessment. The volt is a measure of the energy supplied to each coulomb of charge.

1 volt = 1 joule per coulomb
1 V = 1 J/C

Word bank

- **across a component/in parallel**

the term 'across a component' is the same as saying 'in parallel'. To attach a device in parallel, the wires/connectors must be connected over the device so that it creates a secondary loop. If you trace your finger round a parallel circuit there will be more than one choice of path.

Potential difference (voltage)

We have so far seen that electrons, which have a negative charge, flow around a circuit. This flow of electrons is known as electric current. The electrons need energy to move round the circuit. A lot of the early studies of the energy within electrical circuits were carried out by Alessandro Volta. For this reason, the concepts are named after him and we talk of a voltage in the circuit measured in volts. The flow of electrons requires a 'push' to move around a circuit. Voltage in a circuit, sometimes called electrical potential or potential difference, is often compared to the 'push' that the negative charge electrons require to flow around the circuit. To understand the process of charge flow we need to define voltage more rigorously.

Defining electrical potential (voltage)

Electrical potential is defined as the potential energy carried by each coulomb of charge:

$$V = \frac{W}{Q}$$

where W is energy (work) in joules (J) and Q is the charge in coulombs (C).

You can draw an analogy between electrical potential energy and gravitational potential energy – both are measured in joules as both are types of energy! However, gravitational potential energy is used to make objects with a given mass move towards the Earth, while electrical potential energy causes electric charges to move in a circuit towards the positive terminal.

Potential difference

The potential difference is described as the change in voltage between two points in the circuit. This could be between the positive and negative terminals of a battery. We might say a battery has a voltage of 1·5 V. This means that there is a potential difference of 1·5 V between the positive and negative terminals.

Returning to our definition of the volt above, this also means that the battery supplies each coulomb of charge with 1·5 J of energy. This energy is then carried around the circuit and can be transformed by components in the circuit into other forms of energy.

Potential difference is measured using a voltmeter. The voltmeter is connected **across a component** in a circuit (**in parallel** with the component). The voltmeter therefore measures the difference in potential between one side of the component and the other.

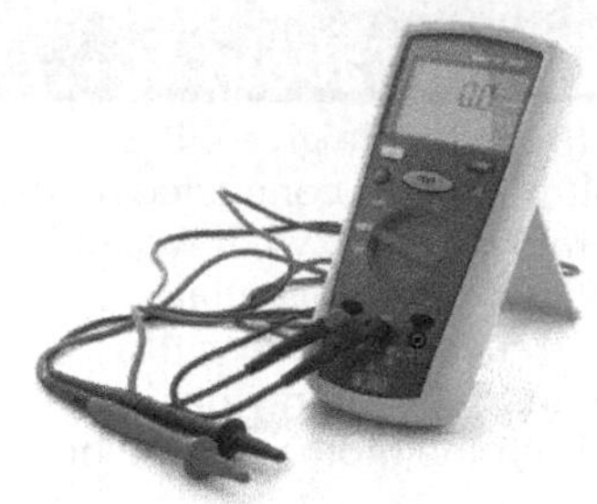

Fig 3.10.2: *Voltmeter*

Consider measuring the potential difference across a light bulb. The voltmeter would be connected across the bulb as shown in the circuit diagram here.

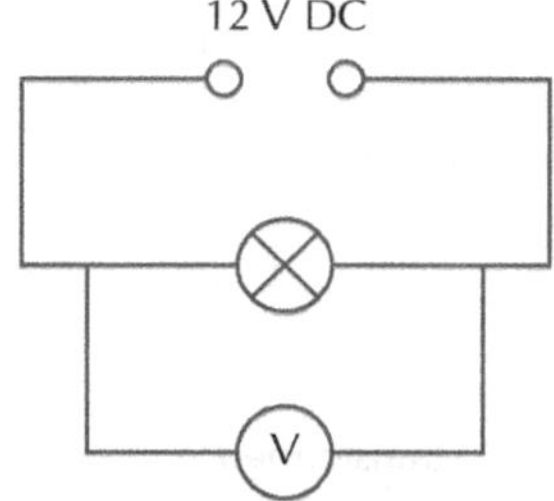

Fig 3.10.3: *A series circuit with a voltmeter across the filament bulb*

In electronics, we talk of the voltage dropping across a component. In the above example of the bulb, if the potential difference measured across the bulb was 12 V, this would mean that 12 V has been dropped going across the bulb: one side of the bulb would be at a potential of 12 V and the other at 0 V which makes the difference equal to 12 V. Linking this to energy, this means that every coulomb of charge passing through the bulb loses 12 J of energy. This energy may be transformed into heat and light as described earlier.

Think about it

Consider a ball being lifted to different heights in a gravitational field. A ball at a given height is said to have a certain gravitational potential energy. If we consider two different heights, there will be a potential difference between these heights. How does this relate to electrical potential energy above?

Energy and voltage in a circuit

The definition for voltage links it to the energy in the circuit. Consider connecting a light bulb and a motor to a battery as shown below. Current (electrons) flows round the circuit in the direction of the arrows. Connecting a voltmeter across one of the components will give a measure of the voltage across it. This is a measure of how much energy is transferred in the component.

Electrons are supplied with electrical energy by the battery. The amount of energy depends on the voltage of the battery – a 10 V battery will supply each coulomb of charge with 10 J of energy.

In going from point A to point B, the electrons lose electrical energy. The energy is transformed to kinetic energy in the motor. This means the voltage at point B (7 V) will be less than it is at point A (10 V). This difference in voltage is the potential difference across the motor (3 V). Connecting a voltmeter to points A and B will measure this potential difference.

In going from point B to point C, the electrons lose electrical energy again. This time the energy is converted to light and heat in the bulb. The voltage at point C (0 V) is less than it is at point B (7 V). The potential difference across the bulb is 7 V.

Key point

The potential difference (voltage) is measured using a voltmeter. The voltmeter must be connected across (in parallel with) the component as shown.

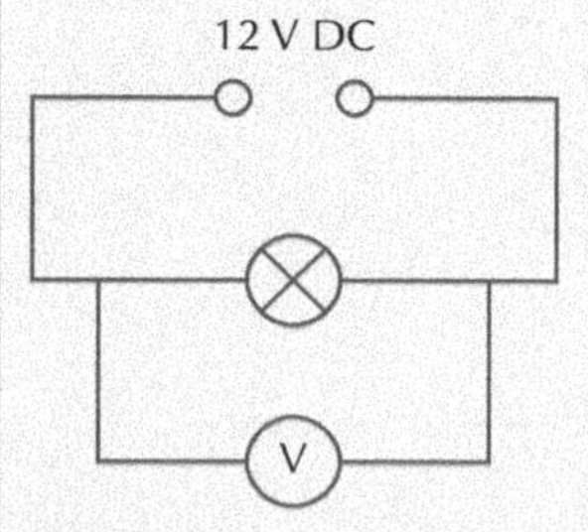

Fig 3.10.4

> **Think about it**
>
> In the example given, if a voltmeter was connected across points A and C, what would be the measured potential difference? How does this relate to the energy dissipated in both the motor and the bulb?

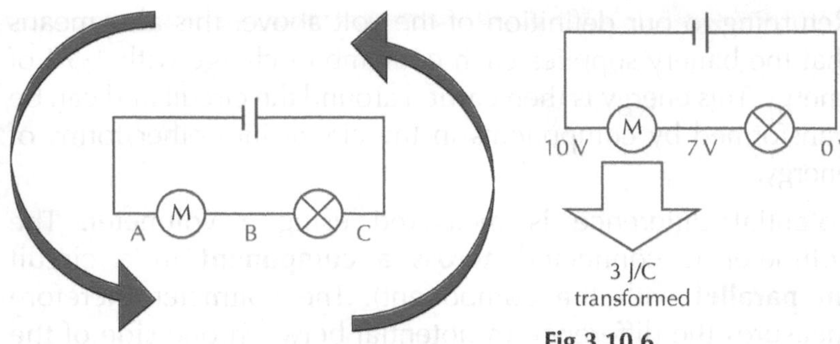

Fig 3.10.5: *Current flows from negative to positive*

Fig 3.10.6

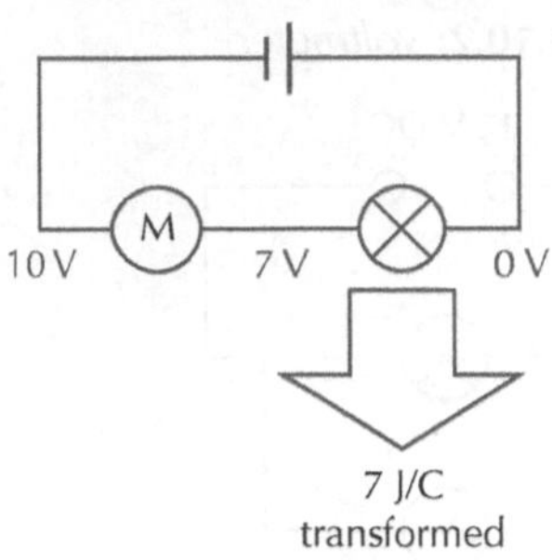

Fig 3.10.7

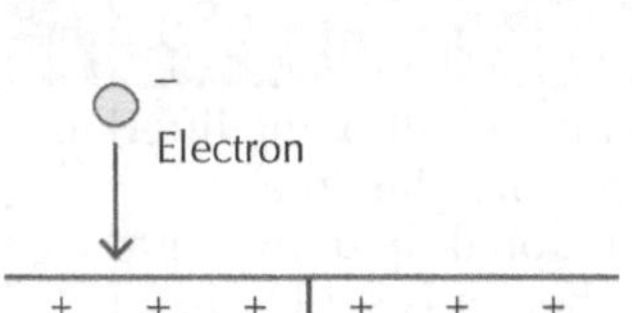

Fig 3.10.8: *An electron is placed between two charged plates*

Potential difference in an electric field

In the same way as we have gravitational potential energy in a gravitational field, we can have electrical potential energy in an electric field. Consider an electron (negative charge) between two plates as shown opposite. The bottom plate is positively charged and the top plate is negatively charged. This means there is a potential difference between the two plates.

The electron will be attracted to the positive plate and repelled from the negative plate. This is similar to a mass being attracted downwards to the ground in a gravitational field. The closer the electron is pushed to the negative plate, the more potential energy it has. This means there will be a potential difference between different positions.

If we want to move the electron towards the negative charge, we have to apply a force to overcome the repulsive force. This means we have to do work to move it. The amount of work (W) required to move a charge depends on the size of the charge (Q) and the potential difference (voltage) over which the charge is moved:

$$W = QV$$

This equation is simply a rearranged version of the equation for voltage above!

> **Technique: Calculations involving potential difference**
>
> 1. Write down what you know from the question, leaving blank what you are trying to find:
>
>   Energy (work), W = ____ J
>
>   Voltage (potential difference), V = ____ V
>
>   Charge, Q = _____ C
> 2. Write down the equation linking the quantities above as you see it on the formula sheet:
>
> $$W = QV$$
> 3. Substitute into the equation what you know.
> 4. Solve for the unknown.

Worked example

In the circuit shown below, the voltmeter measures a potential difference of 6 V across the bulb. If a charge of 0·5 C flows through the bulb, calculate the amount of energy transformed by the bulb.

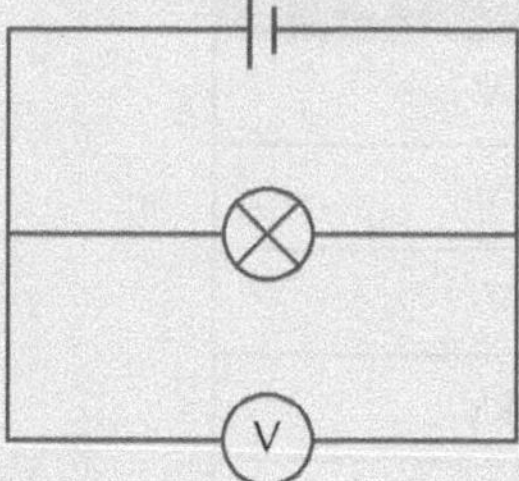

Fig 3.10.9

Write down what we know and what we are looking for:

- *Energy, W = ? J*
- *Charge, Q = 0·5 C*
- *Potential difference, V = 6 V*

We are using the equation that links energy, charge and potential difference:

$$W = QV$$

Substitute what we know:

$$W = 0{\cdot}5 \times 6$$

Solve for energy (work done):

$$W = 3J$$

Worked example

The size of the charge on an electron is $1{\cdot}6 \times 10^{-19}$ C. The electron is moved through an electric field, requiring 0·5 μJ of energy. Calculate the potential difference over which the electron was moved.

Write down what we know and what we are looking for:

- *Energy, W = $0{\cdot}5 \times 10^{-6}$ J*
- *Charge, Q = $1{\cdot}6 \times 10^{-19}$ C*
- *Potential difference, V = ? V*

We are using the equation that links energy, charge and potential difference:

$$W = QV$$

Substitute what we know:

$$0{\cdot}5 \times 10^{-6} = 1{\cdot}6 \times 10^{-19} \times V$$

Solve for charge:

$$V = \frac{0{\cdot}5 \times 10^{-6}}{1{\cdot}6 \times 10^{-19}}$$

$$V = 3{\cdot}1 \times 10^{13} V$$

Think about it

In the example of an electron between two metal plates, you need to do work to move the electron towards the negative plate. This leads to a store of electrical potential energy. If the electron is released, what will the energy transformation be?

Exercise 3.10.1 Potential difference

1 Find the missing variables in the table below.

Energy (work) (W)	Charge (Q)	Potential difference (V)
(a)	4 C	20 V
(b)	0·5 C	17 V
(c)	20 mC	5 V
(d)	7 C	30 kV
20 J	(e)	8 V
15 μJ	(f)	1 kV
10 J	5 C	(g)
100 nJ	$1{\cdot}6 \times 10^{-19}$ C	(h)

Table 3.10.1

2 In an electric circuit, a voltmeter is used to measure the potential difference across a bulb. It is measured to be 8 V. In a time of 1 hour, a charge of 150 C flows through the bulb. Calculate the amount of energy transformed into heat and light by the bulb.

3 A positive test charge is placed between two parallel plates as shown. One plate is held at a positive charge, the other is negative. Describe the motion of the test charge and the energy transformation which is taking place.

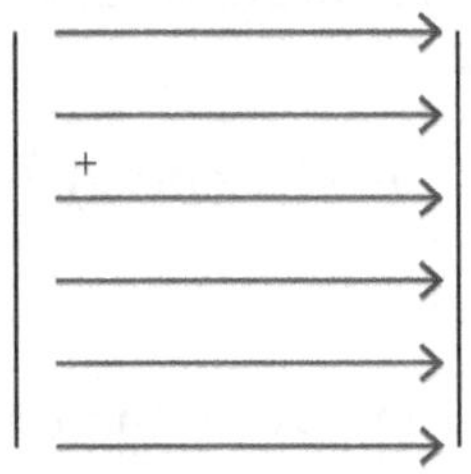

4 An energy of 12 μJ is used to move a charge of 50 nC across a potential difference. Calculate the size of the potential difference.

Fig 3.10.10

5 In an electrical circuit, a current of 4 A flows through a motor for a time of 30 s. The potential difference across the motor is 12 V.

a) Find the total charge that flows through the motor in 30 s.

b) Calculate the amount of energy transformed by the motor in 30 s.

Fig 3.10.11: *Motor*

6 A metal sphere has a charge of 0·5 C. It is moved through a potential difference of 40 V. It has a mass of 0·04 kg.

a) Calculate the work done moving the charge through the potential difference.

b) The charge is released and the stored electrical potential energy is transformed into kinetic energy. Find:

i) The final kinetic energy of the sphere.

ii) The final speed of the sphere.

7 A negative charge with energy of 5 mJ moves along a wire from negative to positive. When the charge reaches the positive plate it has 4 mJ of energy remaining. Describe how this is possible if energy is conserved?

(continued)

8. A Cathode Ray Tube (CRT) for an oscilloscope uses an electron gun to fire electrons at the screen to produce a picture. Electrons with a charge of $1{\cdot}6 \times 10^{-19}$ C and mass of $9{\cdot}11 \times 10^{-31}$ kg are accelerated through a potential difference of 5 kV.

a) How much energy is gained by an electron when it is fired across the potential difference of 5 kV?

b) Find the speed of the electron when it leaves the electron gun.

Force on charged particles

As mentioned earlier, charged particles will experience a force in the presence of an electric field, the direction of this force depends upon the nature of the charge (positive or negative) as well as the shape of the surface which provides the electric field. **Remember that the electric field direction is always that which a positive charge would follow**. Negative charges will move in the opposite direction. Field lines represent 'lines of force' and are often represented by arrows in diagrams. Some examples are shown below:

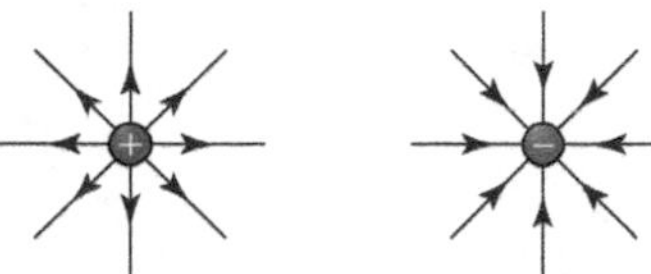

Direction a **positive** charge would move close to two isolated point charges

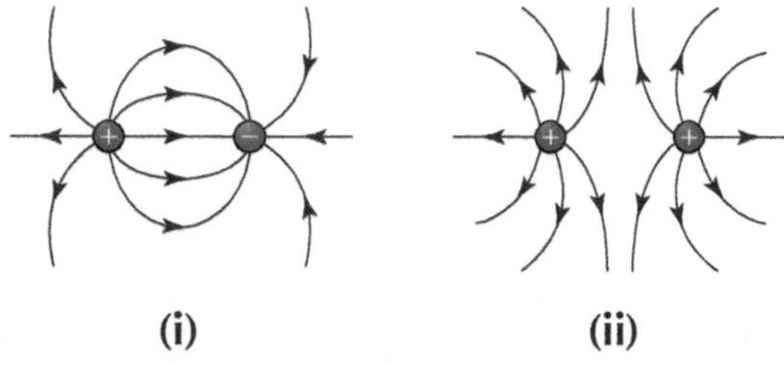

(i) (ii)

Direction a **positive** charge would move close to (i) oppositely charged particles and (ii) like charged particles

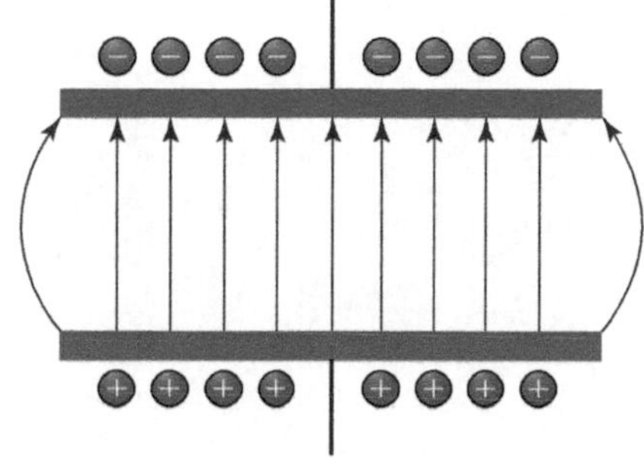

Direction a **positive** charge would move between two parallel plates

Learning checklist

In this section you will have learned:

- The definition of potential difference (voltage)
- That a charged particle experiences a force in an electric field.
- About the path a charged particle would follow between two oppositely charged plates, near a single point charge, between two oppositely charged points and between two like charged points.
- That the potential difference (voltage) of a supply is a measure of the amount of energy given to the charge carriers in a circuit.

11 Ohm's law

In this chapter you will learn about:

- Ohm's law and the relationship between voltage, current and resistance.
- How to evaluate a voltage-current graph.

Ohm's law

Ohm's law was formulated by German physicist Georg Ohm. It describes a relationship between the voltage and the current in an electric circuit. Ohm's experiments of the early 19th century form fundamental laws of electronics that are the basis of many of the circuits we use today.

Fig 3.11.1: *Georg Ohm*

Resistance

Current is able to flow through some materials more easily than others. For example, copper is used for wiring because current can flow through it easily. However, plastic and wood do not allow current to flow through them easily. We classify materials into groups depending on whether they allow electrical current to flow through them: **conductors** which allow current to flow as there are many free electrons; **insulators** which prevent current from flowing as electrons do not flow easily, and semiconductors through which, in some circumstances, current can flow if energy is supplied.

Word bank

- **conductor**
  a material, generally a metal, that allows current to flow through it easily. Conductors are materials that have a large amount of free electrons. As they are free electrons, they can move around more easily, which is why charge flows more easily.

Experiment: Current compared to length of wire

This experiment investigates the current carried by a wire for a given voltage and how this changes with the length of the wire.

You will need:

- Different lengths of wire*
- DC power supply
- Ammeter
- Connecting wires

* A wire board with different lengths of wire already attached could be used.

Instructions

1 Connect the length of wire under test, and the ammeter, to the DC power supply as shown in the circuit diagram below. Crocodile clips can be used to attach the wire under test to leads.

(*continued*)

2 Switch on the DC power supply and use the ammeter to measure the current flowing in the circuit.

3 Keeping the voltage constant, repeat the above experiment for different lengths of wire. Record your results in a table similar to the following.

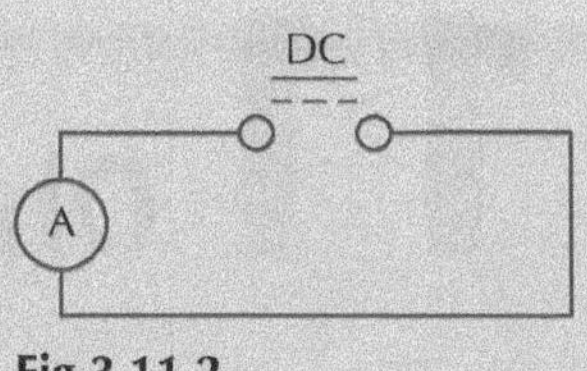

Fig 3.11.2

Length of wire (cm)	Current (A)

Table 3.11.1

4 Plot a graph of the current against the length of the wire.

5 How does the current change with different lengths of wire – what is the relationship between current and length?

Experiment: Current compared to thickness of wire

Now investigate the current carried by a wire for a given voltage and how it changes as the thickness of the wire is changed.

Record your results in a table similar to the following.

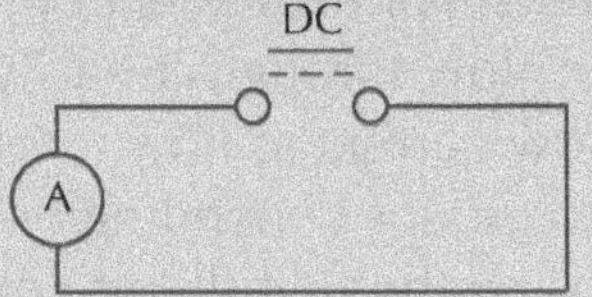

Fig 3.11.3

Thickness of wire (cm)	Current (A)

Table 3.11.2

Plot a graph of the current against the thickness of the wire. How does the current change with different thicknesses of wire – what is the relationship between current and thickness of wire?

Word bank

- **insulator**

a material, like wood, plastic or rubber, that does not allow current to flow easily. Insulators are materials that have electrons which are very tightly bound to their atoms. As there are no (or few) free electrons, there are no charge carriers to flow.

Ohm realised that the longer the wire the harder it was for charge to pass through. He observed similar results to those you have seen in the experiments above. He defined the resistance to current flow as resistance, *R*. This is measured in ohms (Ω).

The resistance of a material depends on the following:

- Type of material
- Length of material
- Thickness of material
- Temperature of material

Hint

The resistance of a conducting material alters depending on its temperature: as temperature increases, resistance increases. It is directly proportional.

According to Ohm, the resistance of a material is its opposition to charge flowing through it. The greater the resistance, the harder it is for charge to flow. The resistance of a component in a circuit must therefore affect the current (I) and the voltage (V).

Make the link

For more information about how the resistance of a thermistor changes with temperature, see Practical electronics.

Experiment: Ohm's law in practice

This experiment investigates how the current through a resistor changes with applied voltage.

You will need:

- A variable DC power supply*
- A resistor
- A voltmeter
- An ammeter
- Connecting wires

* As an alternative to changing the voltage from the power supply, the current values can be changed using a variable resistor.

Instructions

1. Set up the circuit shown in the circuit diagram below, consisting of a resistor and ammeter connected to the DC power supply. A voltmeter is connected across the resistor.
2. Set the DC power supply to 2 V and switch it on. Measure the current in the circuit, and the voltage across the resistor.
3. Change the voltage of the DC power supply to 3 V, and repeat the measurements above. It is important to measure the voltage and not use the voltage setting of the power supply!
4. Repeat for different voltages.

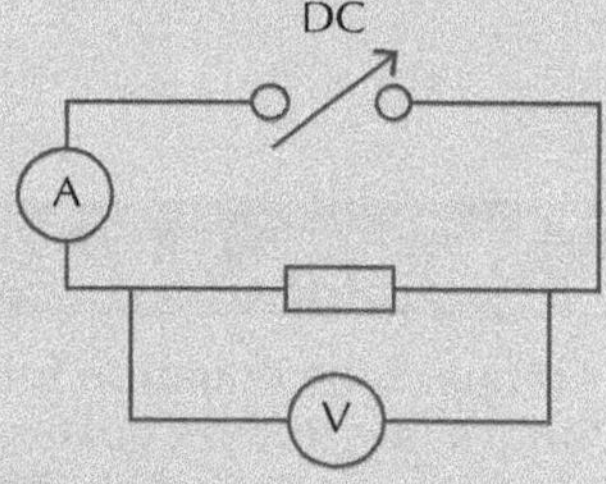

Fig 3.11.4

(*continued*)

5 Record your results in the table below. Include the third column where you calculate the value of voltage divided by current for each pair of readings:

Voltage (V)	Current (I)	Voltage/Current

Table 3.11.3

6 Look at the values you have calculated for voltage divided by current (V/I) – how do they compare to each other? How do they compare to the accepted value of the resistor you have used in your experiment?

7 Plot a graph of voltage against current and use it to determine the relationship between these two quantities (Ohm's law).

Word bank

• constant

a value that generally does not change. A constant is used to turn proportionality into an equation that can be used to solve problems. Examples of constants include: speed of light, Planck's constant and gravitational acceleration of Earth.

Typically, a graph for voltage versus current in a circuit shows a straight line through the origin.

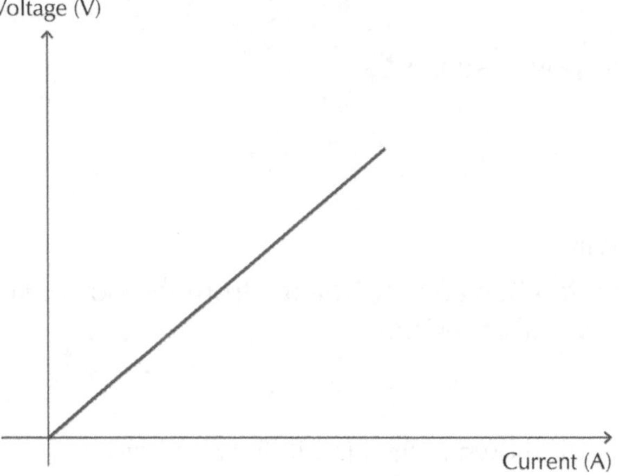

Fig 3.11.5: *A voltage-current graph showing a straight line for an ohmic conductor*

Make the link

Compare the equation for Ohm's law to the equation for a straight line through the origin in Mathematics,

$$y = mx$$

How could you work out the resistance of the resistor using just the graph?

In other words, voltage and current are directly proportional if conditions are constant:

$$V \propto I$$

This means that,

$$V = kI$$

where k is a **constant**.

The constant in the above equation is the resistance. This leads to the following equation, which is Ohm's law:

$$V = IR$$

Technique: Calculations involving Ohm's law

1. Write down what you know from the question, leaving blank what you are trying to find:
   Voltage, V = ____ V
   Current, I = ____ A
   Resistance, R = ______ Ω
2. Write down the equation linking the quantities above as you see it on the formula sheet:
   $$V = IR$$
3. Substitute into the equation what you know.
4. Solve for the unknown.

Worked example

An electric heater is connected to mains voltage (230 V) and draws a current of 7 A.

Calculate the resistance of the heater.

Write down what we know and what we are looking for:

- *Voltage, V = 230 V*
- *Current, I = 7 A*
- *Resistance, R = ? Ω*

We are using the equation that links voltage, current and resistance:

$$V = IR$$

Substitute what we know:

$$230 = 7 \times R$$

Solve for resistance:

$$R = \frac{230}{7}$$

$$R = 32{\cdot}9\ \Omega$$

Exercise 3.11.1 Ohm's law

1 Find the missing variables in the table below.

Voltage (V)	Current (I)	Resistance (R)
10 V	5 A	(a)
230 V	0·4 A	(b)
(c)	11 A	25 Ω
(d)	50 mA	4 kΩ
50 V	(e)	25 kΩ

Table 3.11.4

Fig 3.11.6: *Desk fan*

2 A desk fan is connected to a 230 V mains supply. A current of 1·5 A is found to flow to the desk fan when it is operating normally. Calculate the resistance of the desk fan.

3 A light bulb is designed to operate when connected to a 230 V supply. The resistance of the bulb is 25 Ω. Find the current drawn by the bulb.

4 A torch bulb has a resistance of 50 Ω. When it is connected to a battery of the correct voltage it draws a current of 0.24 A. Calculate the voltage of the battery.

5 Consider the circuit shown in Fig 3.11.7. An ammeter and a voltmeter have been used to measure the current and voltage in the circuit.

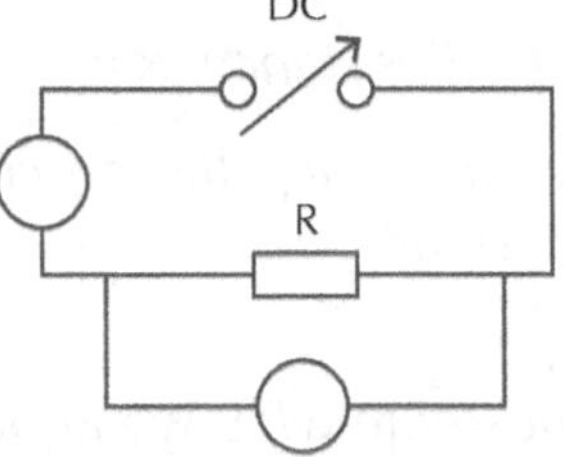

Fig 3.11.7

a) Copy the diagram, indicating which meter is the voltmeter and which is the ammeter.

b) The current was measured to be 2 A. Enter this reading onto your diagram.

c) The voltage measured was 8 V. Enter this reading onto your diagram.

d) Calculate the resistance of the resistor, R.

6 The circuit in question 5 is used to investigate Ohm's law. Students measure the following pairs of readings for voltage and current:

Voltage (V)	1·2	1·8	2·4	3·0	3·6
Current (A)	0·12	0·18	0·24	0·30	0·36

Table 3.11.5

a) Plot a graph of voltage against current and use the graph to derive the relationship between voltage and current.

b) Using your graph, or otherwise, find the resistance of the resistor in this experiment.

Non-ohmic materials

Not all materials obey Ohm's law! Materials in which the current is not proportional to the resistance are known as non-ohmic materials. An example of such a component would be a light emitting diode (LED), which has a voltage-current graph similar to the one for a diode shown in the examples below.

The graphs below show the relationship between voltage and current for both ohmic and non-ohmic materials.

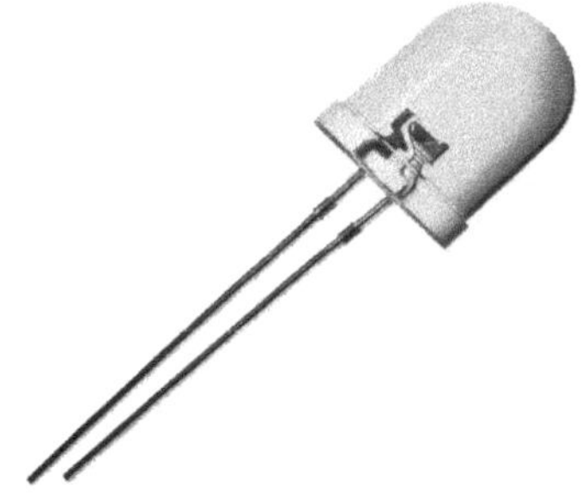

Fig 3.11.8: *Light emitting diode*

Large resistance

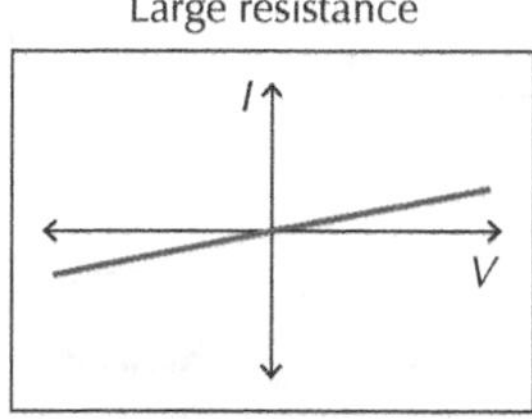

Small resistance

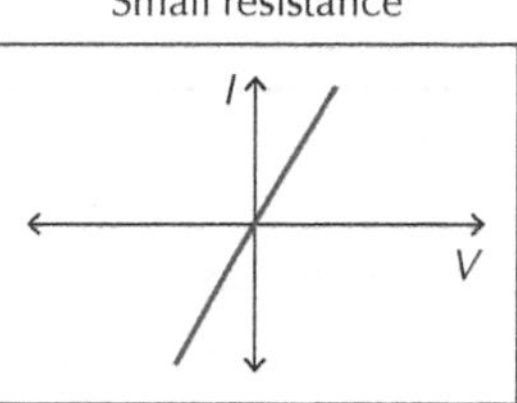

Diode

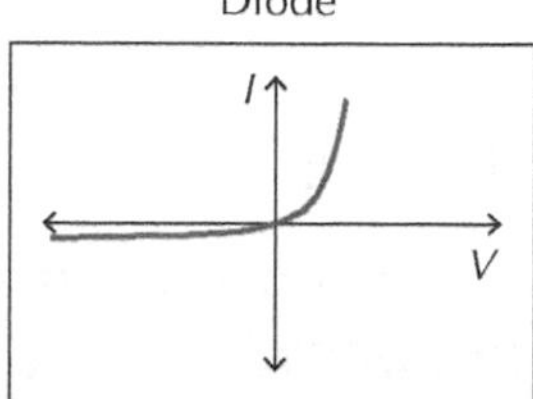

Battery

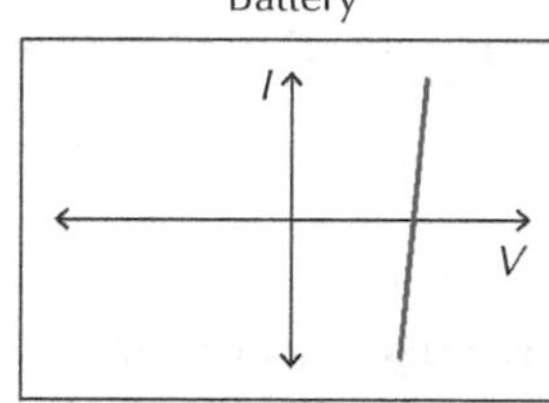

Fig 3.11.9

Experiment: Is a filament bulb ohmic?

This experiment investigates how the resistance of a filament bulb varies with temperature.

You will need:

- A DC power supply
- A filament bulb
- A voltmeter
- An ammeter
- Connecting wires

Fig 3.11.10

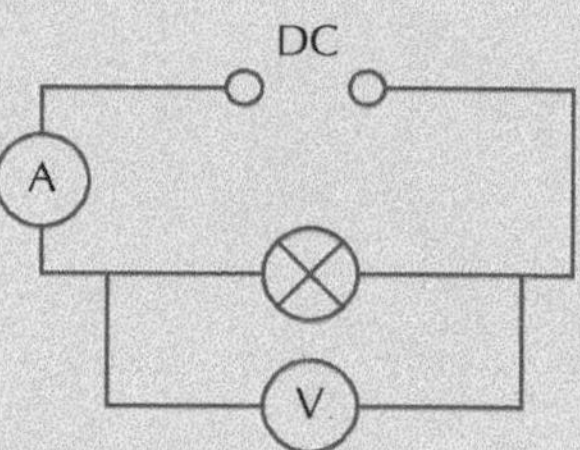

Fig 3.11.11

Instructions

1. Set up the circuit shown in Fig 3.11.11. The bulb is connected to the DC power supply, with an ammeter in series and a voltmeter connected across the bulb.
2. Switch on the DC power supply. At the same instant, note the voltage across the bulb and the current through the bulb using the voltmeter and the ammeter.
3. Wait for 5 minutes to allow the bulb to heat up.
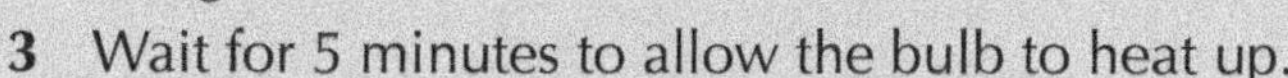
4. Repeat the voltage and current measurements.
5. Use these measurements to calculate the resistance of the bulb when it is cold and when it is hot.

   How does the resistance of the bulb vary with temperature?

Think about it

A light bulb is most likely to blow (fail) when it is first switched on. By considering the current flowing, and the resistance of the bulb when it is cold compared to when it is hot, can you explain why this is the case?

Key point

An ohmic material is one which obeys Ohm's law (current and voltage are directly proportional):

$$V = IR$$

A non-ohmic material is one which does not obey Ohm's law (such as a diode).

Learning checklist

In this section you will have learned:

- That the gradient of the line of best fit on a V-I graph can be used to determine resistance.
- How to use appropriate relationships to solve problems involving potential difference (voltage), current and resistance.
- About the qualitative relationship between the temperature and resistance of a conductor.
- How to describe an experiment to verify Ohm's law.

12 Practical electricity and electronic circuits

In this chapter you will learn about:

- Measuring voltage, current and resistance in series, parallel and complex circuits.
- The function and application of electrical and electronic components.
- Calculating the resistance of resistors in series and in parallel circuits.
- Applications of the rules for potential difference (voltage) and current in series and parallel circuits.
- The impact of adding more resistors to the total resistance of a series or parallel network.
- Transistors acting as electrical switches in simple control circuits.

Practical electrical and electronic circuits

Before you can really start to understand electronic circuits you need to recognise all the symbols used in constructing electronic circuit diagrams. Shown below are most of the symbols you need to recognise:

SYMBOL	NAME/FUNCTION
———	connecting wire
┼	two wires not connected
┼ (dot)	two wires connected
⊣⊢	cell – provides electrical energy
⊣⊢--⊣⊢	battery – provides electrical energy
⊗	lamp – acts as an indicator
—⁄—	switch – opens a circuit preventing current from passing
—▭—	resistor – restricts the passage of current
—▭̸—	variable resistor – restricts the passage of current
Ⓥ	voltmeter – measures p.d. (voltage)
Ⓐ	ammeter – measures current
LED symbol	LED – acts as an indicator
mosfet symbol	mosfet – acts as an electrical switch
Ⓜ	motor – provides rotational motion

SYMBOL	NAME/FUNCTION
microphone symbol	microphone – converts sound energy to electrical energy
loudspeaker symbol	loudspeaker – converts electrical energy to sound energy
photovoltaic cell symbol	photovoltaic cell – converts light to electrical energy
fuse symbol	fuse – protects cables from overheating
diode symbol	diode – restricts the movement of current in a circuit
⊣⊢	capacitor – stores charge/electrical energy
thermistor symbol	thermistor – tempearture dependent resistor
LDR symbol	LDR – resistance varies with light intensity
relay symbol	relay – acts as an electrical switch
transistor symbol	(NPN) transistor – acts as an electrical switch
buzzer symbol	buzzer – converts electrical energy to sound energy

The function of some of these components represented by the symbols above are described in more detail later on.

Series circuits

Think about it

The headlights in a car are not connected in a series circuit. Using the results of the above experiment to help you, can you explain why a series circuit is not used?

Key point

A series circuit consists of only one loop, meaning current can only flow along one path. A break anywhere in the circuit will cause the whole circuit to stop working.

So far we have seen that meters can be connected either in series or in parallel with a component. When connected in series, the meter (an ammeter) is connected in the circuit with the component as shown in the circuit diagram below.

A series circuit is one which has only one loop. Current can only flow along one path in a series circuit. Any break in the circuit will cause the whole circuit to stop working as the only current path has been broken.

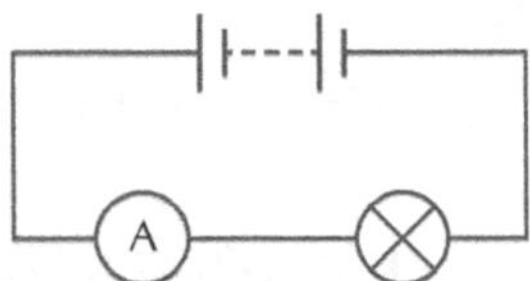

Fig 3.12.1: *A series circuit with a filament bulb and ammeter*

Experiment: Series circuit basics

This experiment investigates the basic function of a series circuit, comparing the effects of adding additional components and putting a break into the circuit.

You will need:

- A DC power supply
- 12 V filament bulbs (at least 3)
- Connecting wires

Instructions

1. Set up the series circuit shown in Fig 3.12.2 with one bulb connected to the DC power supply.
2. Switch on the power supply. How bright is the bulb?
3. Now add a second bulb to the circuit as shown in Fig 3.12.3. The second bulb is connected in series with the first. Make sure you switch the power supply off before making any changes to the circuit!
4. Switch on the power supply. How does the brightness of both bulbs compare to the brightness of one bulb on its own?
5. Repeat the above experiment with three bulbs, commenting on the relative brightness of the bulbs.

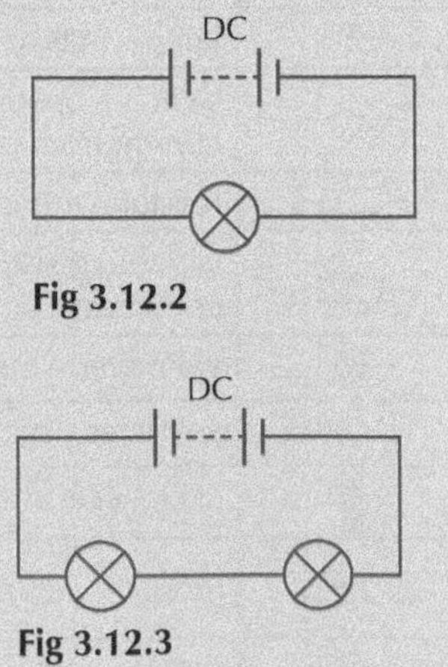

Fig 3.12.2

Fig 3.12.3

(*continued*)

6 How does the brightness of the bulb depend on the number of bulbs in the circuit?
7 Remove one of the bulbs from its holder to produce a break in the circuit – what happens to all of the other bulbs in the circuit?

Experiment: Current in a series circuit

This experiment investigates how current changes in a series circuit.

You will need:

- A DC power supply
- Two 12 V filament bulbs
- An ammeter
- Connecting wires

Instructions

1 Set up the series circuit shown in Fig 3.12.4 with two bulbs, a DC power supply and an ammeter to measure the current.

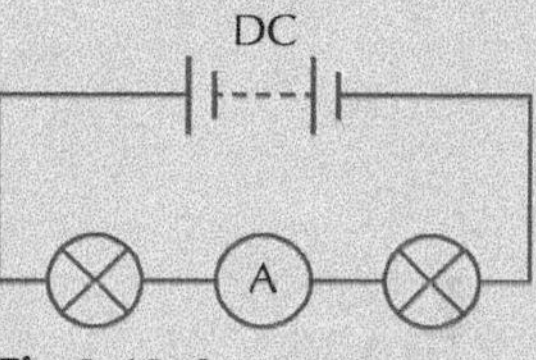

Fig 3.12.4

2 Switch on the DC power supply, and check that the bulbs light. Measure the current in the circuit using the ammeter.
3 Repeat the above experiment with the ammeter at different positions in the circuit as shown by the diagrams below.

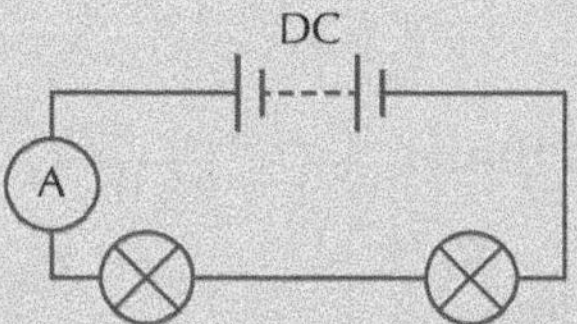

Fig 3.12.5

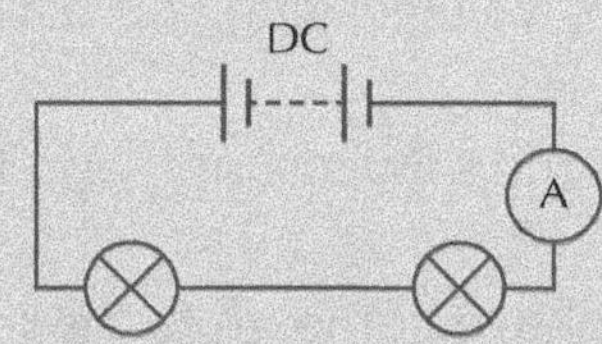

Fig 3.12.6

4 Comment on how the measured value of current depends on the position of the ammeter.
5 Now connect a third bulb in series in the circuit. How does the current with three bulbs compare to that with just two bulbs?
6 Using your results for part 5 above, predict a relationship between the current flowing and the brightness of the bulb.

Experiment: Voltage in a series circuit

This experiment investigates how the voltage changes in a series circuit.
You will need:

- A DC power supply
- Two resistors with different resistances
- Two filament bulbs
- A voltmeter
- Connecting wires

Instructions

1 Set up the series circuit shown in the diagram below with two bulbs, a DC power supply and a voltmeter to measure the voltage.

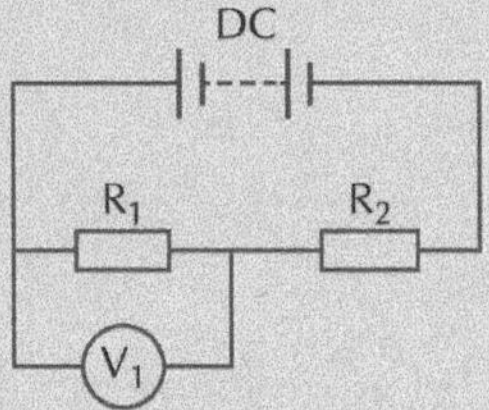

Fig 3.12.7

2 Switch on the DC power supply and measure the voltage across resistor R_1. For convenience we will call this voltage V_1.

3 Now move the voltmeter to be positioned across resistor R_2 as shown below and measure the voltage V_2.

4 Comment on the value of the two voltages that you have measured. Are they the same?

5 Now measure the voltage across the power supply, V_T. This is the total voltage in the circuit.

6 How does the size of the total voltage compare to the individual voltages you measured across the resistors?

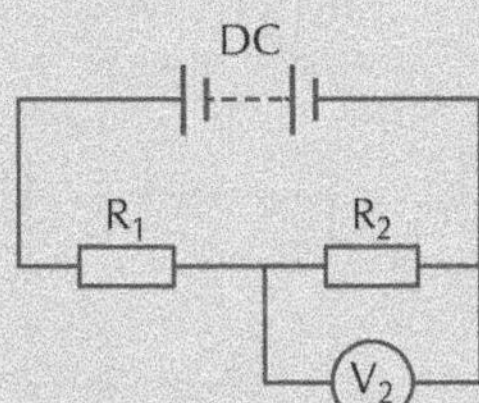

Fig 3.12.8

Key point

The current in a series circuit is the same at all points in the circuit.

This is due to the fact that there is only one path for the current to take, so it is the same everywhere.

Technique: Current and voltage in a series circuit

Current and voltage in a series circuit can be worked out using the two main rules:

Current is the same at all points.

$$A_1 = A_2$$

(continued)

Voltage splits between the individual components.

$$V_T = V_1 + V_2$$

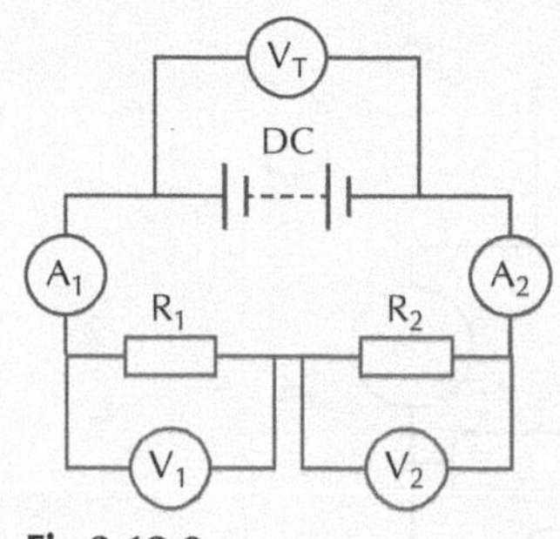

Fig 3.12.9

Key point

The voltage in a series circuit splits between the components. The sum of the individual voltages adds up to the total voltage in the circuit,

$$V_T = V_1 + V_2$$

Worked example

Calculate the missing current and voltage values for the circuit shown.

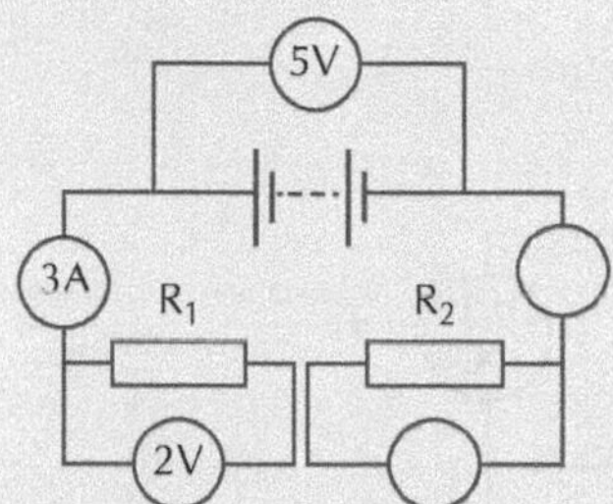

Fig 3.12.10

Starting with the current, we know that current is the same at all points,

$$A_1 = A_2$$

Therefore the missing current must also be 3 A.

$$A_2 = 3A$$

The voltage can be calculated using the relationship that the individual voltages add up to equal the supply voltage,

$$V_T = V_1 + V_2$$

This gives,

$$5 = 2 + V_2$$

Solve for the missing voltage,

$$V_2 = 3\ V$$

Exercise 3.12.1 Current and voltage in a series circuit

1 For each of the circuits shown below, find the missing voltage and current values.

(a)

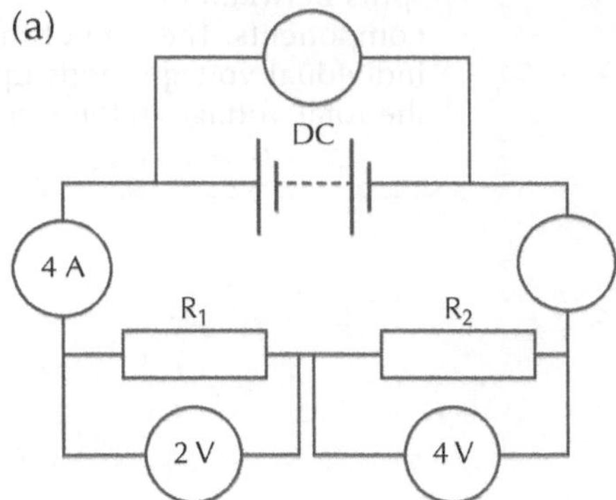

Fig 3.12.11

(b)

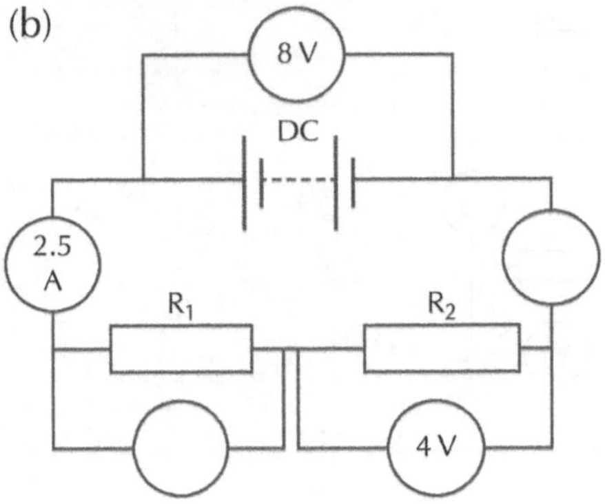

Fig 3.12.12

(c)

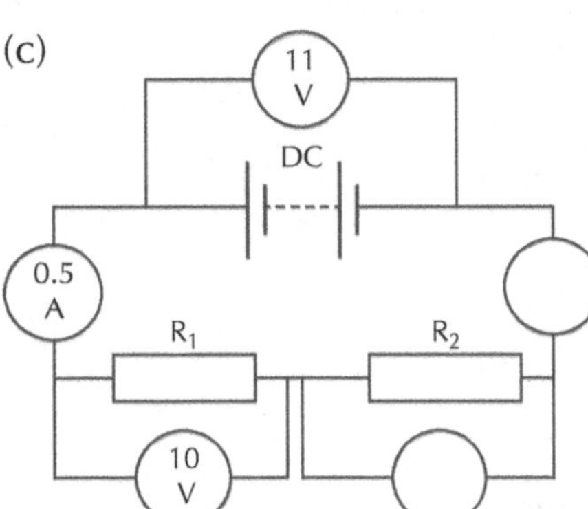

Fig 3.12.13

(d)

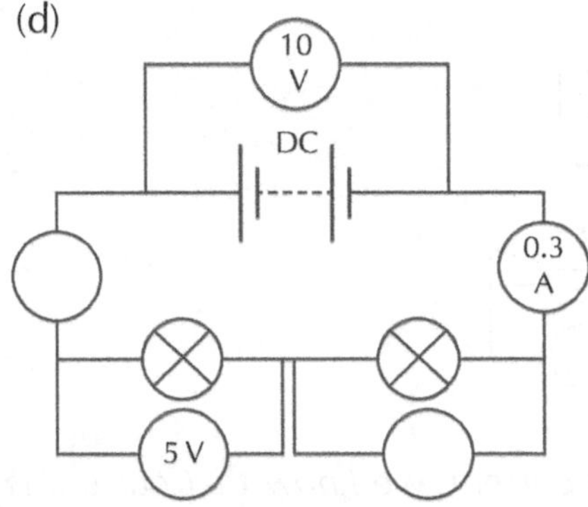

Fig 3.12.14

2 In a decorative ornament, four filament light bulbs are connected in series with each other and a 12 V battery as shown in the circuit diagram below (Fig 3.12.14).

a) One of the bulbs blows. How does this affect the other bulbs in the circuit?

b) An electrician diagnoses the fault with the ornament. She fits a new bulb to replace the faulty one. In checking the appliance, she measures the current flowing through different parts of the circuit – how does the current change at different point in a series circuit?

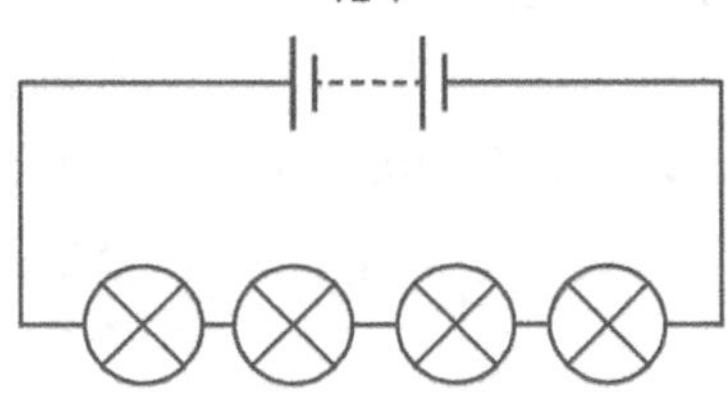

Fig 3.12.15

3 Draw a circuit diagram to show a filament bulb and two resistors (R_1 and R_2) connected in series with a 20 V DC power supply. On your diagram include:

a) the position of the ammeter you would use to measure the current in the circuit;

b) the position of the voltmeter you would use to measure the voltage across the bulb;

c) the position of the voltmeter you would use to measure the voltage across R_2.

Resistance in a series circuit

We have seen that the current is the same at all points in a series circuit, while the voltage splits across the individual components. Each electrical component in a circuit will have a resistance. Combining different components will result in the total resistance, R_T, of the circuit changing.

Fig 3.12.16: *Resistors are coloured to indicate the size of resistance*

GO! Experiment: Resistance in a series circuit

This experiment investigates the effects of combining resistors in series on the total resistance, R_T, in a circuit.

You will need:

- A DC power supply
- Two resistors with different resistances
- An ammeter
- A voltmeter
- Connecting wires

Instructions

1 Set up the series circuit shown below consisting of two resistors connected to a DC power supply, with an ammeter to measure the current and a voltmeter to measure the voltage across the power supply.

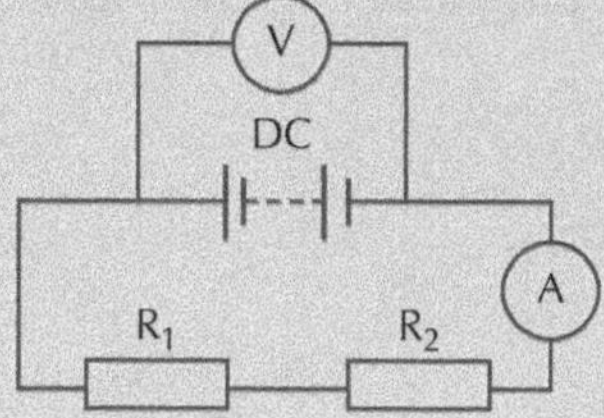

Fig 3.12.17

2 Switch on the DC power supply.

3 Measure the total voltage, V_T, in the circuit using the voltmeter.

4 Measure the current, I, in the circuit using the ammeter.

5 The total resistance can then be calculated using Ohm's law:

$$R_T = \frac{V_T}{I}$$

6 How does the total resistance compare to the individual resistances in the circuit?

Key point

The total resistance in a series circuit is equal to the sum of the individual resistances:

$$R_T = R_1 + R_2 + ...$$

Technique: Current, total voltage and total resistance

1. Work out the total resistance in a series circuit by adding together the individual resistances in the circuit,
$$R_T = R_1 + R_2 + ...$$
2. Write down what you know from the question, leaving blank what you are trying to find:
Total voltage, V_T = ____ V
Current, I = ____ A
Total Resistance, R_T = ______ Ω (calculated in part 1)
3. Write down the equation linking the quantities above as you see it on the formula sheet:
$$V = IR$$
4. Substitute into the equation what you know.
5. Solve for the unknown.

Worked example

Find the total resistance of the circuit shown below. Calculate the current flowing in the circuit given that it is connected to a 12 V DC power supply.

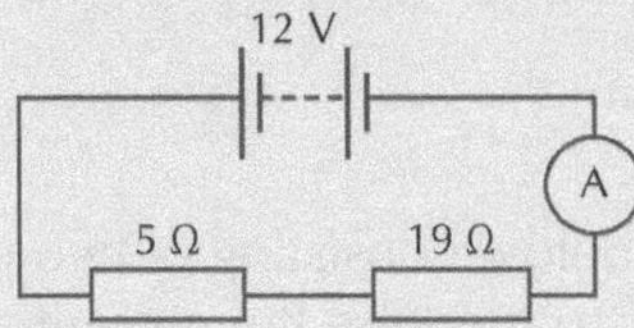

Fig 3.12.18

The total resistance is found by adding together the individual resistances:

$$R_T = R_1 + R_2 + ...$$

Substitute the values of the resistors,

$$R_T = 5 + 19$$
$$R_T = 24\ \Omega$$

The current is then found using Ohm's law. Write down what we know and what we are looking for:

Voltage, V = 12 V

Current, I = ? A

Resistance, R = 24 Ω

We are using the equation that links voltage, current and resistance:

$$V = IR$$

(continued)

Substitute what we know:

$$12 = I \times 24$$

Solve for current:

$$I = \frac{12}{24}$$

$$I = 0{\cdot}5\ A$$

Exercise 3.12.2 Total resistance in a series circuit

1 For each of the circuits shown below, calculate the total resistance.

a)

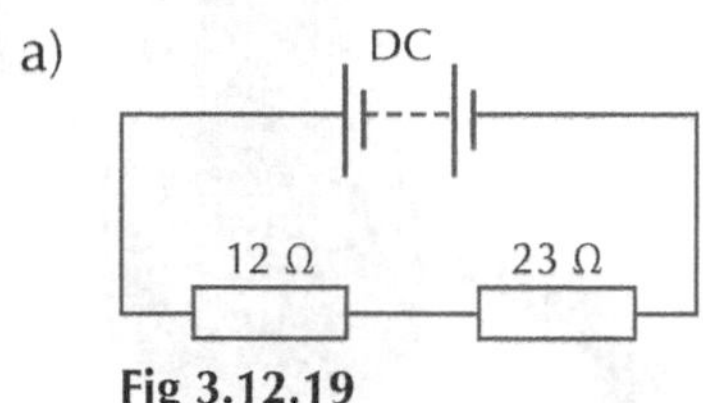

Fig 3.12.19

b)

Fig 3.12.20

c)

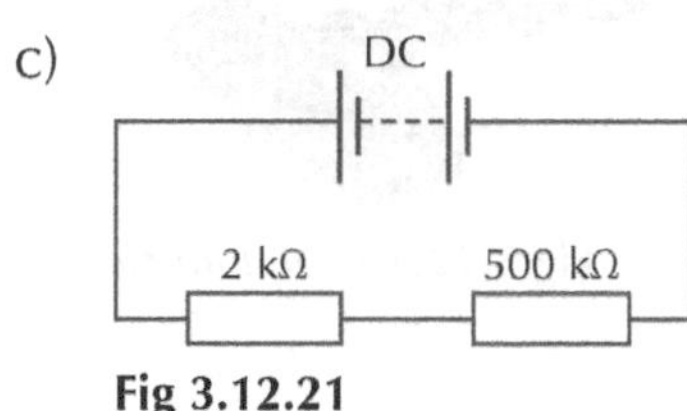

Fig 3.12.21

d)

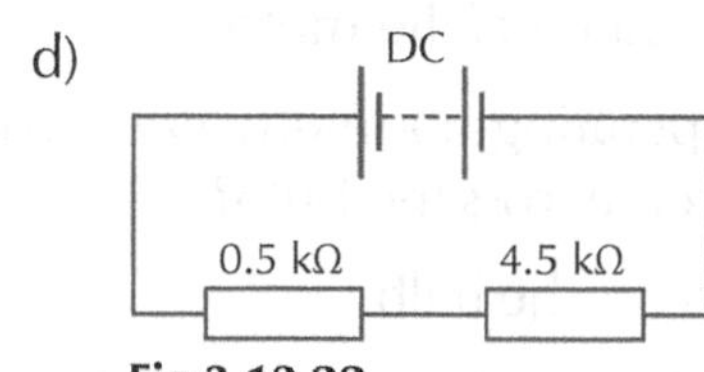

Fig 3.12.22

2 The circuit shown below is connected to a 20 V DC power supply.

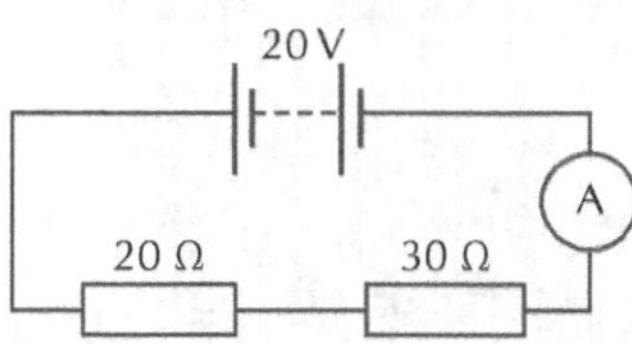

Fig 3.12.23

a) Find the total resistance of the two resistors in series.

b) Calculate the current flowing in the circuit.

3 Christmas tree lights are connected in series. Explain the potential problem with this design if conventional filament light bulbs are used. *Hint – think what would happen if one bulb failed!*

4 An engineer is designing a lighting rig for a display cabinet. He chooses to use four identical filament bulbs. Each bulb is designed to work with a voltage of 12 V and a current of 0·5 A. The engineer connects the four bulbs in series with a voltage supply.

a) Calculate the resistance of one bulb.

b) Draw a circuit diagram showing the engineer's planned circuit, with the four bulbs connected in series with the DC power supply.

Fig 3.12.24: *Bulb*

(*continued*)

c) Calculate the total resistance of the four bulbs in series.

d) What should the voltage of the DC power supply be to ensure the bulbs operate at their correct voltage?

5 The series circuit shown is used in a vacuum cleaner.

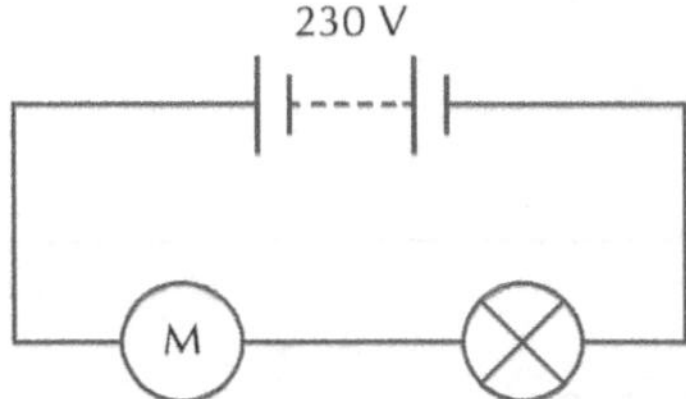

Fig 3.12.25: *A motor and filament bulb connected in series*

The motor and a bulb are connected in series. The bulb lights when the motor is on to give a visual indicator that the motor is switched on. The motor is designed to work with 225 V and a current of 10 A.

a) Calculate the resistance of the motor.

b) If the motor is operating normally, what voltage would be measured across the bulb?

c) Find the resistance of the bulb.

d) Describe why this circuit would not be recommended for connecting the motor and bulb in the vacuum cleaner. Your discussion should include series and parallel circuits, resistors and current flow. A diagram may be helpful.

Fig 3.12.26 *Vacuum cleaner*

6 Resistors are sold in sets of common values. It is therefore not possible to buy a resistor of any value you want. Explain how you could use a series circuit with more than one resistor to obtain the value of resistance that you want.

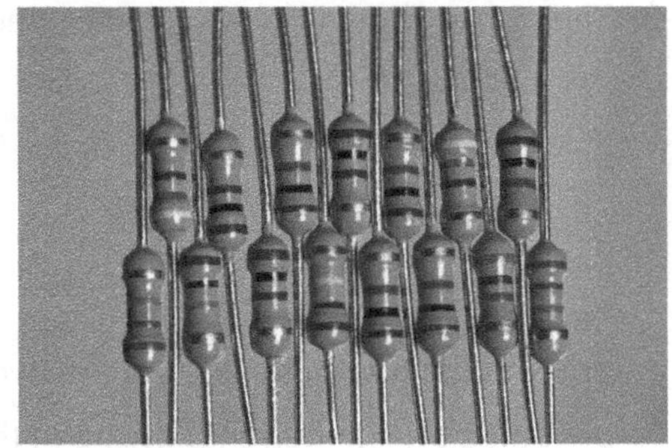

Fig 3.12.27: *Resistors*

Parallel circuits

A voltmeter is not connected in series with a component; instead it is connected across the component. This is an example of connecting a component in parallel and is shown in the diagram on the left.

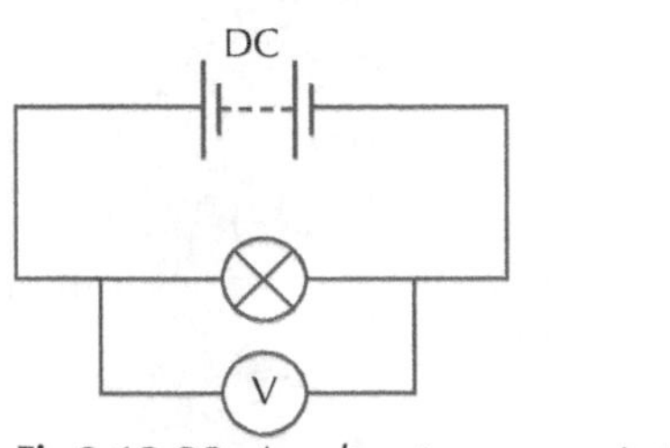

Fig 3.12.28: *A voltmeter connected in parallel across a filament bulb*

A parallel circuit has more than one loop. Current is therefore able to flow along more than one path in the circuit. A break in one of these paths will not prevent other paths from carrying charge. We name the circuit 'parallel' because on a circuit diagram, the different branches of the circuit run parallel to each other.

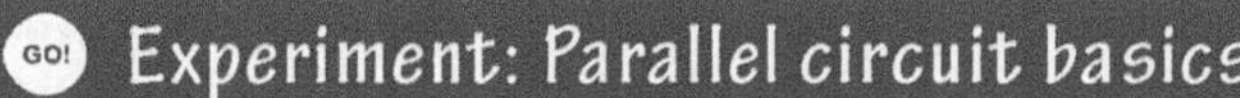

Experiment: Parallel circuit basics

This experiment investigates the basic properties of a parallel circuit, comparing the effects of adding additional components and putting a break in the circuit.

You will need:

- A 12 V DC power supply
- Three identical 12 V filament light bulbs
- Connecting wires

Instructions

1 Build the parallel circuit shown in the diagram below, consisting of two bulbs connected to the DC power supply along separate branches.

2 Use the photograph to help you.

Fig 3.12.30

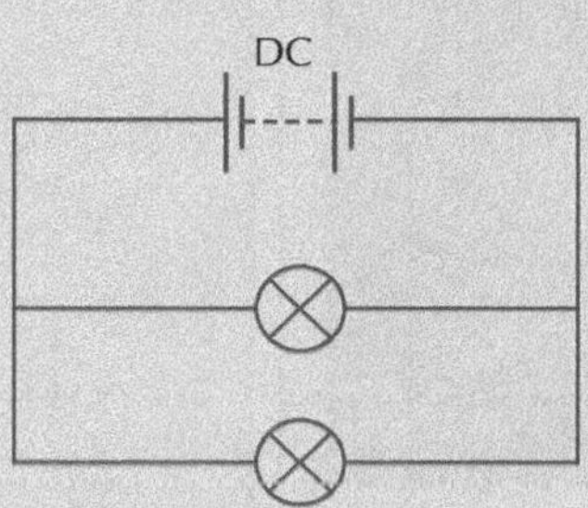

Fig 3.12.29

3 Switch on the DC power supply and note the brightness of the bulbs.

4 Now add a third bulb in parallel with the first two bulbs as shown in the circuit diagram below (Fig 3.12.31). Remember to switch off the power supply before making any changes to the circuit!

5 Comment on the brightness of the three bulbs compared to when there were just two in the circuit.

6 Now remove one of the bulbs from its holder, to form a break in one of the branches in the circuit. What happens to the other bulbs?

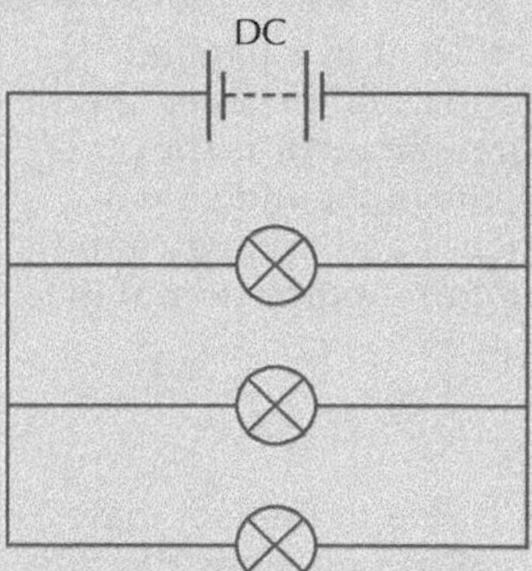

Fig 3.12.31

Key point

A parallel circuit consists of more than one loop, meaning current can flow along more than one path. A break along one of the branches of the circuit does not prevent current from flowing along the other branches. Only components in the failed branch will be affected.

Think about it

Car headlights are connected in parallel. Based on the previous experiment and the experiment with a series circuit, explain why a parallel circuit is most suitable for wiring a car's headlights.

Fig 3.12.32: *Car headlights*

Key point

The current in a parallel circuit is split between the separate branches. The currents in the individual branches add up to give the total current drawn from the supply,

$$I_T = I_1 + I_2 + \ldots$$

This is due to the fact that current can go along different paths, therefore the total current splits.

Experiment: Current in a parallel circuit

This experiment investigates how current changes in a parallel circuit.

You will need:

- A DC power supply
- Two resistors with different resistances
- An ammeter
- Connecting wires

Instructions

1 Set up the parallel circuit shown below consisting of two resistors connected in parallel with a DC power supply. Include the ammeter as shown on the diagram, immediately after the power supply.

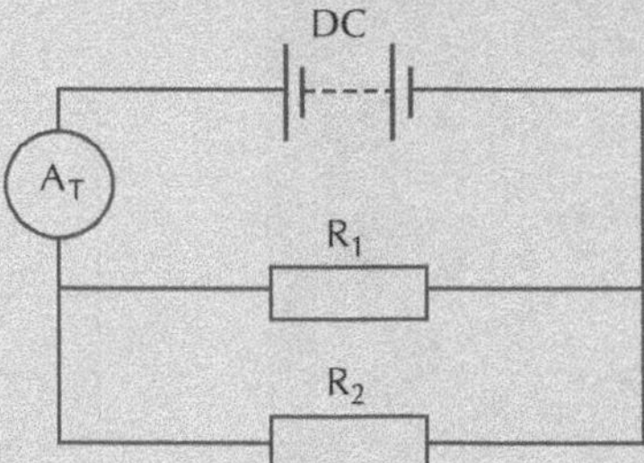

Fig 3.12.33

2 Switch on the DC power supply and use the ammeter to measure the total current, I_T, that is drawn from the power supply.

3 Now connect the ammeter to each branch in turn as shown. Measure the current, I_1 and I_2, through each resistor, R_1 and R_2.

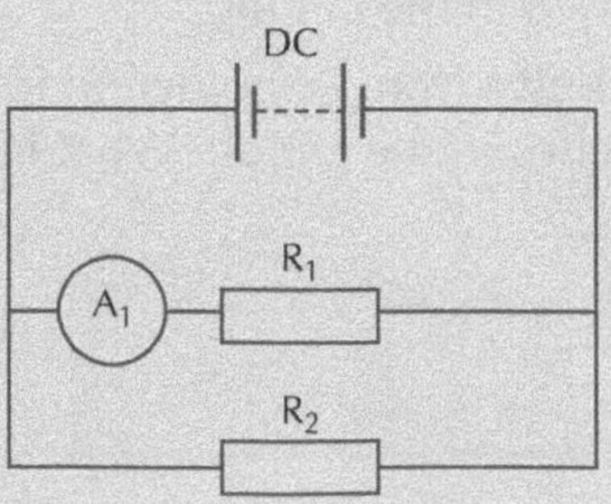

Fig 3.12.34

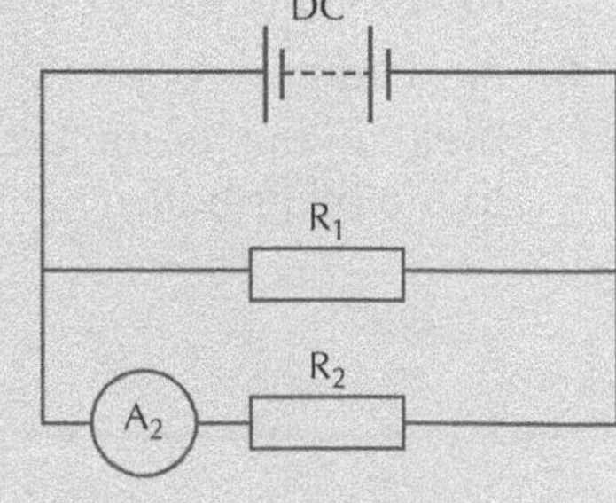

Fig 3.12.35

4 Comment on how the current through each branch compares to the other branch, and the total current drawn from the supply.

Experiment: Voltage in a parallel circuit

This experiment investigates voltage in a parallel circuit.

You will need:

- A DC power supply
- Two resistors with different resistances
- A voltmeter
- Connecting wires

Instructions

1. Connect the parallel circuit shown opposite consisting of two resistors connected in parallel with a DC power supply. Include the voltmeter as shown across the first resistor.
2. Switch on the DC power supply and measure the voltage across R_1 using the voltmeter. How does this voltage compare with the supply voltage?
3. Now measure the voltage across the second resistor, R_2, by moving the voltmeter as shown in the diagram below.
4. Compare the voltage across R_2 with the voltage across R_1 and comment on your findings.

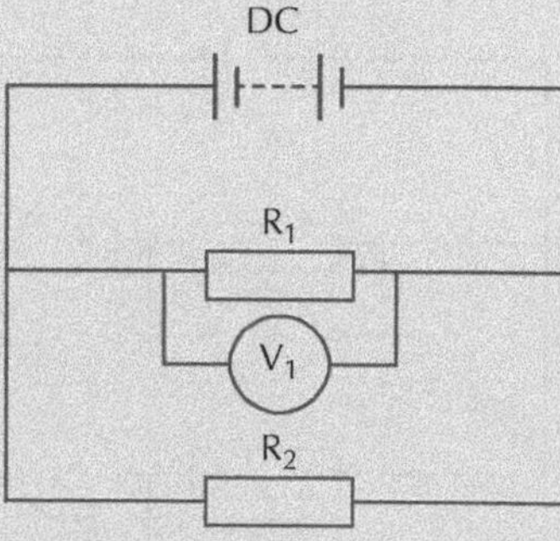

Fig 3.12.36

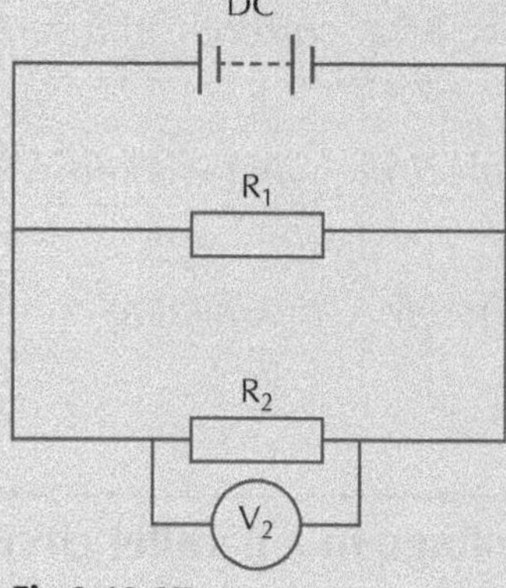

Fig 3.12.37

Worked example

Calculate the missing current and voltage values for the circuit shown below.

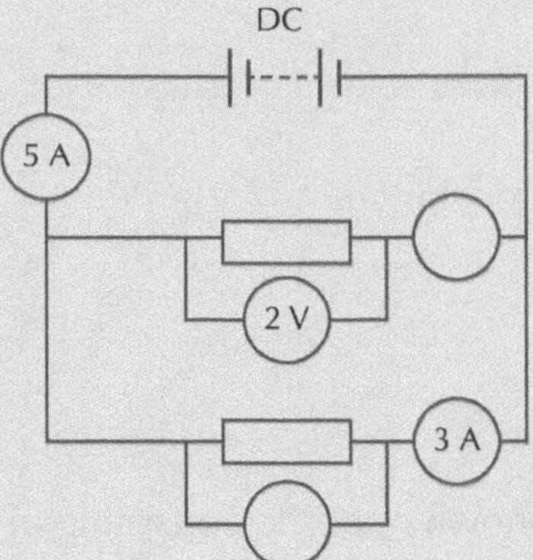

Fig 3.12.38: *A parallel circuit with 2 branches. Voltmeters are connected in parallel but do not change the circuit structure*

Start with the voltage. We know that the voltage across each branch is the same:

$$V_T = V_1 = V_2$$

Therefore the missing voltage is 2 V,

$$V_2 = 2\ V$$

The current in the circuit splits across the different branches. The total current is the sum of the individual currents,

$$I_T = I_1 + I_2$$

Substitute what we know,

$$5 = I_1 + 3$$

Therefore the missing current is:

$$I_1 = 2\ A$$

Key point

The voltage in a parallel circuit is the same across each branch:

$$V_T = V_1 = V_2$$

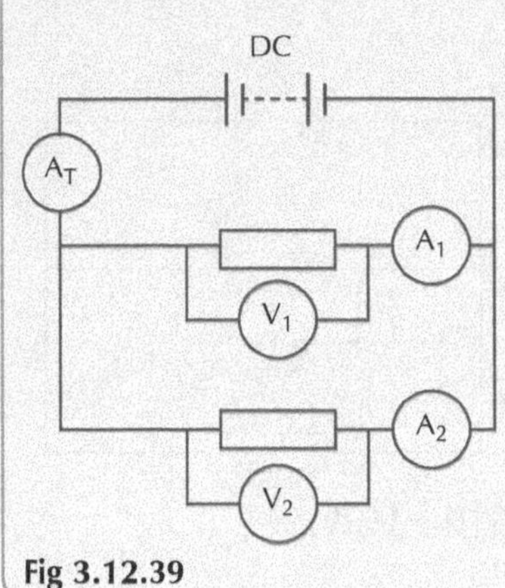

Fig 3.12.39

Exercise 3.12.3 Current and voltage in a parallel circuit

1 For each of the circuits below, find the missing current and voltage values.

(a)

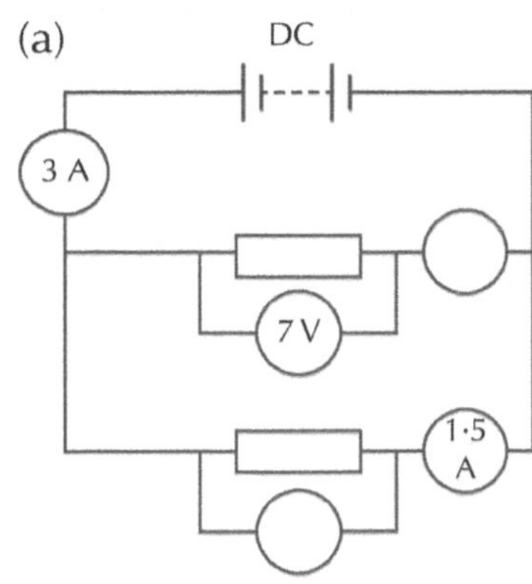

Fig 3.12.40

(b)

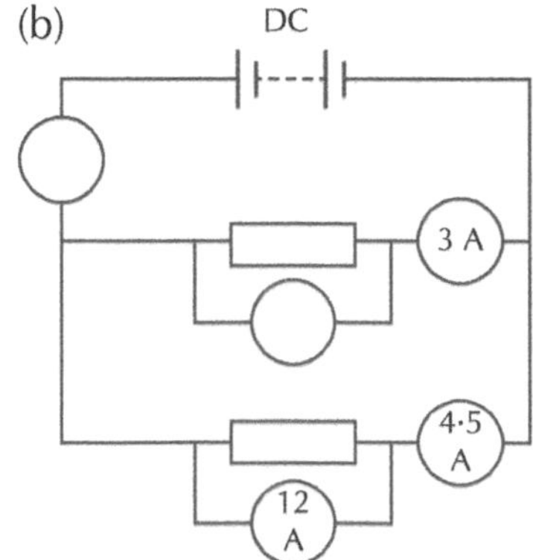

Fig 3.12.41

(continued)

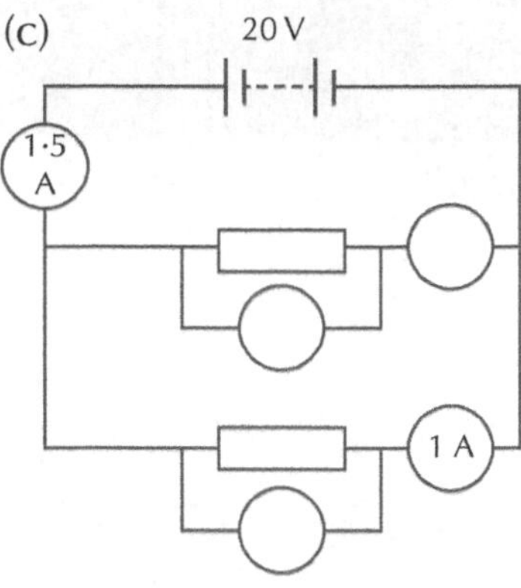

Fig 3.12.42

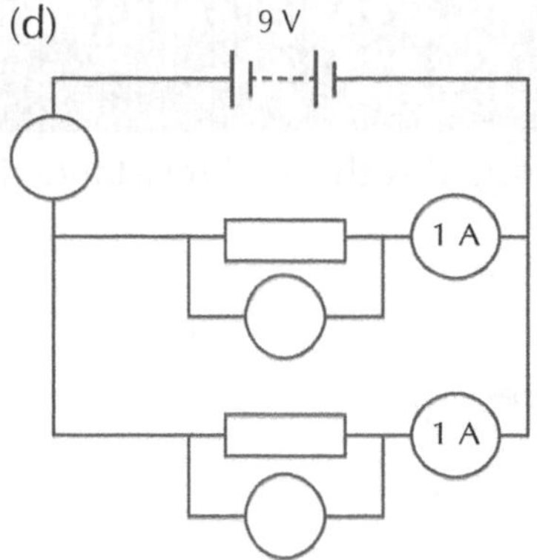

Fig 3.12.43

2 A desk lamp consists of two filament bulbs both connected to the 230 V mains supply.

The bulbs are connected in parallel.

a) Explain why both bulbs can use a voltage of 230 V.

b) Each bulb draws a current of 2·4 A. Calculate the total current drawn from the mains.

c) One of the bulbs fails. State and explain the effect this has on the other bulb.

Fig 3.12.44: *Desk lamp with two bulbs connected to mains voltage*

3 A student builds a circuit in a lab consisting of three resistors, R_1, R_2 and R_3 in parallel with a 4 V DC supply. Draw a circuit diagram to show this circuit and include on it:

a) a meter to measure the total current drawn from the supply.

b) a meter to measure the current through R_2.

c) a meter to measure the voltage across R_3.

Resistance in a parallel circuit

We have seen that the voltage is the same across each branch in a parallel circuit, while the current splits through the individual components. Each electrical component in a circuit will have a resistance. Combining different components in parallel will result in the total resistance, R_T, of the circuit changing.

Experiment: Total resistance in a parallel circuit (identical resistors)

This experiment investigates the total resistance in a parallel circuit.

You will need:

- DC power supply
- Three identical resistors
- An ammeter
- A voltmeter
- Connecting wires

Instructions

1 Set up the parallel circuit shown in the diagram consisting of two identical resistors, an ammeter to measure the total current and a voltmeter to measure the voltage across the power supply (total voltage).

2 Switch on the DC power supply.

3 Measure the total voltage, V_T, in the circuit using the voltmeter.

4 Measure the current, I, in the circuit using the ammeter.

5 The total resistance can then be calculated using Ohm's law:

$$R_T = \frac{V_T}{I}$$

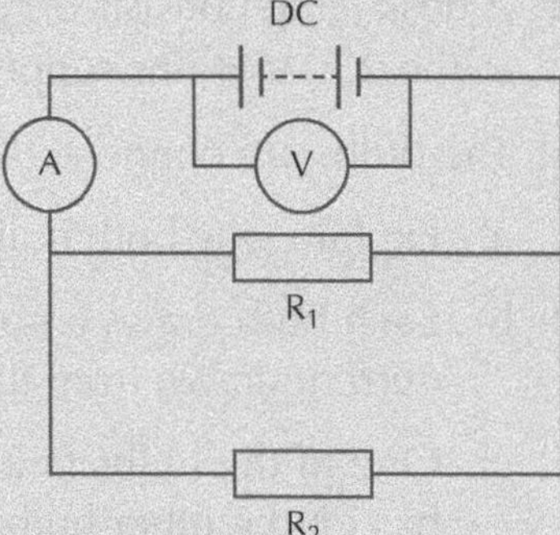

Fig 3.12.45

6 How does the total resistance compare to the resistance of a single resistor in the circuit?

7 Repeat the experiment using a third identical resistor in parallel. How does the total resistance compare now to the resistance of a single resistor?

Experiment: Total resistance in a parallel circuit (different resistors)

This experiment investigates the total resistance in a parallel circuit.

You will need:

- DC power supply
- Three different resistors
- An ammeter
- A voltmeter
- Connecting wires

(continued)

Instructions

1 Carry out the above experiment again, repeating steps 1 to 6, but this time add a second resistor with a different value.

2 Repeat the experiment using a third resistor with a different resistance in parallel. How does the total resistance now compare to the resistance of a single resistor?

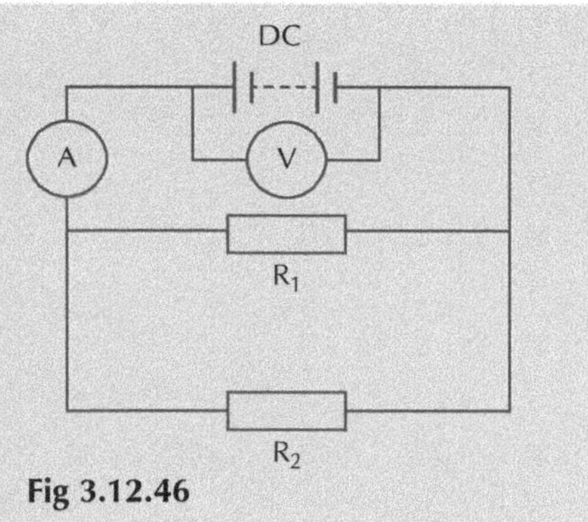

Fig 3.12.46

Technique: Current, total voltage and total resistance

1. Work out the total resistance in a parallel circuit as follows:

$$\frac{1}{R_T} = \frac{1}{R_1} + \frac{1}{R_2} + \ldots$$

Remember to invert your answer at the end of the resistance calculation to find R_T.

2. Write down what you know from the question, leaving blank what you are trying to find:

Total voltage, V = ____ V

Current, I = ____ A

Total resistance, R = ______ Ω (calculated in part 1)

3. Write down the equation linking the quantities above as you see it on the formula sheet:

$$V = IR$$

4. Substitute into the equation what you know.
5. Solve for the unknown.

Key point

The total resistance in a parallel circuit is worked out using the following equation:

$$\frac{1}{R_T} = \frac{1}{R_1} + \frac{1}{R_2} + \ldots$$

Worked example

Find the total resistance for the circuit shown below, and calculate the total current drawn from the 9 V battery.

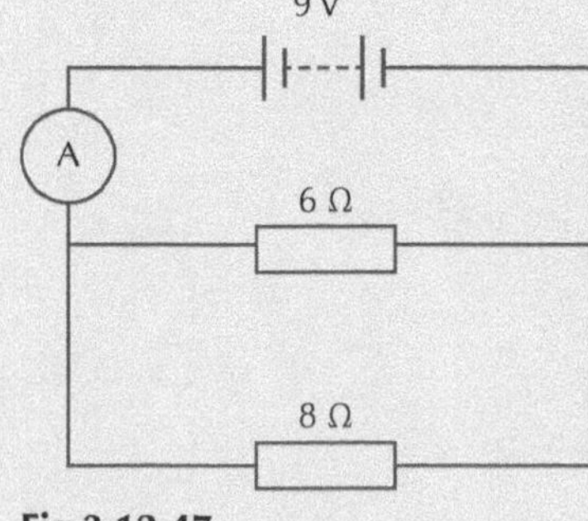

Fig 3.12.47

(continued)

The total resistance is found by using the following equation:

$$\frac{1}{R_T} = \frac{1}{R_1} + \frac{1}{R_2}$$

Substitute the values of the resistors,

$$\frac{1}{R_T} = \frac{1}{6} + \frac{1}{8}$$

Evaluate the right hand side of this equation in decimal form using a calculator,

$$\frac{1}{R_T} = 0{\cdot}17 + 0{\cdot}13$$

$$\frac{1}{R_T} = 0{\cdot}30$$

Solve for the total resistance by dividing 1 by the right hand side:

$$R_T = \frac{1}{0{\cdot}30}$$

$$R_T = 3{\cdot}3\ \Omega$$

The current is then found using Ohm's law. Write down what we know and what we are looking for:

Voltage, $V = 9$ V

Current, $I = ?$ A

Resistance, $R = 3{\cdot}3\ \Omega$

We are using the equation that links voltage, current and resistance:

$$V = IR$$

Substitute what we know:

$$9 = I \times 3{\cdot}3$$

Solve for current:

$$I = \frac{9}{3{\cdot}3}$$

$$I = 2{\cdot}7\,A$$

Exercise 3.12.4 Total resistance in a parallel circuit

1 Calculate the total resistance for each of the circuits below.

(a)

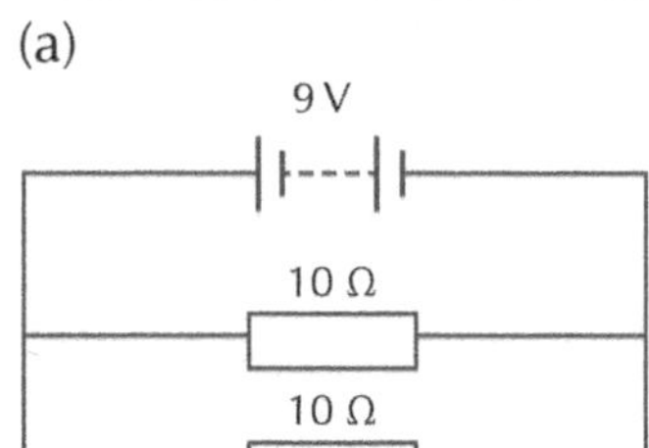

Fig 3.12.48

(b)

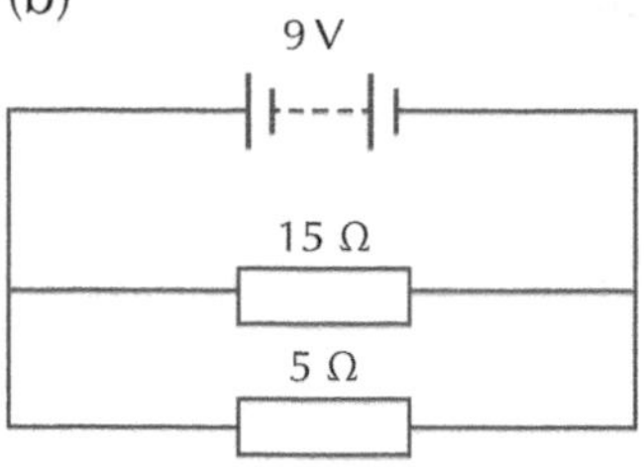

9 V

15 Ω

5 Ω

Fig 3.12.49

(c)

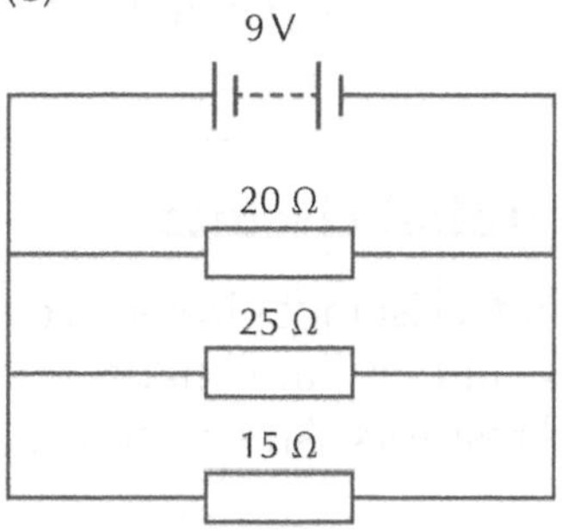

9 V

20 Ω

25 Ω

15 Ω

Fig 3.12.50

(d)

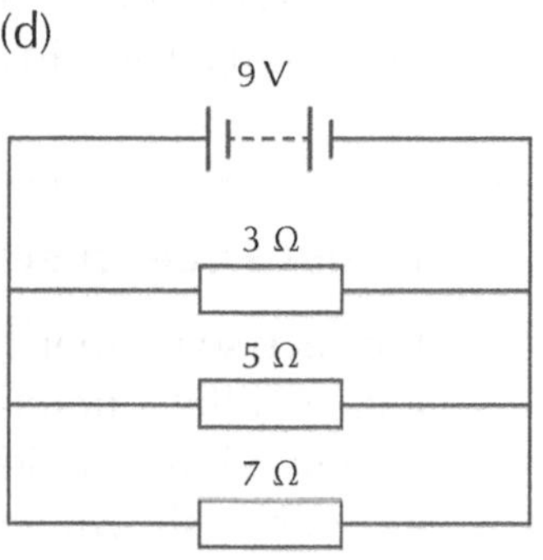

Fig 3.12.51

2 The circuit shown is connected to a 10 V DC power supply.

a) Calculate the total resistance of the three resistors in parallel.

b) Calculate the current drawn from the 10 V battery.

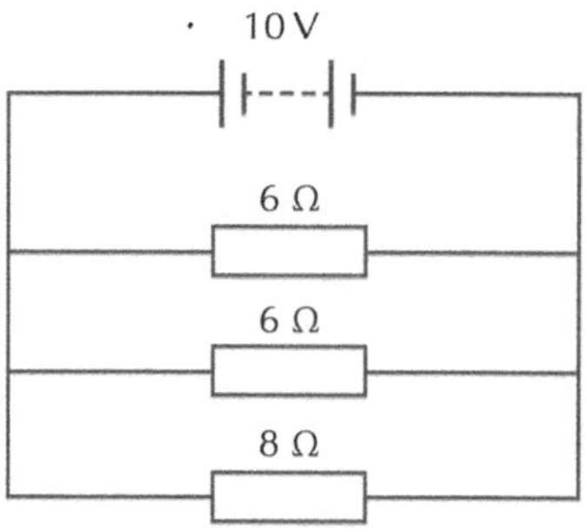

10 V

6 Ω

6 Ω

8 Ω

Fig 3.12.52

3 A car's headlights are connected in parallel with the 12 V car battery.

a) Explain why connecting the bulbs in parallel allows all of the 12 V bulbs to operate at their correct brightness.

b) Explain why the failure of one bulb does not result in all of the other bulbs going off.

4 A cooling system for a room consists of two cooling fans driven by motors connected to the 230 V mains supply. The cooling fans are connected in parallel. They are rated at 230 V, 8 A.

a) Draw a circuit diagram showing the 230 V power supply and the two motors connected in parallel.

b) Find the resistance of one of the motors when it is operating at its rated voltage and current.

c) Find the total resistance of the two motors connected in parallel.

(continued)

5 Resistors are only sold in certain values. However, you can use different resistors in series or parallel to get the resistance you want. You have the following resistors to choose from: 10 Ω; 25 Ω; 50 Ω; 100 Ω; 500 Ω; 1000 Ω. Describe how you would connect resistors to get the following resistances.

a) 60 Ω

b) 20 Ω

c) 32·26 Ω

d) 55·5 Ω

6 When more than one appliance is connected to a mains socket, the appliances are connected in parallel. Explain, in terms of the total current drawn, why connecting too many appliances to one socket is dangerous. Include the current rule for parallel circuits in your explanation.

Resistance in series and parallel circuits

We have so far seen how to combine resistances in series circuits and parallel circuits. Sometimes, circuits are made up of a combination of series and parallel resistors. An example of such a circuit is shown here.

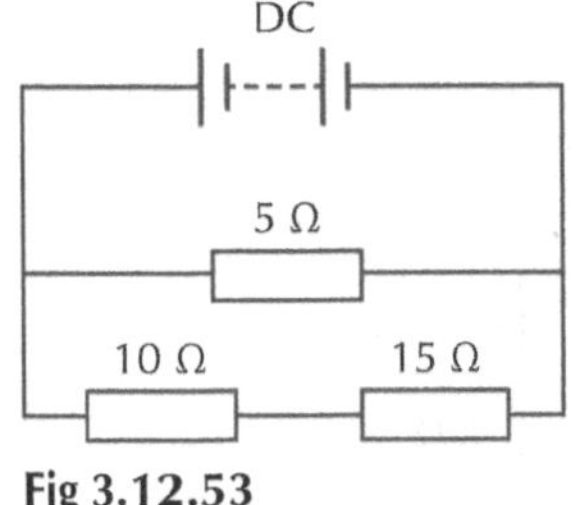

Fig 3.12.53

In order to work out the total resistance of the resistors shown, you need to:

- work out the series resistance of the series branch first
- then work out the parallel combination.

It is important that any branch calculations are carried out first. The total from the branches can be summed to find the total equivalent resistance. In other circuits it may be necessary to work out the parallel combination first and then the series combination. See the worked examples below.

Worked example

Calculate the total resistance of the circuit shown below.

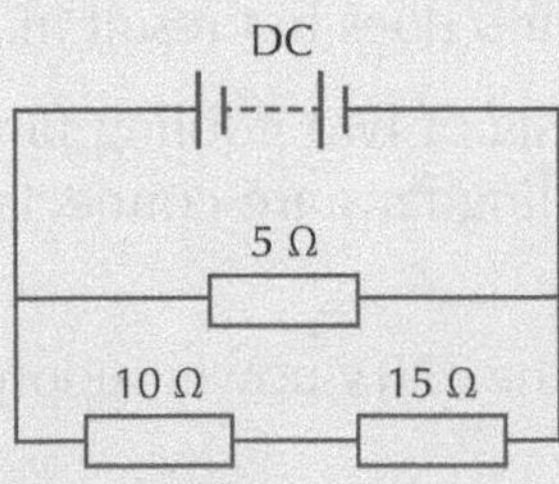

Fig 3.12.54

Calculate the total resistance of the two resistors in series first:

$$R_T = R_1 + R_2$$
$$R_T = 10 + 15$$
$$R_T = 25\ \Omega$$

(continued)

We can now re-draw the circuit as shown below,

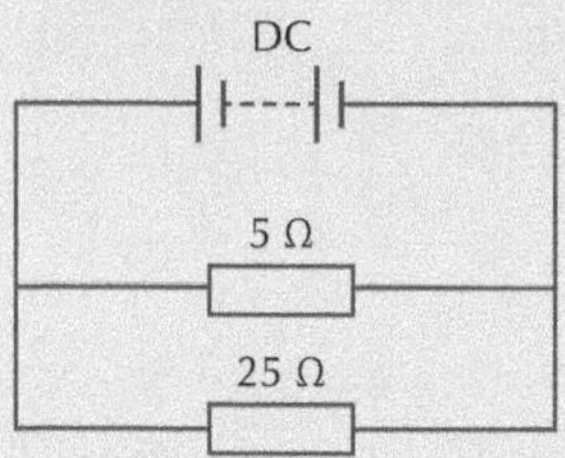

Fig 3.12.55

Now use the rule for a parallel circuit to find the total resistance,

$$\frac{1}{R_T} = \frac{1}{R_1} + \frac{1}{R_2}$$

$$\frac{1}{R_T} = \frac{1}{5} + \frac{1}{25}$$

$$\frac{1}{R_T} = 0{\cdot}2 + 0{\cdot}04$$

$$\frac{1}{R_T} = 0{\cdot}24$$

$$R_T = \frac{1}{0{\cdot}24}$$

$$R_T = 4{\cdot}17\Omega$$

Worked example

Calculate the total resistance of the circuit below.

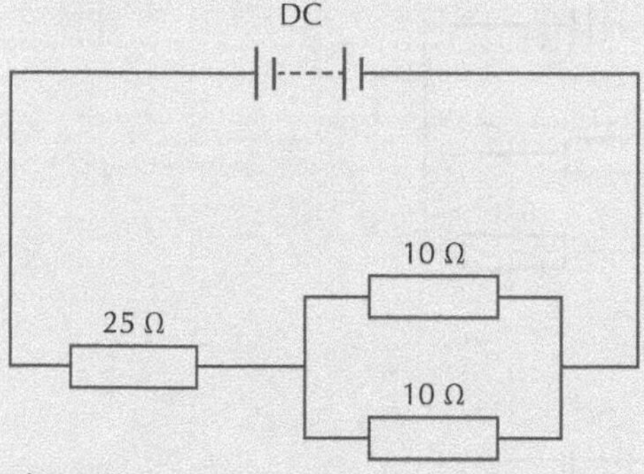

Fig 3.12.56

Start off by working out the parallel resistors,

$$\frac{1}{R_T} = \frac{1}{R_1} + \frac{1}{R_2}$$

$$\frac{1}{R_T} = \frac{1}{10} + \frac{1}{10}$$

$$\frac{1}{R_I} = 0{\cdot}1 + 0{\cdot}1$$

$$\frac{1}{R_T} = 0{\cdot}2$$

$$R_T = \frac{1}{0{\cdot}2}$$

$$R_T = 5\Omega$$

(continued)

The above circuit can now be represented as shown below – a simple series circuit!

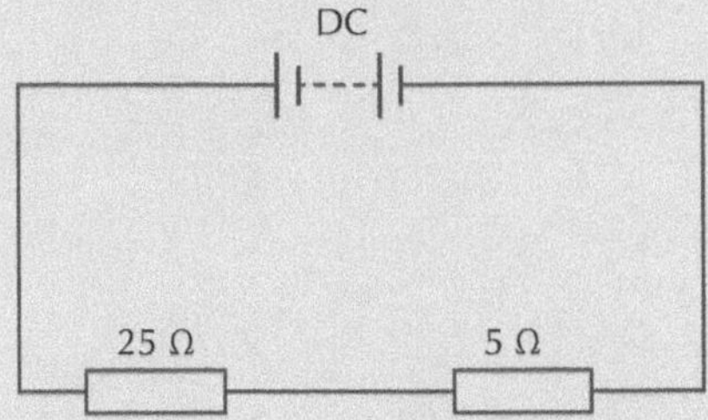

Fig 3.13.57

The total resistance can then be found using the rule for a series circuit,

$$R_T = R_1 + R_2$$
$$R_T = 25 + 5$$
$$R_T = 30\Omega$$

Exercise 3.12.5 Resistance in series and parallel circuits

1 For each of the circuits shown below calculate the total resistance.

(a)

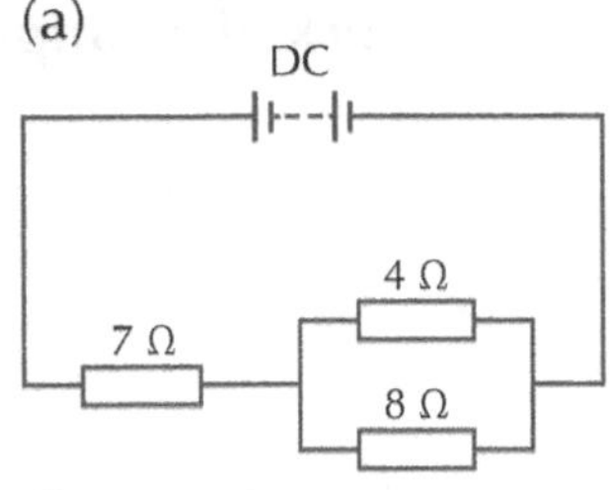

Fig 3.12.58

(b)

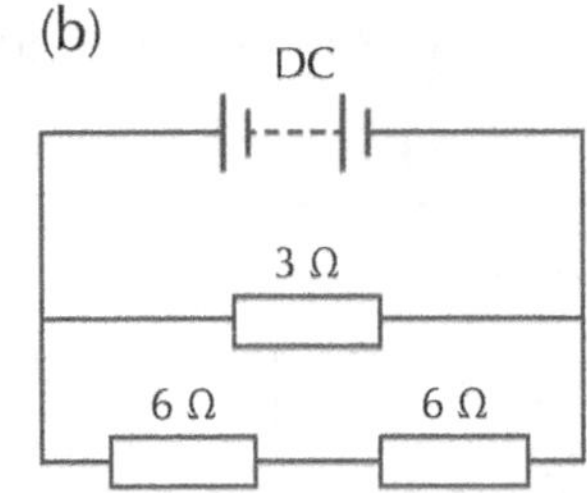

Fig 3.12.59

(c)

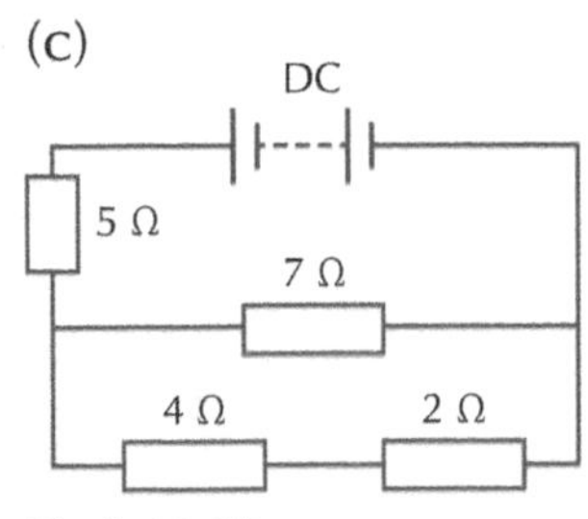

Fig 3.12.60

(d)

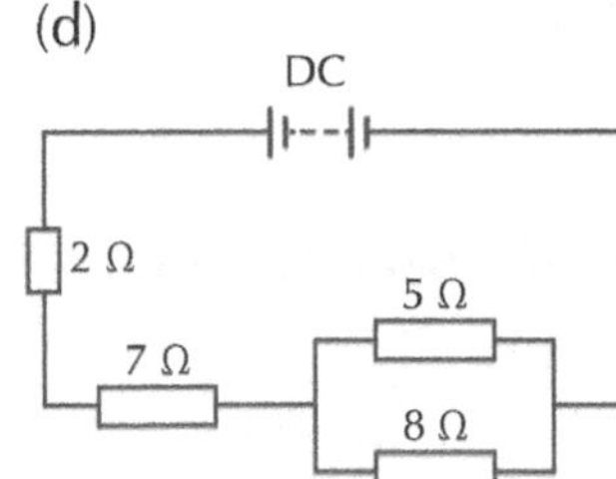

Fig 3.12.61

Useful circuits and components

Electrical components are used in many circuits that we encounter in our day to day lives, from our alarm clock in the morning to the climate control system in the car. In the latter, electrical sensors measure the temperature inside the car. When it is too hot, an electrical circuit switches on a cooling fan to cool the car down. When it is too cold, the electrical system switches on a heater to heat up the interior.

Fig 3.12.62: *Climate control system in a car*

The potential divider

We have already seen that the voltage splits across components in a series circuit. We give this type of circuit a specific name – potential divider. A potential divider takes the supply voltage (potential) and divides it between the components in the circuit.

Experiment: The potential divider

This experiment investigates how the voltage changes in a potential divider circuit (series circuit).

You will need:

- A DC power supply and battery connector*
- Potential divider alpha board*
- A variety of resistors
- A voltmeter

* These components can be substituted with resistors and connecting wires.

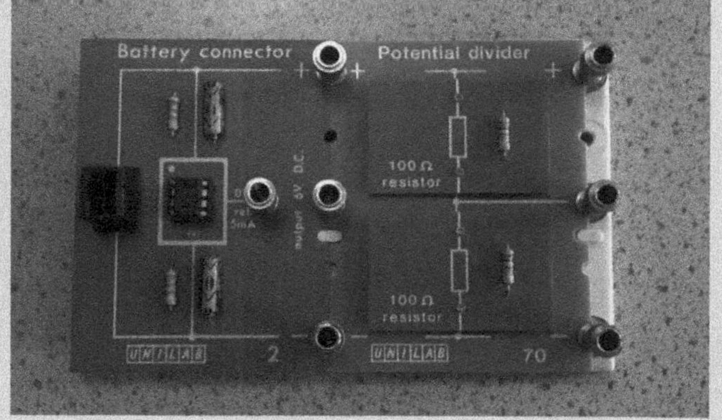

Fig 3.12.63: *Circuit board*

Instructions

1 Connect the potential divider circuit to a DC power supply. Choose equal resistance resistors to start with, for example, two 100 Ω resistors.

2 Switch on the DC power supply and use a voltmeter to measure the voltage across each of the resistors (V_1 and V_2).

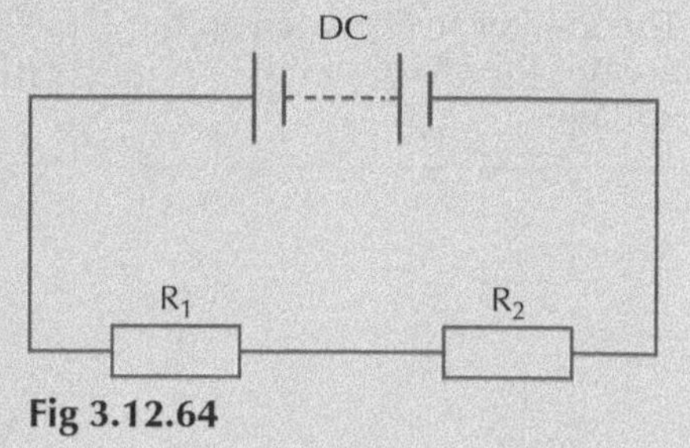

Fig 3.12.64

(continued)

3 Change the resistors and measure the voltages V_1 and V_2 for each combination of resistors. Make sure you use combinations of different resistors. Record your results in a table similar to the following:

Resistor 1 (Ω)	Resistor 2 (Ω)	Voltage 1 (V)	Voltage 2 (V)	Total voltage (V)

Table 3.12.1

4 Answer the following questions:

a) If you know the voltages V_1 and V_2, how do you calculate the total voltage?

b) What happens to the voltages across the resistors (V_1 and V_2) when you increase the resistance of resistor R_1?

c) How does the size of the voltage across a resistor depend on the resistance of the resistor in a potential divider?

In a potential divider, when the resistance of one of the resistors is increased it causes the voltage across that resistor to increase as well. If the voltage across one resistor increases, the voltage across the other resistor decreases.

When the sizes of the resistors are equal (the resistors are balanced), the size of the voltage across each resistor is the same.

Key point

In a series circuit the voltage across each of the resistors adds up to equal the supply voltage.
The greater the resistance, the greater the share of the voltage.

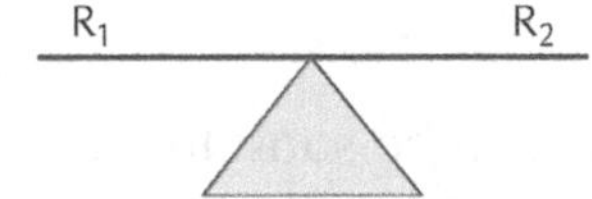

Fig 3.12.65: *Resistors of equal sizes*

When R_1 is greater than R_2, the voltage across R_1 is greater than R_2.

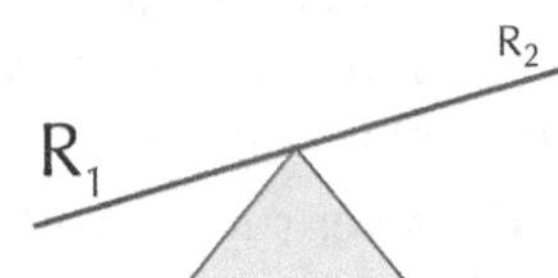

Fig 3.12.66: *Resistor 1 is greater than resistor 2*

When R_1 is less than R_2, the voltage across R_1 is less than R_2.

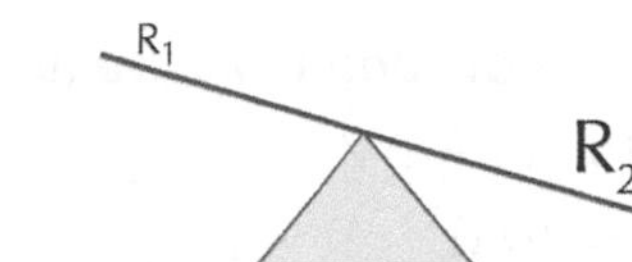

Fig 3.12.67: *Resistor 2 is greater than resistor 1*

Remember – the voltages across both the resistors always add up to the supply voltage.

Calculating the voltage

To calculate the voltage across resistor R_1 in the circuit, the following equation can be used:

$$V_1 = \left(\frac{R_1}{R_1 + R_2}\right)V_S$$

This equation shows the ratio in which the voltages are split.

Technique: Calculating the voltage in a potential divider

1. Write down what you know from the question, leaving blank what you are trying to find:
   Voltage across resistor, V_1 = ____ V
   Resistance of R_1, R_1 = ____ Ω
   Resistance of R_2, R_2 = ____ Ω
   Supply voltage, V_S = ____ V
2. Write down the equation linking the quantities above as you see it on the formula sheet:

$$V_1 = \left(\frac{R_1}{R_1 + R_2}\right)V_S$$

3. Substitute into the equation what you know.
4. Solve for the unknown.

Worked example

Calculate the voltage across the 10 Ω resistor in the circuit shown below.

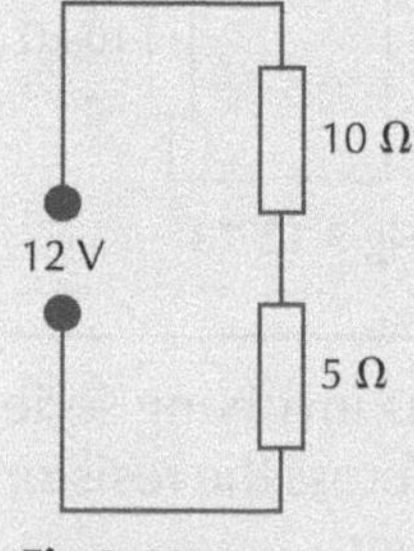

Fig 3.12.68

(continued)

Write down what we know and what we are looking for:
Voltage across resistor, V_1 *= ? V*
Resistance of R_1*,* $R_1 = 10\ \Omega$
Resistance of R_2*,* $R_2 = 5\ \Omega$
Supply voltage, $V_S = 12\ V$
We are using the equation that links charge, current and time:

$$V_1 = \left(\frac{R_1}{R_1 + R_2}\right)V_S$$

Substitute what we know:

$$V_1 = \left(\frac{10}{10 + 5}\right) \times 12$$

Solve for the voltage:

$$V_1 = \left(\frac{10}{15}\right) \times 12$$

$$V_1 = 8\ V$$

Exercise 3.12.6 The potential divider

1 Calculate the voltage across each resistor for the circuits below.

(a)

(b)

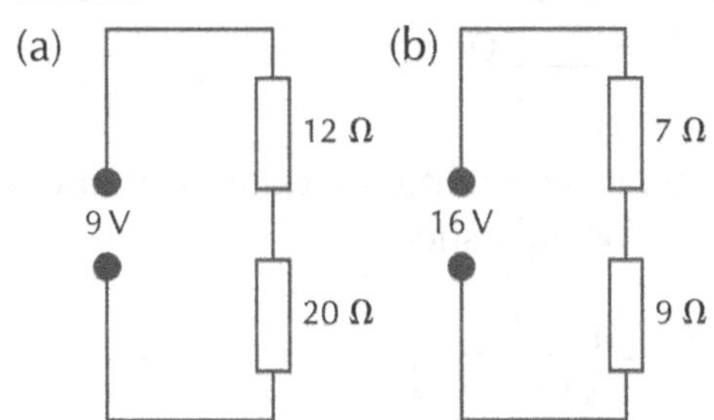

Fig 3.12.69 **Fig 3.12.70**

(c)

(d)

(e)

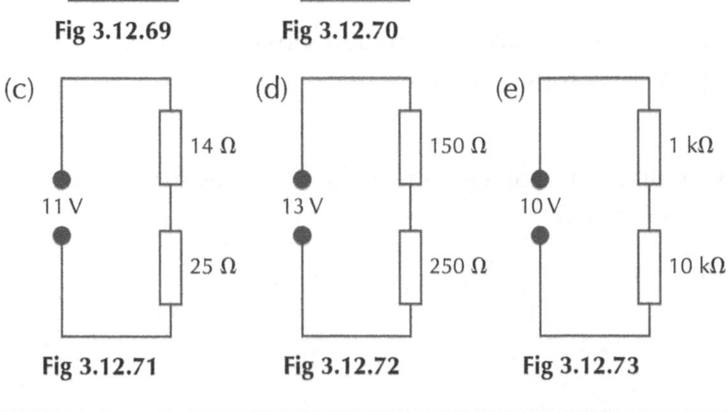

Fig 3.12.71 **Fig 3.12.72** **Fig 3.12.73**

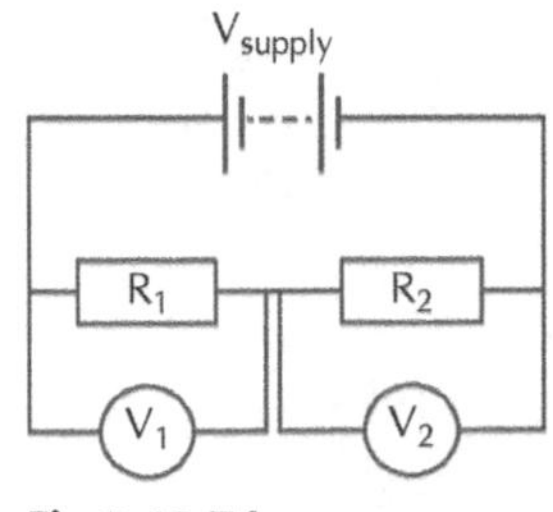

Fig 3.12.74

Another useful fact that helps us solve problems involving series circuits is that the ratio of the voltage dropped across the resistors is equal to the ratio of the resistances themselves,

$$\text{i.e. } \frac{V_1}{V_2} = \frac{R_1}{R_2}$$

Together with the fact that $V_{supply} = V_1 + V_2$ this can be a very useful relationship.

Worked example

Calculate the voltage V across the 10 Ω resistor and hence the total current leaving the battery.

Substituting values into

$$\frac{V_1}{V_2} = \frac{R_1}{R_2} \text{ gives } \frac{4}{V?} = \frac{20}{10} \text{ hence } V = 2\text{ V}$$

and since the supply voltage, $V_{supply} = V_1 + V_2$, $V_{supply} = 6\text{ V}$

Using Ohm's law $V = IR$

$6 = I \times 30$

$I = 0.2\text{ A}$

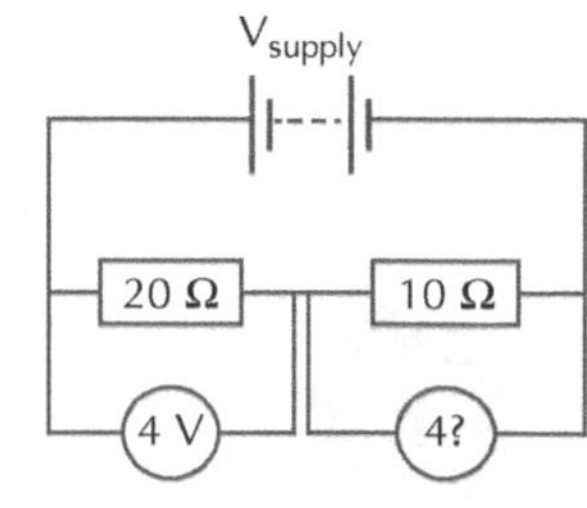

Fig 3.12.75

The bulb and light emitting diode (LED)

There are two main electrical components we will consider that transform electrical energy into light energy the light bulb and the LED.

Experiment: Bulb vs. LED

This experiment compares the function of a filament light bulb to a light emitting diode.

You will need:

- A 12 V light bulb
- An LED with series resistor*
- Connecting wires
- DC power supply

* Most LEDs you will find in the lab have a series resistor connected to them to protect the LED from too high a current.

Instructions

1. Connect a series circuit including the DC power supply and filament bulb as shown in the diagram.
2. Switch on the power supply and make sure that the bulb lights.
3. Reverse the connections on the power supply. Is the bulb still lit?
4. Replace the bulb in the circuit with the LED (notice the circuit symbol for the LED), as shown.
5. Switch on the power supply. Does the LED light up?
6. Reverse the connections on the power supply. Is the LED still lit?
7. Based on your observations, state the differences between a filament bulb and an LED.

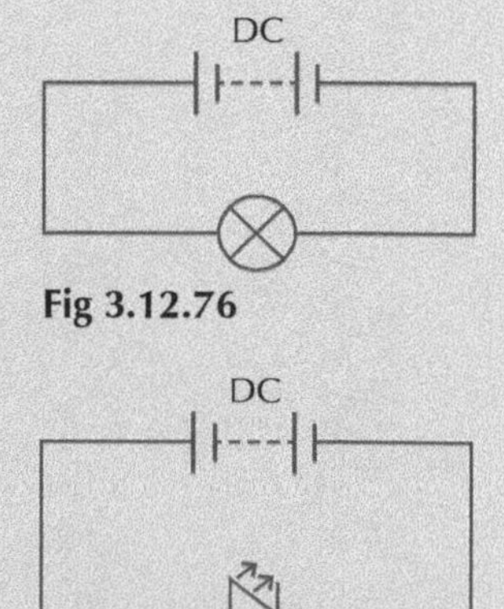

Fig 3.12.76

Fig 3.12.77

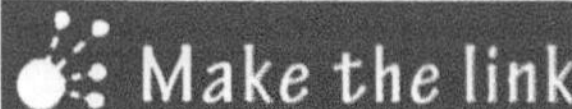

For more information on power, see Chapter 5.

The light bulb

The traditional filament light bulb is an output device that is used to transform electrical energy into light energy. Filament bulbs come in many shapes and sizes, as well as different brightnesses.

A filament light bulb works with both AC and DC.

The power of a light bulb is linked to its brightness – the more powerful the bulb, the brighter it is.

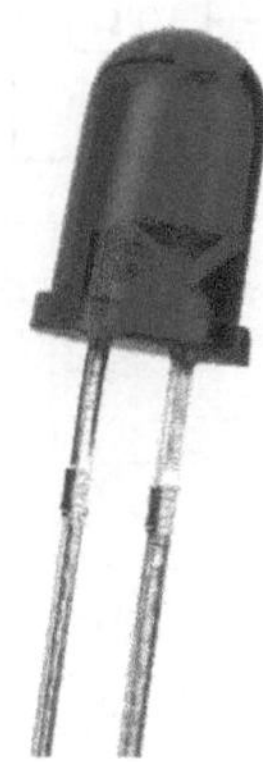

Fig 3.12.78: *Light emitting diode – the LED uses a fraction of the power of a filament bulb and has less wasted energy*

The LED

Light emitting diodes are semi-conducting diodes that typically emit light of only one colour. They last longer than filament bulbs.

An LED will only work with DC when connected the correct way round.

A series resistor is used to protect an LED from too high a current.

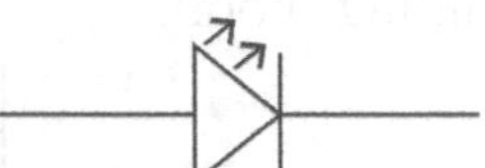

Fig 3.12.79: *Circuit symbol for LED*

Choosing a series resistor for an LED

An LED is usually connected in series with a resistor. The job of the resistor is to protect the LED from too high a current. It ensures the correct voltage across the LED. A typical LED circuit therefore looks like this:

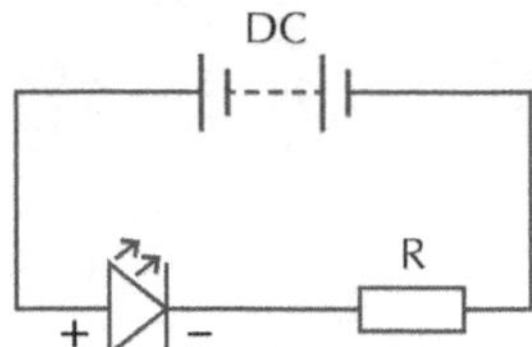

Fig 3.12.80: *Notice how the arrow head is pointing to the negative terminal of the battery*

The LED will be specified to operate at a certain voltage and current. The voltage across the series resistor, V_R, is therefore the remainder of the voltage from the supply, i.e.

$$V_R = V_S - V_{LED}$$

The current flowing through the series resistor, I, must be the same as the current flowing through the LED. This is because current is the same at all points in a series circuit. Both of these factors combined will specify the resistance of the series resistor to be:

$$R = \frac{V_R}{I}$$

Technique: Calculating the resistance of the LED series resistor

1. Calculate the voltage across the series resistor using the following equation:

$$V_R = V_S - V_{LED}$$

where V_S is the supply voltage and V_{LED} is the voltage required across the LED.

2. Find the current flowing through the series resistor using the specification for the LED – the current flowing must be equal to the current specified for the LED.
3. Write down what you know from the above, and what you are trying to find out:
Voltage across resistor, V_R = ____ V
Current through resistor, I = ____ A
Resistance of series resistor, R = ____ Ω
4. Use the following equation to find the resistance of the series resistor,

$$R = \frac{V_R}{I}$$

Worked example

A red LED is to be used as an indication that the power is on for a radio receiver. The LED is rated to operate at 2 V and 0·05 A. If the radio uses a 12 V battery as a power supply, calculate the resistance of the series resistor that must be used in series with the LED.

First of all, we must find the voltage across the series resistor. That is, the difference between the supply voltage and the LED voltage:

$$V_R = V_S - V_{LED}$$
$$V_R = 12 - 2$$
$$V_R = 10V$$

Now use Ohm's law to find the resistance of the series resistor, writing down what we know:

Voltage across resistor, V_R = 10 V

Current through resistor, I = 0·05 A

Resistance of resistor, R = ?

Using Ohm's law, we find:

$$R = \frac{V_R}{I}$$

$$R = \frac{10}{0{\cdot}05}$$

$$R = 200\ \Omega$$

Exercise 3.12.7 Bulb vs. LED

1 Explain the key differences between a filament light bulb and a light emitting diode (LED).

2 A student is investigating light bulbs and LEDs. She finds that when she connects a bulb to a battery, it does not matter which way round the battery connections are made. Is the same true for the LED?

3 Car dashboards use LEDs to indicate potential faults to the driver. Each LED is connected to the 12 V car battery. A series resistor is used to ensure the LED is operating at the correct voltage and current of 2·5 V and 0·04 A. Calculate the resistance of the series resistor required for each LED.

Fig 3.12.81: *Car dashboard indicators*

Transistor control circuits

Electrical components are used in automatic control circuits. A good example of this is the automatic lighting circuit in a car which turns on the headlights when it gets dark.

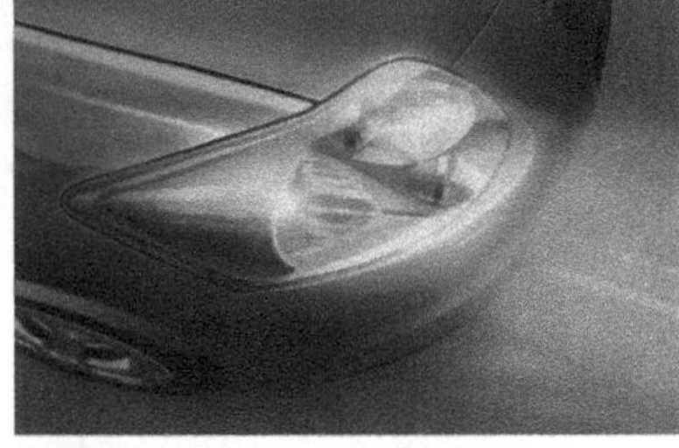

Fig 3.12.82: *Car headlight*

At the heart of many control circuits is a component called a transistor. Transistors come in many guises but they serve the same basic purpose – they are an electronic switch. A transistor is designed to switch on at a certain voltage. This allows the transistor to be switched on automatically by other electrical components.

A transistor is represented by the following circuit symbol:

Fig 3.12.83: *A transistor*

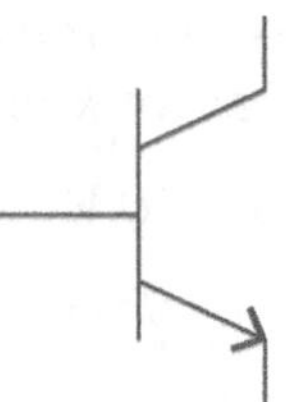

Fig 3.12.84a: *Transistor circuit symbol*

In many cases, a transistor is controlled by a potential divider circuit. We have already seen that the voltages in a potential divider circuit depend on the resistances of the resistors. If we fit a component that changes its resistance in response to certain conditions, then we can have varying voltages for the potential divider.

Another type of transistor used in electronic circuits is the MOSFET (Metal-Oxide semiconductor Field-Effect Transistor). The symbol used is shown below.

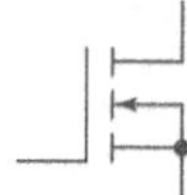

Fig 3.12.84b: *Mosfet circuit symbol*

This transistor operates as a switch in a similar fashion to the transistor above in that it switches on at a certain voltage which often comes from the output of a potential divider circuit (see later).

Light dependent resistors

If we use a light dependent resistor to control the voltage in a circuit controlling automatic headlights the transistor can be switched on and off by changing light levels. This is because a light dependent resistor changes its resistance as the light level changes. It is represented in circuits by the following symbol:

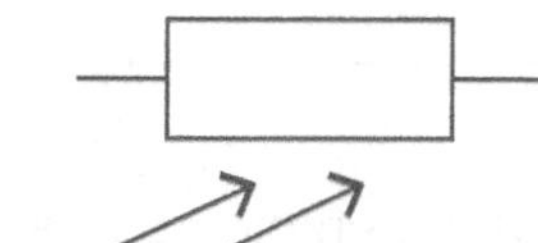

Fig 3.12.85: *LDR circuit symbol*

Experiment: The LDR

This experiment investigates how the resistance of an LDR changes with a changing light level and how this can control the voltage in a potential divider.

You will need:

- A DC power supply and battery connector*
- Potential divider alpha board
- 1 kΩ resistor
- Light dependent resistor
- A voltmeter

* These components can be substituted for resistors and connecting wires.

(*continued*)

Fig 3.12.86: *Circuit board*

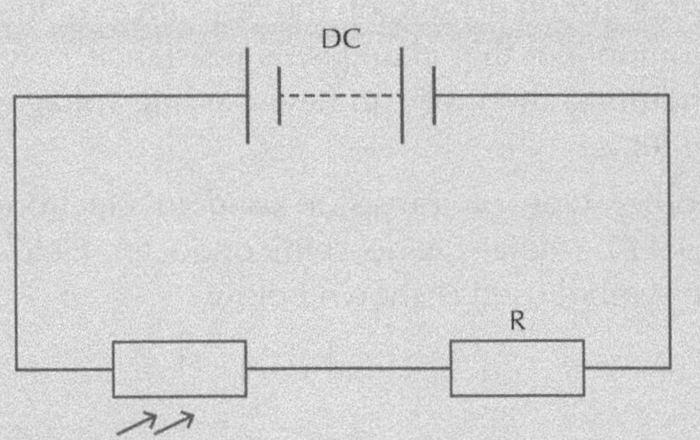

Fig 3.12.87

Instructions

1 Connect the potential divider circuit consisting of an LDR and 1 kΩ resistor to a DC power supply as shown in the circuit diagram, with the alpha board set up as shown in the photograph.

2 Use a voltmeter to measure the voltage across the LDR and the resistor (separately) when the LDR is exposed to high light levels (in a well lit lab).

3 Now cover the LDR with your finger and repeat the voltage measurements.

4 Use your results above to complete the following table:

LDR	Voltage across LDR (V)	Voltage across resistor (V)
Uncovered		
Covered		

Table 3.12.2

5 Describe what happens to the resistance of an LDR when the light level is increased or decreased.

Fig 3.12.88

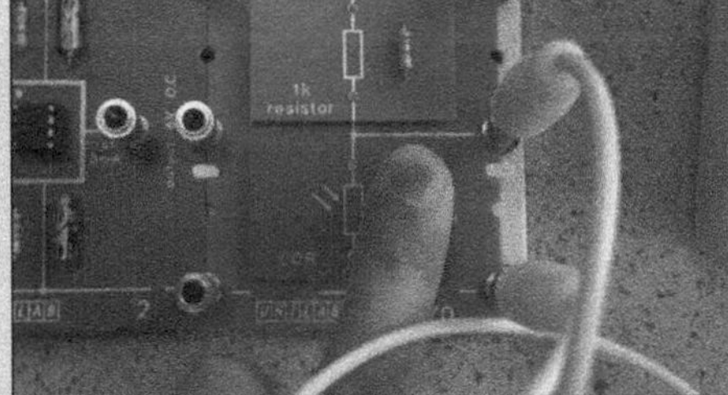

Fig 3.12.89

Thermistor

A thermistor changes its resistance as the temperature changes. It is represented in circuits by the following circuit symbol:

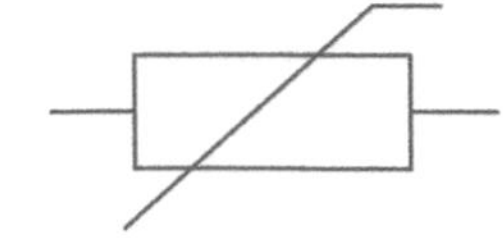

Fig 3.12.90: *Thermistor circuit symbol*

Experiment: Thermistor

This experiment investigates how the resistance of a thermistor changes with temperature and how this can control the voltage in a potential divider.

You will need:

- A DC power supply and battery connector*
- Potential divider alpha board*
- 1 kΩ resistor
- Thermistor
- A voltmeter

*These components can be substituted for resistors and connecting wires.

Fig 3.12.91

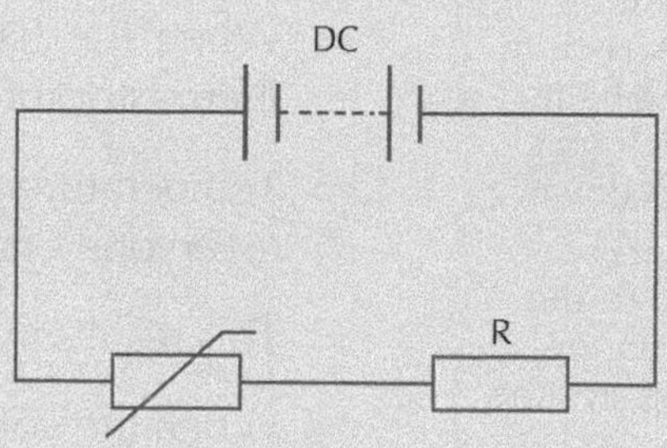

Fig 3.12.92

Instructions

1. Connect the potential divider circuit consisting of a thermistor and 1 kΩ resistor to a DC power supply as shown in the circuit diagram above (alpha board set-up shown in the photograph).
2. Use a voltmeter to measure the voltage across the thermistor and the resistor (separately) when the thermistor is at room temperature.
3. Now cover the thermistor with your finger and repeat the voltage measurements to get the voltages when the thermistor is at hand-temperature (warm).

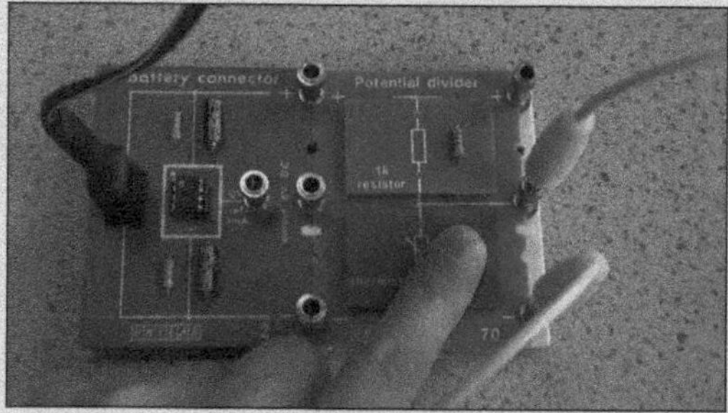

Fig 3.12.93

4. Use an ice cube wrapped in a paper towel to cool the thermistor and measure the voltages when the thermistor is cold.

(continued)

5 Use your results above to complete the following table.

Temperature	Voltage across thermistor (V)	Voltage across resistor (V)
Room temperature		
Hand temperature		
Ice temperature		

Table 3.12.3

6 Describe what happens to the resistance of a thermistor when the temperature is increased or decreased.

Key point

A light dependent resistor (LDR) changes its resistance in response to the varying light levels. As the light level goes up, the resistance goes down.

A thermistor changes its resistance in response to the varying temperature. As the temperature level goes up, the resistance goes down.

When describing the circuit in the thermistor experiment, it is always best to start with the input device and work across the circuit:

⇒ When the temperature decreases, the resistance of the thermistor increases.

⇒ **T**emperature goes **U**p so **R**esistance goes **D**own (TURD) (everyone's favourite!)

⇒ This results in the voltage across the thermistor increasing.

⇒ When the voltage across the thermistor reaches 0·7 V, the transistor switches on.

⇒ This allows current to flow through.

Automatic lights

So far we have seen that an LDR changes its resistance in response to varying light levels. This change in resistance causes a change in the voltages in a potential divider circuit. We can use these changes to control a transistor – in other words, turn a transistor on or off depending on the light level. This can allow us to build a circuit that turns on the lights in a room or the headlights of a car when it gets dark.

Fig 3.12.94: *An LDR*

Experiment: Automatic lighting circuit

A transistor can be used as part of a control circuit to switch on headlights when it gets dark. This experiment investigates such a circuit.

You will need:

- A DC power supply and battery connector*
- Potential divider alpha board*
- 10 kΩ resistor
- Light dependent resistor
- A voltmeter

(*continued*)

- Transistor switch/indicator alpha board**

*These components can be substituted for resistors and connecting wires.

**This component can be replaced with a transistor and an LED with a series resistor to protect it from too high a current.

Fig 3.12.95

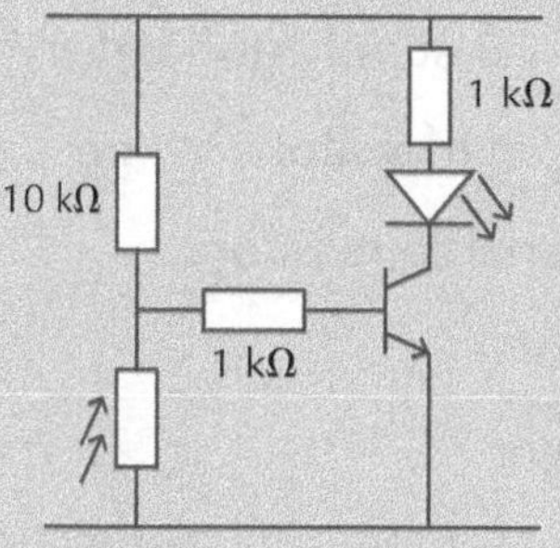

Fig 3.12.96

Instructions

1. Build the circuit shown in the circuit diagram consisting of a potential divider connected to a transistor with an LED as the output device. The circuit is also shown in the photograph above.
2. With the circuit switched on (connected to the battery), measure the voltage across the LDR when the LED is uncovered and when it is covered and complete the table below.

LDR	Voltage (V)	Transistor (on or off)	LED (on or off)
Uncovered			
Covered			

Table 3.12.4

3. Using these results combined with the results of the previous experiment, describe how the above circuit switched on an LED when it got dark. Include in your explanation how the resistance of the LDR changes with light level, how this affects the voltage and what this voltage will do to the transistor.

Thermostat control

A transistor control circuit can be used to monitor and control the temperature of a room or greenhouse, for example. A thermistor will change its resistance with temperature. These changes can be used to control the base voltage of a transistor, switching it on or off depending on temperature.

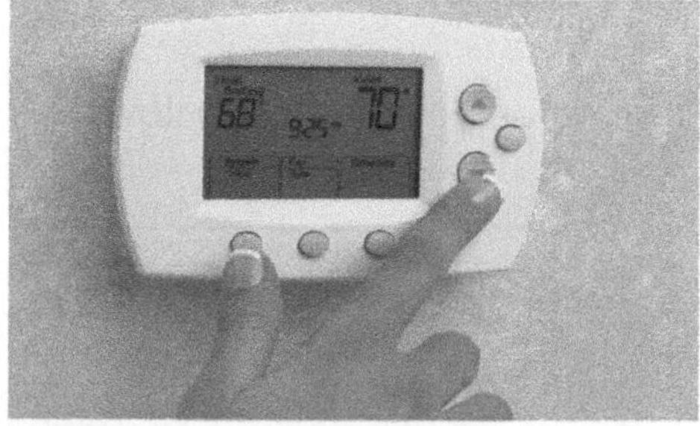

Fig 3.12.97: *Central heating control panels rely on thermistors to regulate temperature*

Think about it

The circuit described in the experiment above can turn on an LED when it gets dark. A similar design could also be used to turn on car headlights when it gets dark.

How could the circuit be altered to allow the user to change the light level at which the LED lights?

What alteration could be made to the circuit to turn on the LED in bright conditions?

Experiment: Thermostat circuit

A transistor can be used as part of a control circuit to switch on a heater when it gets cold. This experiment investigates such a circuit.

You will need:

- A DC power supply and battery connector*
- Potential divider alpha board*
- 4·7 kΩ variable resistor
- Thermistor
- A voltmeter
- Transistor switch/indicator alpha board

*These components can be substituted for resistors and connecting wires.

Fig 3.12.98

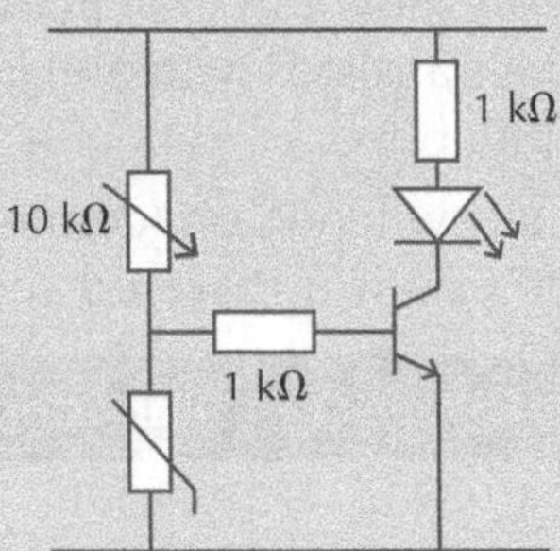

Fig 3.12.98A

Instructions

1. Build a circuit which will control an LED based on temperature as shown in the photograph.
2. Adjust the variable resistor so that at room temperature the base voltage is set to 0·7 V. You can measure the base voltage using a voltmeter.
3. Using a block of ice wrapped in tissue paper, cool the thermistor and note how the voltage across the variable resistor changes in the table below.

Temperature	Base voltage (V)	Transistor (on or off)	LED (on or off)
Room temperature	0·7	On	On
Ice temperature			

Table 3.12.5

4. Describe how this circuit works to turn on an LED when the temperature falls below a certain level.

How a control circuit works

The above experiments investigate two transistor control circuits. They rely on an input device which changes its resistance with brightness such as an LDR. A transistor works as the processing device and switches on or off depending on voltage. This switches on or off an output device such as an LED. Consider the automatic lighting circuit shown in fig 3.12.99.

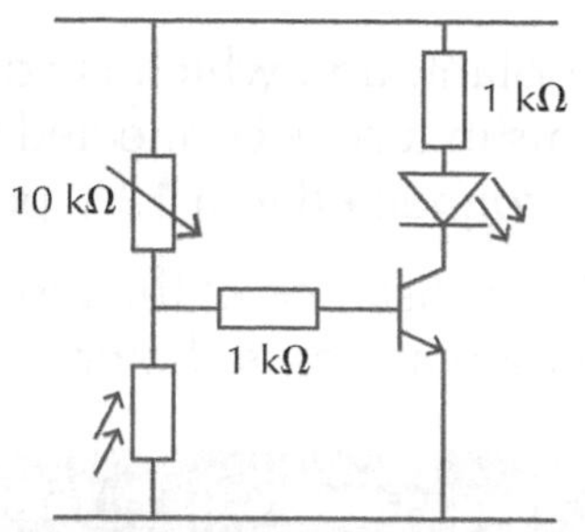

Fig 3.12.99: *Automatic lighting circuit*

When describing this circuit, it is always best to start with the input device and work across the circuit:

- When the light level decreases, the resistance of the LDR increases.
- In an LDR, as **L**ight level goes **U**p, **R**esistance goes **D**own (LURD).
- This results in the voltage across the LDR increasing.
- When the voltage across the LDR reaches 0·7 volts, the transistor switches on.
- This allows current to flow through the LED, causing the LED to light.

Key point

Swapping the position of the input device and resistor in the potential divider will cause the circuit to work in the opposite direction, i.e. switch on an LDR when it is bright rather than when it is dark.
A variable resistor can be used to change the conditions at which the transistor switches on – for example, the light level or temperature.

Exercise 3.12.8 Transistor control circuits

1 An engineer designs the circuit shown to switch on an LED to indicate when it is getting dark. The resistance of the LDR in bright and dark conditions is shown in the table below. It is connected to a 5 V power supply.

Light condition	LDR resistance (kΩ)
Dark	10
Light	0·5

Table 3.12.6

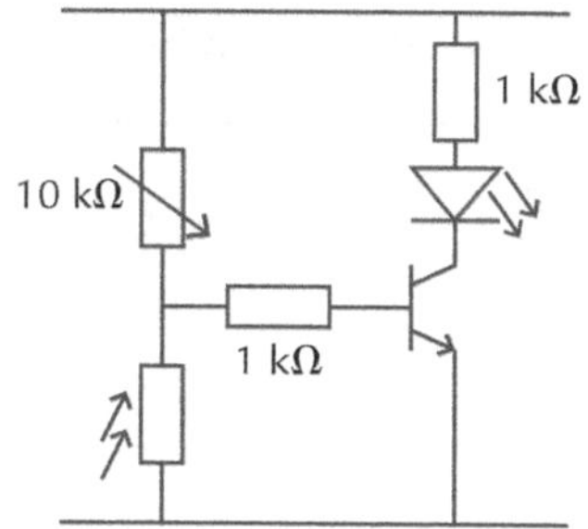

Fig 3.12.100

a) Calculate the voltage across the LDR in light conditions.

b) Calculate the voltage across the LDR in dark conditions.

c) Explain how the above circuit operates to switch on the LED in dark conditions.

d) How could the circuit be modified to allow a user to adjust the brightness at which the LED illuminated?

2 The cooling system for a car engine uses an electric fan to reduce the temperature of the engine coolant when it gets too high. The circuit shown is used to switch on the cooling fan at a certain temperature. It consists of a thermistor which is placed in the engine

(*continued*)

coolant, and which is connected to an LED indicator. This in turn is connected to the cooling fan. The circuit is connected to a 12 V power supply.

The resistance of the thermistor for different temperatures is shown in the table below.

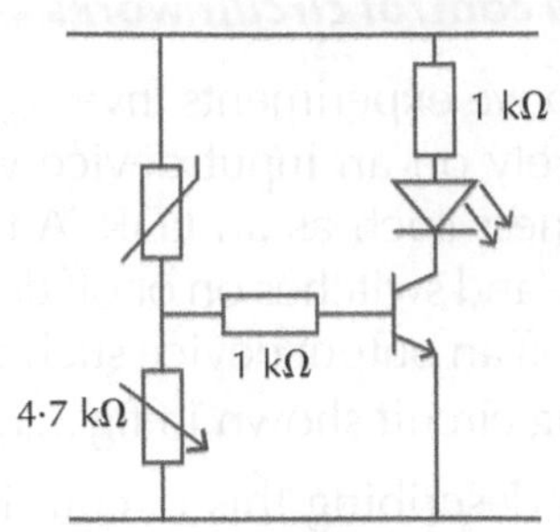

Fig 3.12.101: *Control circuit*

Temperature (°C)	Thermistor resistance (kΩ)
50	100
60	90
70	80
80	70
90	60
100	50
110	40

Table 3.12.7

a) By working out the voltage across the 4·7 kΩ resistor for the different temperatures above, predict the temperature at which the cooling fan will operate.

b) Explain how the circuit above operates the cooling fan when the engine coolant gets too hot.

c) Describe a modification that can be made to the above circuit so that the cooling fan operates at a lower temperature.

d) Describe a modification that can be made to the above circuit to allow the fan to switch on when the coolant temperature falls below a set value.

The capacitor

A capacitor is a device which stores electric charge. Initially, capacitors were designed as two parallel plates. This gave rise to the circuit symbol for a capacitor which is shown below.

Fig 3.12.102: *Circuit symbol for a capacitor*

Fig 3.12.103: *Rolled capacitors*

Modern capacitors now use an electrolytic design which looks like a Swiss roll when opened. The principle of the capacitor is based on there being two conducting plates – one positively charged and the other negatively charged – which leads to energy being stored between the plates.

A thunderstorm – nature's capacitor!

Capacitance refers to the ability to store separated charge, and because of this, store energy. This occurs in nature in a thunderstorm.

In the cloud, a large negative charge builds. This is separated from the positive charge on the ground by air which is a good insulator. Only when a large voltage builds up will the current flow from cloud to ground – this is when we get a lightning strike. The amount of energy stored is very large.

A capacitor is similar to a scaled down thunderstorm! There are two plates (one positive and one negative) and energy is stored.

Fig 3.12.104: *Thunderstorm*

The capacitor in a circuit

When connected to a DC power supply, such as a battery, a capacitor will charge up. Shown here is a capacitor connected to a battery.

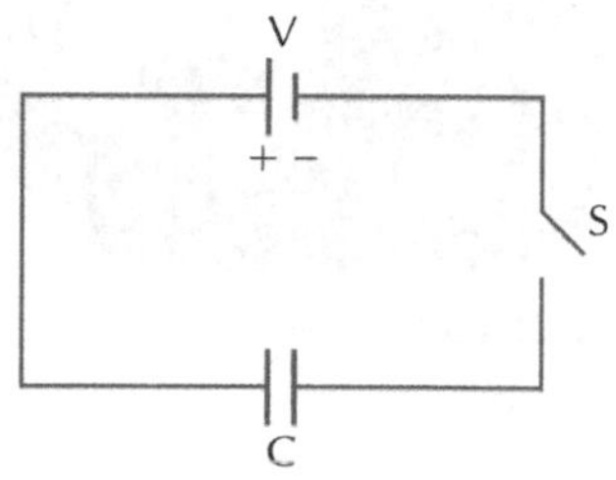

Fig 3.12.105

When the switch S is closed, the capacitor charges up. Electrons flow from the negative terminal of the battery to the negative plate of the capacitor. This repels electrons from the positive plate of the capacitor and creates a potential difference, *V*, between the plates. Energy is stored in the electric field between the plates.

It takes time to charge up a capacitor – this makes the capacitor a very useful component in circuits requiring a time delay. An example of such a circuit is the time delayed pedestrian crossing. When you push the button to cross the road, there is a time delay before the lights change. This time delay can be created by using a capacitor.

Fig 3.12.106: *Pedestrian crossing*

A capacitor can also be used for car interior lights. When the car door is open, the capacitor fully charges and the lights are on brightly. When the door is shut, the capacitor discharges powering a timing mechanism that controls how long the lights remain on (the lights draw power from the car battery).

Fig 3.12.107: *Car interior lights*

The electric motor

An output device that converts electrical energy into kinetic energy is an electric motor. A motor is similar to an electric generator. Motors come in all shapes and sizes and have been put to use in a huge variety of applications – from powering golf carts and electric cars to operating lifts.

Electrical safety: the fuse

A fuse is a safety device that you would find in a plug. It is designed to protect the appliance and the cable connecting it to the mains (flex) from too high a current. Fuses are also used in other applications – e.g. in a car to protect the electrical systems.

The circuit symbol for a fuse is shown below.

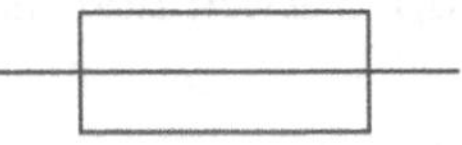

Fig 3.12.109: *Fuse circuit symbol*

A fuse that you would find inside a plug in the home is made from a piece of wire inside a casing as shown in the photograph. The wire is designed to melt when the current flowing through it is too high. When the wire melts, the fuse trips and leaves a break in the circuit. This break isolates the appliance from the electricity supply. A typical fuse you would find in a car is also shown below – you can see the wire inside the plastic case that melts when the current is too high.

Fig 3.12.110: *Fuses typically found in the home*

Fig 3.12.111: *Fuses typically found in plugs in a car*

Think about it

Can you think of other devices which use a capacitor? These could be devices where a time delay is required before switching on, where a large amount of energy is required quickly or where a device needs to remain on for a given time after an event, then switch off.

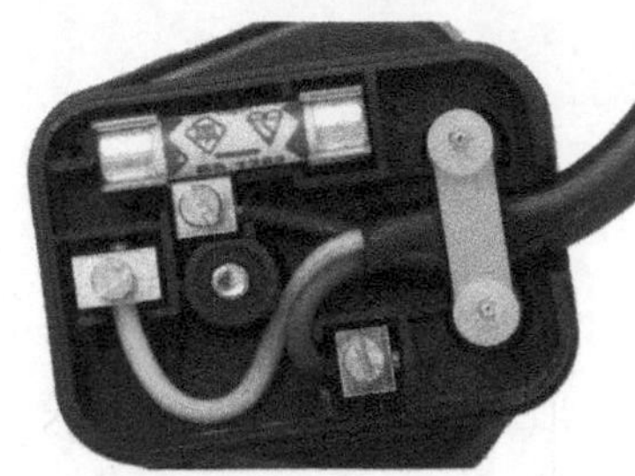

Fig 3.12.108: *Fuse in a plug*

Learning checklist

In this section you will have learned:

- How the measurement of potential difference (voltage), current and resistance can be carried out using appropriate meters in simple and complex circuits.
- The circuit symbols and function of a variety of standard electrical and electronic components.
- How to apply the rules for potential difference (voltage) and current in series and parallel circuits.
- Transistors acting as electrical switches.
- The impact of adding more resistors in series and parallel to the total resistance of a network.
- How to use appropriate relationships to solve problems involving the total resistance of resistors in series and parallel circuits, and in circuits with a combination of series and parallel resistors.
- How to explain the operation of simple control circuits using transistors as electrical switches.

13 Electrical power

In this chapter you will learn about:

- The relationship between power, energy and time.
- The relationship between power, current, potential difference (voltage) and resistance.

Electrical power

We have seen that electrical devices such as a filament light bulb convert electrical energy into other useful forms of energy. Some devices convert more energy in a given time than other devices, e.g. some light bulbs are brighter than others. The brighter light bulbs are converting more electrical energy to light energy in a given time. This is related to the power of the light bulb. Power is a measure of how quickly energy is converted from one form to another. We call this the rate of energy transfer. Power (P) is measured in watts (W):

$$P = \frac{E}{t}$$

Where E = energy transferred (J) and t = time (s).

Experiment: Measuring power and energy transfer

This experiment compares the amount of energy transferred from electricity to heat and light by low and high power bulbs.

You will need:

- Bulbs of various power ratings
- Power supply (DC)
- Wires
- Joulemeter
- Stop clock

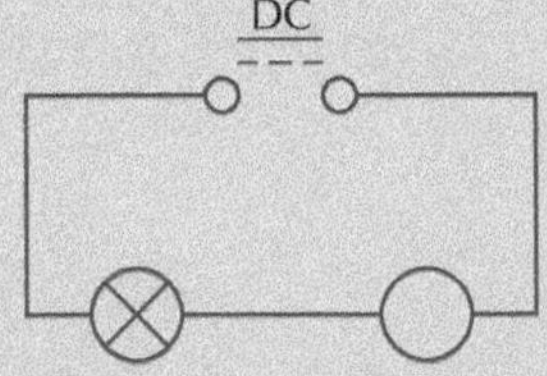

Fig 3.13.1

Instructions

1 Connect a series circuit containing a bulb, joulemeter and DC power supply as shown in the circuit diagram above.

2 Set the DC power supply to the operating voltage for the bulb (usually 6 V or 12 V). Ensure this is kept constant throughout the experiment.

(continued)

3 Switch on the bulb and use the joulemeter to measure the amount of energy transferred in the circuit in one minute (measure the time using the stop clock).

4 Repeat for bulbs with different powers, noting your results in a table similar to the one shown below. You can also note the relative brightness of each of the bulbs.

Power (W)	Energy transferred (J)	Energy transferred per second (J/s)	Relative brightness

Table 3.13.1

5 Compare the power and the energy transferred – what is the link between these two quantities?

6 Compare the power and the energy transferred per second – what is the link between these two quantities?

Technique: Power, energy and time

1. Write down what you know from the question, leaving blank what you are trying to find:

   Energy transformed, E = ____ J

   Power, P = ____ W

   Time, t = ____ s

2. Write down the equation linking the quantities above as you see it on the formula sheet:

$$P = \frac{E}{t}$$

3. Substitute into the equation what you know.
4. Solve for the unknown.

Key point

Power is a measure of the amount of energy transformed per second. That is, power is the rate of change of energy from one form to another. It is measured in watts where 1 watt is equal to 1 joule per second:

$$P = \frac{E}{t}$$

where E is the energy transferred (J) and t is the time in seconds (s).

STEP BACK IN TIME: HORSEPOWER

Power was initially measured in horsepower (hp). The term was adopted in the late 18th century by Scottish engineer James Watt to compare the output of steam engines with the power of draft horses. It was later expanded to include the output power of piston engines, turbines, electric motors and other machinery. 1 hp = 746 watts

Worked example

An electric fire converts 25 000 J of electrical energy into heat and light energy in a time of 30 s. Calculate the power of the electric fire.

Fig 3.13.2: *Electric fire*

We are using the equation that links power, energy and time:

$$P = \frac{E}{t}$$

Substitute what we know, and solve for power:

$$P = \frac{25000}{30}$$

$$P = 833\,W$$

Exercise 3.13.1 How much power?

1 Find the missing values in the table below.

Power (W)	Energy (J)	Time (s)
(a)	2400	12
(b)	95 000	100
450	(c)	12
60	(d)	360
2500	47 000	(e)

Table 3.13.2

2 A light bulb has a power rating of 60 W. It is switched on for a time of 4 minutes. Calculate the amount of electrical energy converted to light and heat during this time.

3 An electric motor has a power rating of 2500 W. How long does it take the motor to convert 4900 J of electrical energy into movement energy? You may assume the motor is 100% efficient.

4 An electric heater converts 45 000 J of energy from electricity to light and heat in a time of 2 minutes. Calculate the power of the electric heater.

Linking power to current and voltage

We have seen in Chapter 3 that an electric current flows round a circuit. Increasing the current increases the rate of flow of charge, and so increases the flow of energy around a circuit. Increasing the potential difference means each unit of charge has more energy to move around the circuit and so must also increase the energy flow around a circuit.

Electrical power given by P = IV means a larger current increases the transport of energy around the circuit by increasing the flow of charge. Increasing V means each carrier has more energy available to be transferred into other forms.

Experiment: Power, current and voltage

This experiment examines the link between power, current and voltage.

You will need:

- Various bulbs
- DC power supply
- Voltmeter
- Ammeter

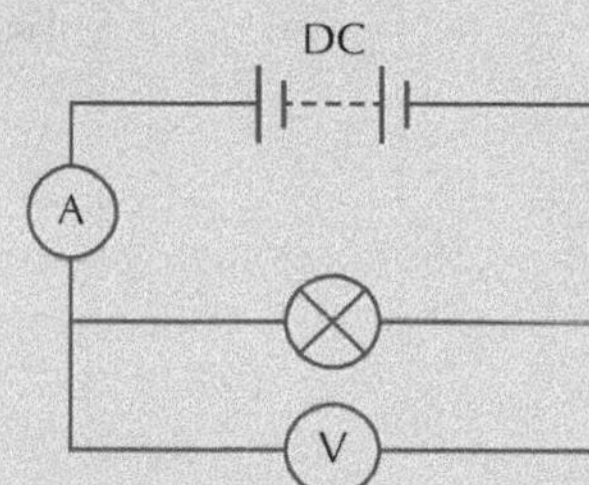

Fig 3.13.3

Instructions

1 Set up a series circuit consisting of a DC power supply, a bulb and an ammeter as shown in the circuit diagram above. Connect a voltmeter across the bulb to measure the voltage across the bulb.

2 Set the DC power supply to the operating voltage of the bulb (usually 6 V or 12 V).

3 Switch on the circuit and measure the current and the voltage in the circuit.

4 Repeat with different bulbs, noting your results in the table below.

Bulb power (W)	Current (A)	Voltage (V)	Current x Voltage

Table 3.13.3

5 How does the value of current multiplied by voltage compare to the rated power of the bulb?

Key point

The power in an electrical circuit, P, is linked to the current, I, and voltage, V, in the circuit by the following equation:

$$P = IV$$

In Chapter 3, we saw that current and voltage in a circuit are linked by Ohm's law,

$$V = IR$$

This can be combined with the equation for power to obtain two new equations.

Consider the equation for power,

$$P = IV$$

We can substitute for voltage based on Ohm's law:

$$P = I(IR)$$

Multiplying out the bracket, we yield:

$$P = I^2R$$

We can also substitute for current in the power equation. According to Ohm's law, current is given by:

$$I = \frac{V}{R}$$

Substituting into the equation for power

$$P = IV$$

$$P = \left(\frac{V}{R}\right)V$$

$$P = \frac{V^2}{R}$$

Technique: Electrical power

1. Write down what you know from the question, leaving blank what you are trying to find:

   Power, P = ____ W
   Current, I = ____ A
   Voltage, V = ____ V
   Resistance, R = ____ Ω
2. Choose the required equation from the following list:

   $$P = IV \quad P = I^2R \quad P = \frac{V^2}{R}$$
3. Substitute into the equation what you know.
4. Solve for the unknown.

Worked example

A vacuum cleaner motor has a power of 2200 W. It draws a current of 6 A. Calculate the resistance of the motor.

Write down what we know and what we are trying to find out from the question:

Power, P = 2200 W

Current, I = 6 A

Resistance, R = ? Ω

Choose the equation that links the above quantities. In this case, we will use:

$$P = I^2R$$

Substitute what we know and solve for the resistance,

$$2200 = 6^2 \times R$$

$$2200 = 36R$$

$$R = \frac{2200}{36}$$

$$R = 61{\cdot}1\ \Omega$$

Exercise 3.13.2 Power, current and voltage

1 Fill in the blanks in the table below:

(a) P = ?	R = 60 Ω	I = 1 A
(b) P = ?	R = 30 Ω	I = 3 A
(c) P = ?	R= 60 Ω	I = 1 mA
(d) P = ?	R = 2 Ω	I = 200 mA
(e) P = 960 W	R = ?	I = 3 A
(f) P = 1200 W	R = 60 Ω	I = ?
(g) P = 230 W	R = ?	I = 0·5 A
(h) P = 1 kW	R = ?	I = 5 A
(i) P = ?	R = 90 μΩ	I = 180 mA
(j) P = 1·75 mW	R = 500 mΩ	I = ?

Table 3.13.4

(*continued*)

2 Fill in the blanks in the table below:

(a) P = ?	R = 60 Ω	V = 1 V
(b) P = ?	R = 30 Ω	V = 3 V
(c) P = ?	R= 60 Ω	V = 1 mV
(d) P = ?	R = 2 Ω	V = 200 mV
(e) P = 960 W	R = ?	V = 30 V
(f) P = 1200 W	R = 60 Ω	V = ?
(g) P = 230 W	R = ?	V = 5 V
(h) P = 1 kW	R = ?	V = 5 V
(i) P = ?	R = 90 μΩ	V = 180 mV
(j) P = 1·75 mW	R = 500 mΩ	V = ?
(k) P = 2300 W	R = 1 kΩ	V = ?

Table 3.13.5

3 A loudspeaker has a resistance of 6 Ω and draws a current of 5 A. Calculate the power of the loudspeaker.

Fig 3.13.4: *Loudspeaker*

4 A television has a resistance of 360 Ω. It is connected to the mains voltage supply. What is the power of the television?

5 How much power is **dissipated** in an iron if it draws a current of 5·6 A and has a resistance of 90 Ω while operating from the mains?

6 Calculate the voltage of a buzzer which has a power rating of 9 W and resistance of 220 Ω.

7 Calculate the current drawn from a power supply by a radio if it has a power rating of 20 W and a resistance of 6 Ω?

Fuse ratings

One major use of these power equations is in calculating the current handling capacity of mains operated circuits and hence the size of fuse required to protect the flex.

Worked example

A mains operated 2 kW kettle is used to boil water. What size of fuse is required to protect the flex from overheating?

Since this is mains operated the voltage used will be 230 V.

Using $P = VI$, $2000 = 230 \times I$ therefore $I = 8.7$ A - This will require a 13 A fuse

A general rule of thumb is that for most appliances rated up to 720 W, a 3 A fuse is used and appliances with a power handling capacity in excess of 720 W will use a 13 A fuse.

Learning checklist

In this section you will have learned:

- How to define power in terms of electrical energy and time.
- How to use an appropriate relationship to solve problems involving energy, power and time.
- How the potential difference (voltage) and resistance affect the current in and the power developed across components in a circuit.
- How to use appropriate relationships to solve problems involving power, potential difference (voltage), current and resistance in electrical circuits.
- How to select an appropriate fuse rating for a given electrical appliance given its power rating.

14 Specific heat capacity

- The relationship between temperature, mass, heat energy and specific heat capacity
- The temperature of a substance and the mean kinetic energy of its particles
- Using the principle of conservation of energy to determine heat transfer

15 Specific latent heat

- Quantity of heat required to change the state of unit mass
- Latent heat of fusion and latent heat of vaporisation

16 Gas laws and the kinetic model

- Definition of pressure in terms of force and area
- Kinetic model and pressure of a gas
- Relationship between Kelvin and degrees Celsius
- Explanation of the gas laws in terms of the kinetic model
- Experiments to verify the gas laws

AREA 4

Properties of matter

14 Specific heat capacity

In this chapter you will learn about:

- How to define heat and temperature.
- How to define specific heat capacity.
- How to solve problems involving specific heat capacity and heat energy.
- Heat transfer and energy conversion.

Defining heat and temperature

We use the words heat and temperature every day. Often we use these words to mean the same thing, but in physics they have two different meanings!

- Heat is a form of energy, measured in joules (J). It is a measure of how fast the particles making up an object are moving – the greater the heat energy, the faster the particles of an object are moving. The greater the heat energy of an object, the greater the kinetic energy of the particles making up the object.
- Temperature is a measure of the hotness of an object. It is linked to heat energy, but should not be confused with it. It is not the same thing! Temperature is measured in degrees Celsius (°C) or degrees Kelvin (K). When the heat energy of an object is increased, its temperature will rise – the size of the rise in temperature depends on the mass and type of the material being heated. Temperature can also be defined as a measure of the mean kinetic energy of the particles of the substance.

Fig 4.14.1: *A Thermos flask*

Heat transfer

Heat energy always flows from a hot place to a cold place. Heat can travel by three different methods: conduction, convection and radiation. Any object which is designed to keep things hot or cold, such as a Thermos flask, must act to counteract these three methods of heat transfer.

Conduction is where heat travels through a solid. The solid must be made from a material that conducts heat, such as a metal. A Thermos flask has a vacuum layer to prevent heat travelling by conduction.

Convection is where heat travels through a fluid. Hot fluids (e.g. air) rise, cold fluids fall. This forms a **convection current**. A Thermos flask uses a lid to prevent heat from escaping by convection.

Radiation is where heat travels as a wave. It does not need anything to travel through. Radiation is the method by which heat travels from the sun to Earth. A Thermos flask uses a reflective inside to stop heat escaping by radiation.

Word bank

- **convection current**

the motion of air or liquid due to heating and cooling. When air or liquid is heated it rises. As it rises it cools down by transferring heat to its surroundings. As it cools it falls back down. It is then re-heated and the cycle continues. Convection does not occur in a solid.

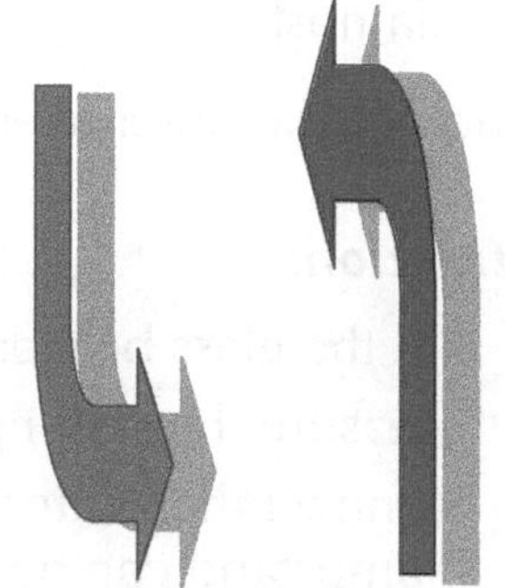

Fig 4.14.2: *Convection current*

Changing temperature

When you heat an object, for example by placing it in the flame of a Bunsen burner, you are transferring heat energy to the object which will result in a rise in temperature. An example of this is the water in a kettle. Heat energy is supplied to the water by the heating element in the kettle. This causes the water's temperature to rise until it reaches its boiling point.

Specific heat capacity

Supplying heat energy to an object will result in a rise in temperature. Similarly, if heat energy is removed from the object its temperature will fall. The temperature rise or fall will depend on the mass of the object (how much there is to heat up) and the type of material. Some materials are more difficult to heat than others.

Different materials require different amounts of heat energy to change their temperature by one degree Celsius (1 °C). To raise the temperature of 1 kg of the following materials by 1 °C the amounts of energy shown are required:

- 1 kg of water – 4180 J
- 1 kg of iron – 470 J
- 1 kg of air – 900 J

The amount of energy that is required to increase the temperature of 1 kg of a particular material by 1°C is a constant known as specific heat capacity, *c*.

$$c = \frac{E_h}{m\Delta T}$$

where E_h is the energy, m is the mass (in kilograms) and ΔT is the change in temperature. For the purposes of this chapter, specific heat capacity has the units J/kg°C. However, specific heat capacity can also be quoted using degrees Kelvin: J/kg K.

Fig 4.14.3: *Kettle*

Think about it

What are the main energy transformations in the process of using a kettle to heat up water?

Experiment: Specific heat capacity of water

This experiment investigates the specific heat capacity of water.

You will need:

- A glass beaker
- A 50 W water heater, e.g. a fish tank heater
- 12 V DC power supply
- Connecting wires
- Thermometer
- Stop clock
- Clampstand

Optionally, a joulemeter can also be used to measure the energy supplied to the heater.

Instructions

1. Fill the glass beaker with 200 ml of cold water. 200 ml of water has a mass 0·2 of kg.
2. Measure the starting temperature of the water using the thermometer.
3. Connect the heater to the DC power supply and lower the heater into the water. A clampstand can be used to hold the heater in place.
4. Switch on the heater and allow it to heat the water for 5 minutes (300 s).
5. Calculate the energy supplied to the heater using the following equation:
$$E = Pt$$
where P is the heater power and t is the time in seconds.
6. Measure the temperature of the water after heating and calculate the temperature rise, ΔT.
7. Use the equation below to find the specific heat capacity of water.
$$c = \frac{E_h}{m\Delta T}$$

Think about it

The accepted value for the specific heat capacity of water is 4180 J/kg°C. Compare this to the value you obtained in your experiment. Can you account for the difference?

The energy required to heat a material up by a certain temperature depends on three variables:

- The specific heat capacity, c, of the material (J/kg°C).
- The mass, m, of the material (kg).
- The change in temperature, ΔT, (°C)

$$E_h = cm\Delta T$$

Technique: Heat energy calculations (temperature change)

1. Write down what you know from the question, leaving blank what you are trying to find:

   Energy, E = ____ J

   Specific heat capacity, c = ____ J/kg°C

   Mass, m = ____ kg

   Temperature change, $\Delta T = T_{end} - T_{start}$ = ______ °C
2. Write down the equation linking the quantities above as you see it on the formula sheet:

   $$E_h = cm\Delta T$$
3. Substitute into the equation what you know.
4. Solve for the unknown.

Worked example

A student is carrying out an experiment to work out the specific heat capacity of a new material. She finds that it requires 200 J of energy to change the temperature of a mass of 2 kg by 5°C. Calculate the specific heat capacity of the new material.

Fig 4.14.4: *Applying heat energy to a new material*

Write down what we know and what we are looking for:

Energy, $E_h = 200\ J$

Mass, $m = 2\ kg$

Temperature change, $\Delta T = 5°C$

Specific heat capacity, $c = ?\ J/kg$

(continued)

We are using the equation that links energy, specific heat capacity, mass and temperature change:

$$E_h = cm\Delta T$$

Substitute what we know:

$$200 = c \times 2 \times 5$$

Solve for heat energy:

$$200 = c \times 10$$

$$c = \frac{200}{10}$$

$$c = 20\ J/kg°C$$

Worked example

The specific heat capacity of a certain liquid is 2430 J/kg°C. What energy is required to heat up 40 kg of the liquid from a starting temperature of 15°C to a finishing temperature of 60°C?

Write down what we know and what we are looking for:

Energy, E_h = ? J

Mass, m = 40 kg

Temperature change, ΔT = 60 − 15 = 45°C

Specific heat capacity, c = 2430 J/kg

We are using the equation that links energy, specific heat capacity, mass and temperature change:

$$E_h = cm\Delta T$$

Substitute what we know:

$$E_h = 2430 \times 40 \times 45$$

Solve for heat energy:

$$E_h = 4\ 374\ 000\ J$$

$$E_h = 4{\cdot}4\ MJ$$

Exercise 4.14.1 Specific heat capacity

1 Calculate the specific heat capacity of a material which needs 2760 J of energy to heat up 1 kg by 20°C

2 The specific heat capacity of copper is 0·39 J/kg°C. Calculate the energy required to heat 5 kg of copper from 20°C to 50°C.

3 A car's brakes work by an abrasive pad being pushed into contact with a rotating disc. This generates heat energy, causing the brake discs to get hot. Car manufacturers try to

(*continued*)

minimise the temperature rise of the brake discs to ensure they work at their optimum. For this reason, different materials are trialled for the construction of brake discs.

a) Traditional brake discs are made using cast iron which has a specific heat capacity of 480 J/kg°C. Calculate the temperature rise of a 4·5 kg cast iron brake disc when it absorbs 150 000 J of kinetic energy (converted to heat energy).

b) Some car manufacturers have trialled carbon-ceramic brake discs with a specific heat capacity of 800 J/kg°C. Find the temperature rise of a 4·5 kg carbon-ceramic brake disc when it absorbs 150 000 J of movement energy.

4 Asphalt is used in the building of a new bridge over the river Forth. Calculate the heat energy generated by a 20 kg section on a day with a temperature change from 5°C to 19°C. The specific heat capacity of asphalt is 920 J/kg °C.

5 In a wood burning stove, how much wood would be required to produce 3500 J of heat energy if wood has a specific heat capacity of 1700 J/kg °C and has a $\Delta T = 22°C$?

6 An architect is planning to build a house from granite rock to assist with it retaining heat. However, as the consultant physicist you must advise whether this is a good idea. Calculate the heat energy stored in 25 kg granite rock during a change in temperature of 8°C. Granite has a specific heat capacity of 790 J/kg °C.

Heat transfer and energy conservation

We have already seen that when an object absorbs heat energy, its temperature will rise. The rise in temperature depends on the type of material, the mass and the amount of heat energy supplied. However, we have not considered where the heat energy comes from.

If you want to heat up water to make a hot drink, you can use a kettle which converts electrical energy into heat energy. Remember – we cannot create or destroy energy, we can only change its form.

If the kettle is 100% efficient, all of this energy is converted to heat energy and supplied to the water and not transferred to the surroundings, i.e. it does not heat the kettle itself. To find the heat energy of the water we use:

$$E_h = cm\Delta T$$

However, in reality no kettle (or other appliance) is 100% efficient. When solving problems where heat energy is transferred we may be required to find the efficiency (%) so that we can calculate a more accurate value of energy transfer.

Experiment: Kettle efficiency

This experiment investigates the efficiency of a kettle.

You will need:

- A kettle
- 1 litre (1 kg) of water
- A stop clock
- A thermometer

Instructions

1 Fill the kettle with 1 litre of cold water and measure the starting temperature of the water using a thermometer.

2 Switch on the kettle for a time of two minutes (120 s).

3 Switch off the kettle and then carefully measure the finishing temperature with the thermometer.

4 Calculate the amount of heat energy gained by the water using the following equation:

$$E_h = cm\Delta T$$

5 Calculate the electrical energy supplied to the kettle using the equation:

$$E = Pt$$

6 Work out the efficiency of the kettle using the equation:

$$Efficiency = \left(\frac{Useful\ energy\ out}{Energy\ in}\right) \times 100\%$$

Worked example

A kettle is used to boil water for making a cup of coffee. The water needs to reach a temperature of 100°C in order to boil. The starting temperature of the water is 20°C.

Fig 4.14.5: *Cup of hot coffee*

(*continued*)

The mass of water in the kettle is 2 kg, and the water has a specific heat capacity of 4180 J/kg°C. If the kettle has a power of 2200 W, find the length of time taken for the water to boil. You may assume the kettle is 100% efficient and that all the heat energy goes to the water.

First we must calculate the amount of heat energy that needs to be supplied to the water to make it boil. Write down what we know and what we are trying to find out:

Energy, $E = ?$ J

Mass, $m = 2$ kg

Temperature change, $\Delta T = 100 - 20 = 80$°C

Specific heat capacity, $c = 4180$ J/kg°C

We are using the equation that links energy, specific heat capacity, mass and temperature change:

$$E_h = cm\Delta T$$

$$E_h = 4180 \times 2 \times 80$$

$$E_h = 668\,800 \ J$$

Now, we can work out the time taken for the water to boil using the information we have about the power of the kettle. Write down what we know and what we are trying to find out:

Energy, $E = 668\,800$ J

Power, $P = 2200$ W

Time, $t = ?$ s

We are using the equation that links energy, power and time:

$$E = Pt$$

$$668\,800 = 2200t$$

$$t = \frac{668\,800}{2200}$$

$$t = 304 \ s$$

Exercise 4.14.2 Heat transfer and energy conservation

1 Students are carrying out the following experiment to find the specific heat capacity of water.

Fig 4.14.6: *Specific heat capacity experiment*

A 12 V heater is connected to a power supply and used to supply heat energy to the water. A joulemeter measures the amount of energy supplied to the water. 200 ml of water, with a mass of 0·2 kg, is used in the experiment.

It is found that in a time of 3 minutes, 10 000 J of heat energy is supplied to the water. This results in a temperature rise of 10°C.

a) Calculate the power of the heater.

b) Find the specific heat capacity of water, based on the experimental results above.

c) The accepted value for the specific heat capacity of water is 4180 J/kg°C. Explain why the result calculated in part b) is different from the accepted value.

2 An electric kettle is used to heat a mass of 1·5 kg of water. The power of the kettle is 1700 W. You should assume the kettle is 100% efficient and c = 4180 J/kg°C.

a) How much energy is supplied to the kettle in a time of 2 minutes?

b) How much heat energy is supplied to the water in a time of 2 minutes?

c) The starting temperature of the water is 7°C. Calculate the temperature of the water after heating.

3 A kettle is used to heat 2 kg of water in a time of 5 minutes. The water starts at a temperature of 20°C and is heated until it is boiling at 100°C. Assuming the kettle is 100% efficient and c = 4180 J/kg°C, calculate the power of the kettle.

4 A pan of water is being heated using a Bunsen burner. There is 4 kg of water in the pan. A student finds that is takes 9 minutes to heat the water from 20°C to 75°C.

a) Calculate the amount of energy supplied by the Bunsen burner to heat the water.

b) What is the average heating power of the Bunsen burner?

Learning checklist

In this section you will have learned:

- That different materials require different quantities of heat to raise the temperature of unit mass by one degree Celsius.
- How to use an appropriate relationship to solve problems involving mass, heat energy, temperature change and specific heat capacity.
- That the temperature of a substance is a measure of the mean kinetic energy of its particles.
- How to use the principle of conservation of energy to determine heat transfer.

15 Specific latent heat

In this chapter you will learn about:

- How to define specific latent heat.
- How to solve problems involving mass, specific latent heat and heat energy.

Fig 4.15.1: *Melting ice*

Word bank

- **state**
  the state of an object refers to whether it is solid, liquid or gas

Latent heat

Increasing the heat energy of an object can raise its temperature as we have seen. If you supply heat energy to an object, you can also change the **state** of the object. For example, if you supply enough heat energy to solid water (ice), you will melt the ice and change its state into a liquid.

In the same way as it requires a certain amount of energy to change the temperature of an object, it also requires a certain amount of energy to change the state of a material. This energy depends on:

- the material
- its mass

In a similar way to the specific heat capacity, we define the quantity **specific latent heat**, *L*, which is the amount of energy required to change the state of 1 kg of the substance at its melting point or its boiling point:

$$L = \frac{E_h}{m}$$

where L is specific latent heat, measured in J/kg, E_h is heat energy, measured in J, and m is the mass of the substance, measured in kg.

Every material has two values for specific latent heat – it depends on whether that state is being changed from solid to liquid (melting) or from liquid to gas (boiling). These processes require different amounts of energy. We refer to the processes as:

- **Specific latent heat of fusion**: solid ⇔ liquid
- **Specific latent heat of vaporisation**: liquid ⇔ gas

A material can also change from a gas back into a liquid (condensing), or from a liquid back into a solid (freezing). In these situations, the material loses heat energy.

Re-arranging the above equation, we get an equation for the energy required to change the state of a specific mass of a certain material:

$$E_h = mL$$

Experiment: Latent heat of vaporisation of water

This experiment determines the latent heat of vaporisation of water.

You will need:

- A kettle
- 1 litre measuring cylinder
- A stop clock

Fig 4.15.2: *Kettle*

Instructions

1 Using the measuring cylinder, measure 1 litre of water (which has a mass of 1 kg) and pour it into the kettle.
2 Remove the lid from the kettle for the duration of the experiment.
3 Switch the kettle on and allow the water to boil.
4 Once the water starts boiling, start the stop clock.
5 Use the stop clock to time the water boiling for two minutes (120 seconds).
6 Switch off the kettle after two minutes and allow the water to cool.
7 Use the measuring cylinder to find the new volume of water.
8 Work out the volume of water that turned to steam.
9 Convert the *volume* of water lost to steam to *mass* using: 1 litre = 1 kg
10 Work out the energy supplied to the water by the kettle using the formula,

$$E = Pt$$

where P is the power of the kettle and t is the time in seconds (120 s).

11 Calculate the specific latent heat of water using the equation:

$$L = \frac{E_h}{m}$$

Worked example

The specific latent heat of fusion for water is 330 kJ/kg. Calculate the energy that must be supplied to change 3 kg of water from solid ice into liquid.

If we study a solid changing to liquid, it is the specific latent heat of fusion that we use in the following equation:

$$E_h = mL$$

Substitute what we know and solve for the energy:

$$E_h = 3 \times 330\,000$$

$$E_h = 990\,000\ J$$

Exercise 4.15.1 Latent heat

1 The specific latent heat of fusion for iron is 209 kJ/kg. Calculate the energy required to melt 25 kg of iron.

2 The specific latent heat of fusion of a material is 3 kJ/kg and the specific latent heat of vaporisation of the material is 18 kJ/kg.

a) Calculate the energy that must be supplied to change 12 kg of the material from a solid to a liquid.

b) How much energy must be supplied to change 12 kg of the material from a liquid into a gas?

Fig 4.15.3: *Molten iron*

3 A student is investigating the specific latent heat of fusion for water. She sets up a beaker with a mass of 400 g of ice and gently heats it. She discovers that she requires 66 020 J of energy to melt the ice completely. What would she calculate for the specific latent heat of water?

4 The specific latent heat of vaporisation of a material is 150 kJ/kg. The material loses 250 kJ of energy to fully change from a gas to a liquid. What is the mass of the material?

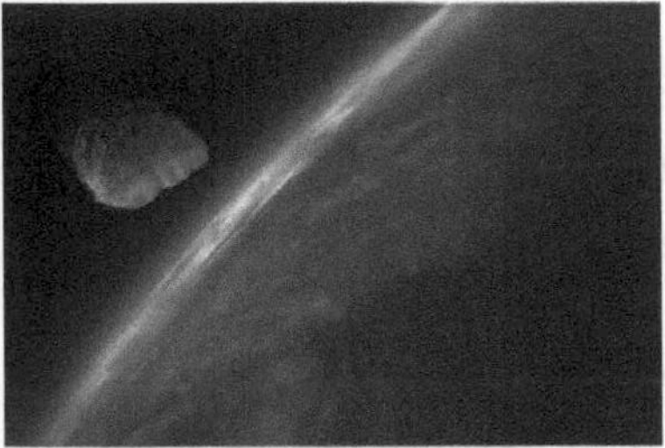

Fig 4.15.4: *Meteorite burning up in Earth's atmosphere*

Re-entry

One of the biggest considerations around rockets is the amount of heat they need to absorb when they re-enter the Earth's (or another planet's) atmosphere. The friction force acting against the rocket slows it down – it loses kinetic energy. This loss of kinetic energy is converted to heat energy, causing the surface of the rocket to heat up. We see this effect in action with a shooting star. A shooting star is a meteorite heating up as it travels through the Earth's atmosphere and burns up.

Understanding heat energy and temperature (covered in Chapter 6) is vital to the design process of a rocket to ensure safe re-entry.

Change temperature or change state?

Heating a material changes either its temperature or its state – only one will be affected at a time. How do you know whether it is the temperature or the state that will change?

Consider heating up a block of ice that has a starting temperature of –50°C.

As the starting temperature is below the melting point (0°C), the temperature of the ice will increase. The temperature will increase until it reaches the melting point of the ice – it will increase by 50°C (ΔT = 50°C)

↓

Once the temperature has reached the melting point, the ice will change state from a solid to a liquid (melt). During this time, the temperature of the ice will not change.

When the ice has fully melted, the temperature of the water will then start to increase again and the state will remain the same (i.e. water). The temperature will increase until it reaches the boiling point of water, at which point the temperature will remain the same but the water will change state from liquid to gas.

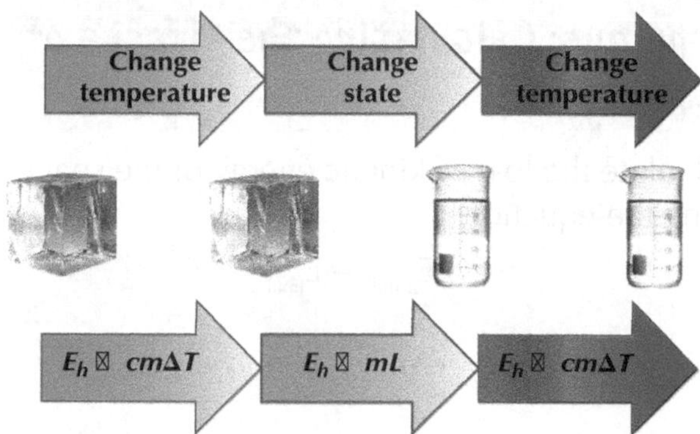

Fig 4.15.5: *State and temperature changes of H_2O heated from −50°C*

When you calculate the amount of energy required to change the temperature you use the equation:

$$E_h = cm\Delta T$$

where E_h is heat energy in J; c is specific heat capacity measured in J/kg°C; m is mass, measured in kg; and ΔT is the change in temperature, measured in °C.

To calculate the amount of energy required to change the state, the following equation is used:

$$E_h = mL$$

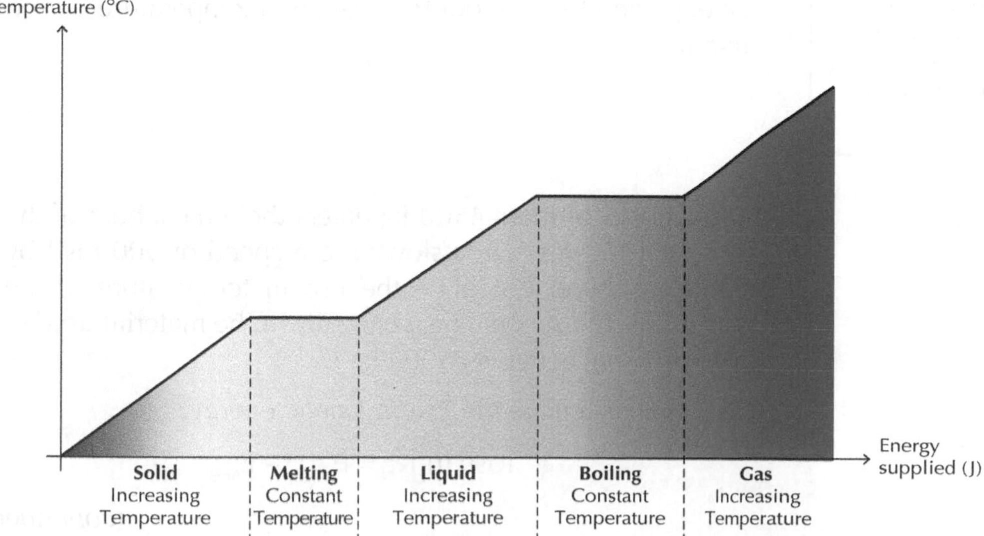

Fig 4.15.6

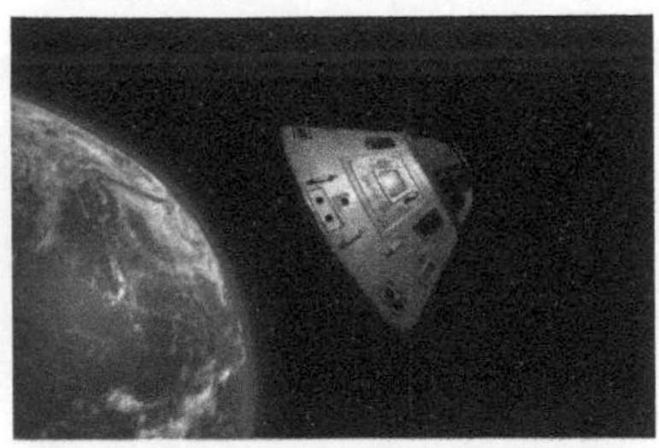

Fig 4.15.7: *Space capsule*

Key point

When a spacecraft enters a planet's atmosphere it is slowed down by air resistance. Kinetic energy is converted into heat energy.

If the heat energy results in a rise in temperature, we have:

$$\frac{1}{2}mv^2 = cm\Delta T$$

If the heat energy results in a change in state, we have:

$$\frac{1}{2}mv^2 = mL$$

Think about it

When Felix Baumgartner tried to break the record for the fastest free fall, he started his jump high in the stratosphere, 39 km above the surface of the Earth. Given that the atmosphere is much thinner at a greater height, explain why Baumgartner was able to reach record breaking speeds. Explain the dangers related to the jump based on the heating effects of being slowed down by the atmosphere.

Re-entry: kinetic energy → heat energy

On re-entry, the kinetic energy from the velocity of the rocket is converted to heat energy as the rocket slows down. The greater the resistive forces, the more the rocket is slowed. Therefore the greater the resistive forces, the greater the amount of heat energy absorbed. This can result in a rise in temperature of the shuttle, or a change in the state of the materials. The latter could have catastrophic results!

Technique: Calculating the effects of re-entry

1. Calculate the loss of kinetic energy of the space craft using the equation

$$E_{initial} - E_{final}$$

$$= \frac{1}{2}mv_i^2 - \frac{1}{2}mv_f^2$$

2. Calculate the rise in temperature of the space craft using the equation, $E_h = cm\Delta T$
3. Substitute what we know into the equation above and solve for the change in temperature, ΔT.
4. If the temperature increase of the space craft causes the material to reach its melting point then the state of the material will begin to change. In this case, use the equation,

$$E_h = mL$$

to find the mass of the material that will melt.*

*In all problems you will meet at National 5 you will only be expected to work out the change in temperature during re-entry.

Worked example

A spacecraft of mass 4500 kg enters the atmosphere of the Earth at 1200 ms^{-1}. It is slowed to a speed of 900 ms^{-1} by the atmosphere. Calculate the rise in temperature of the spacecraft. The specific heat capacity of the material used to make the spacecraft is 7500 J/kg°C.

First of all, calculate the loss in kinetic energy:

$$\text{loss in KE} = E_{initial} - E_{final}$$

(continued)

$$= \frac{1}{2}mv_i^2 - \frac{1}{2}mv_f^2$$
$$= \frac{1}{2}m(1200)^2 - \frac{1}{2}m(900)^2$$
$$= \frac{1}{2}4500(1200^2) - \frac{1}{2}4500(900)^2$$
$$= 3.24 \times 10^9 - 1.825 \times 10^9$$
$$= 1.42 \times 10^9\ J$$

The rise in temperature can be calculated using the equation:

$$E_h = cm\Delta T$$

Substitute what we know and solve for the change in temperature:

$$1{\cdot}42 \times 10^9 = 7500 \times 4500 \times \Delta T$$
$$1{\cdot}42 \times 10^9 = 3{\cdot}38 \times 10^7 \times \Delta T$$
$$\Delta T = \frac{1{\cdot}42 \times 10^9}{3{\cdot}38 \times 10^7}$$
$$\Delta T = 42°C$$

Fig 4.15.8: *Spacecraft*

Exercise 4.15.2 Re-entry

1 During a skydive, a stuntman with a mass of 90 kg enters the Earth's atmosphere at a speed of 340 ms^{-1}. He is slowed to a speed of 100 ms^{-1} by the atmosphere of the Earth. The average specific heat capacity of the stuntman is 4900 J/kg°C

a) Calculate the amount of kinetic energy lost.

b) Calculate the rise in temperature of the stuntman.

2 A space capsule returned to planet Earth following a mission to Mars. The capsule had a mass of 490 kg and entered the atmosphere with a speed of 800 ms^{-1}. The atmosphere slowed the capsule down to a speed of 250 ms^{-1}.

a) Calculate the change in velocity of the space capsule.

b) How much kinetic energy was transferred to heat energy during re-entry?

Fig 4.15.9: *Space capsule*

c) If the temperature rise of the capsule was 600°C, calculate the specific heat capacity of the space capsule.

(continued)

3 When a meteorite of mass 50 kg enters the Earth's atmosphere, it is slowed down from 950 ms^{-1} to just 150 ms^{-1}. Its initial temperature is –70°C. The specific heat capacity of the meteorite is 210 J/kg°C. The melting point of the meteorite is 1700°C.

a) Calculate the amount of kinetic energy transformed into heat energy as the meteorite is slowed by the Earth's atmosphere.

b) Calculate the temperature rise of the meteorite.

c) Does the meteorite remain solid throughout re-entry?

d) Expain, based on your answers to the above, why a meteorite travelling through the Earth's atmosphere is seen as a shooting star.

Fig 4.15.10: *A Shooting Star*

4 Designers for a new space shuttle need to choose an appropriate material for the nose cone to prevent it melting during re-entry. The shuttle will enter the Earth's atmosphere with a velocity of 750 ms^{-1} and be slowed to just 50 ms^{-1} for landing.

The nose cone must have a mass of 800 kg, and should not melt! The following materials are available for the construction of the heat shield.

Material	Melting point (°C)	Specific heat capacity (J/kg°C)
A	175	800
B	90	1100
C	120	1050
D	200	500

Table 3.15.1

The initial temperature of the space shuttle shield on entry to the Earth's atmosphere is –150°C. Choose a suitable material from the list above to construct the heat shield for the shuttle.

Learning checklist

In this section you will have learned:

- That different materials require different quantities of heat to change the state of unit mass.

- That the same material requires different quantities of heat to change the state of unit mass from solid to liquid (fusion) and to change the state of unit mass from liquid to gas (vaporisation).

- How to use an appropriate relationship to solve problems involving mass, heat energy and specific latent heat.

16 Gas laws and the kinetic model

In this chapter you will learn about:

- Pressure, and the relationship between pressure, force and area.
- Gas laws including the kinetic theory of gases, Boyle's law, Charles' law and Guy-Lussac's law.
- The Kelvin temperature scale.
- The combined gas equation relating pressure, volume and temperature from the start of a system to the end position of a system.
- The relationship between Kelvin, degrees Celsius and absolute zero.

Pressure

Pressure is the **force** which acts at 90 degrees to a surface. It is affected by the size of force acting and the area it is acting over. The theory of pressure can be used to explain why a drawing pin can be forced into a brick wall by simply pushing with your thumb, and why a ballet dancer can do more damage to a surface than a Scorpion tank!

Defining pressure

Pressure is defined as the amount of force that an object applies normal to the surface (that is, at right angles to a surface).

When an object is sitting on a table, the force of gravity acts downwards – this is known as the object's weight. The force is spread over a certain area – the contact area between the object and the table. Pressure depends on both the force applied and the contact area.

Compare an object with a small mass to one with a large mass, e.g. an elephant and a mouse. According to the equation

$$W = mg$$

where W = weight (measured in N or kg ms^{-2}), m = mass (measured in kg) and g = gravitational acceleration (9·8 ms^{-2})

The elephant, with the greater mass, will have a greater weight. Therefore the elephant will have a greater downwards force acting on the ground than the mouse. This is why an elephant does more damage to the ground than a mouse – the damage is linked to the pressure applied to the ground.

Think about it

Consider the following example. If you have to walk across snow or soft sand, which of the following should you choose to do it in and why?

Fig 4.16.1: *High heeled shoes*

Fig 4.16.2: *Snow/ice grips*

Word bank

- **force**

an action on an object that can change its direction or shape. Weight is a special force that is linked directly to gravity and is dependent on the mass of the object.

Now consider the effect of the contact area. The downwards force,

$$W = mg$$

is spread across the contact area, A. If the contact area is small, then the force is concentrated over only a small area. This does a lot of damage to the small area – the pressure is large. If the contact area is big, the force is spread across a big area and the damage done to that area is less. In other words, increasing the contact area reduces the pressure.

Fig4.16.3: *The mass of an object affects its weight on a planet*

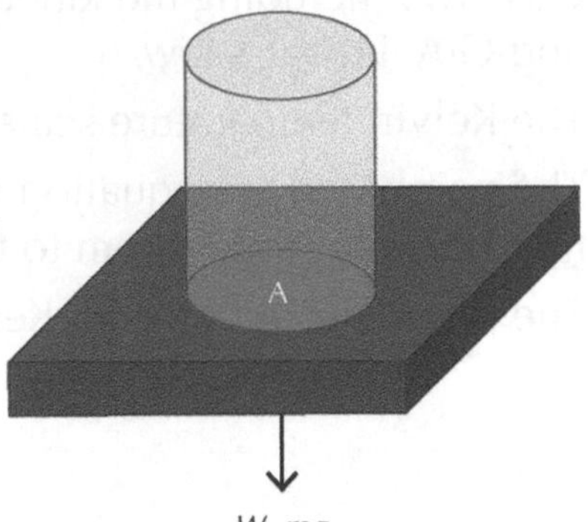

Fig4.16.4: *Pressure is affected by the size of the contact area*

This can be summarised by the equation:

$$P = \frac{F}{A}$$

Pressure is measured in pascals (Pa), where 1 Pa is equal to 1 N/m^2.

Key point

If the applied force is increased, the pressure is increased.

If the force is spread over a larger area, then the pressure is decreased.

Technique: Calculation involving pressure, force and area

1. Write down what you know from the question, leaving blank what you are trying to find:
   Pressure, P = ______ Pa
   Force, F = ______ N (you may need to use $W = mg$ to get the force if you require the weight)
   Area, A = ______ m^2 (ensure the area is measured in square metres!)
2. Write down the equation linking the quantities above as you see it on the formula sheet:
   $$P = \frac{F}{A}$$
3. Substitute into the equation what you know.
4. Solve for the unknown.

Worked example

A crate of mass 25 kg is resting on a flat surface. The crate has a surface area of 2 m^2 in contact with the ground. Calculate the pressure the crate exerts on the ground.

First of all we need to calculate the force, which will be the weight of the crate:

$$W = mg$$
$$W = 25 \times 9{\cdot}8$$
$$W = 245 N$$

Now we can use the equation that links force, pressure and area to find the pressure:

$$P = \frac{F}{A}$$

We know that:

force, $F = 245$ N

area, $A = 2$ m^2

So we have,

$$P = \frac{245}{2}$$
$$P = 122{\cdot}5 Pa$$

Worked example

A pick-up truck has a mass of 2600 kg. Each tyre has a contact area of 0.03 m^2 with the road. Calculate the pressure exerted on the road by the truck.

Fig 4.16.5: *Pick-up truck*

First of all we need to calculate the total contact area:

$$A = 4 \times 0{\cdot}03$$
$$A = 0{\cdot}12 m^2$$

Now work out the pressure:

$$P = \frac{F}{A}$$

(continued)

The force is given by the weight of the truck, so:

$$P = \frac{mg}{A}$$

$$P = \frac{2600 \times 9{\cdot}8}{0{\cdot}12}$$

$$P = 212\,333{\cdot}3\,Pa$$

Exercise 4.16.1 Defining pressure

1 Use the equation that links pressure, force and area to calculate the missing values in the table below.

Pressure (Pa)	Force (N)	Area (m^2)
(a)	200	5
(b)	120	5
(c)	5	0·5
1000	(d)	6
250	(e)	2·3
90	9	(f)
85	6·5	(g)
3000	2×10^5	(h)
(i)	$5{\cdot}4\times10^3$	5·2

Table 4.16.1

2 The mass of a crate is 25 kg. It has a surface area of 4 m^2. Calculate the pressure the crate exerts on the ground.

3 Atmospheric pressure is 1×10^5 Pa. What force does the air exert on the floor of a classroom which measures 3 m wide by 5 m long?

4 Calculate the required surface area of a tyre of the A380 Airbus if the force applied to each tyre is 600 000 N and each tyre requires a design load of 217 PSI. (1 PSI = 6894 Pa)

Fig 4.16.6: *Airbus A380*

Pressure in action

We have already seen the concept of reducing the contact area to increase the pressure – a drawing pin relies on this to allow you to push it through the wall. The small surface area of the point of the pin means that the pressure is very large.

Large tyres

If you increase the contact area, then the pressure will be reduced for a given force. This is put into practice with off-road vehicles such as tractors which use wide tyres to prevent them from sinking into soft mud. The weight of the tractor is spread across a wide area which will reduce the pressure. Tanks and all-terrain vehicles use caterpillar tracks to increase the contact area with the ground to prevent them from sinking.

Fig 4.16.7: *The tractor has large tyres to spread its weight over a large area*

Narrow tyres

Rally cars racing on snow and ice use the opposite effect. Narrow tyres are fitted to reduce the contact area and increase the pressure. This allows the tyres to 'bite in' to the slippy surface in order to get more grip. The effect is increased by using studs on the tyre which act like pins, forcing their way into the ice because of the high pressure. The same effect is applied to studded boots used for football and rugby.

Fig 4.16.8: *Rally cars have narrow tyres to increase the pressure on the surface and 'dig in' more*

Sharp knife

Consider the cutting edge of a sharp knife. It has a very small area. The small area means there is a large pressure applied by the knife. This allows it to cut through objects. The sharper the knife, the smaller the cutting area.

Fig 4.16.9: *A knife has a narrow edge to cut through food*

Fig 4.16.10: *Drawing pin*

Exercise 4.16.2 Pressure in action

1 Using your understanding of pressure, describe why it is more painful to stand on the sharp end of a drawing pin than the blunt side.

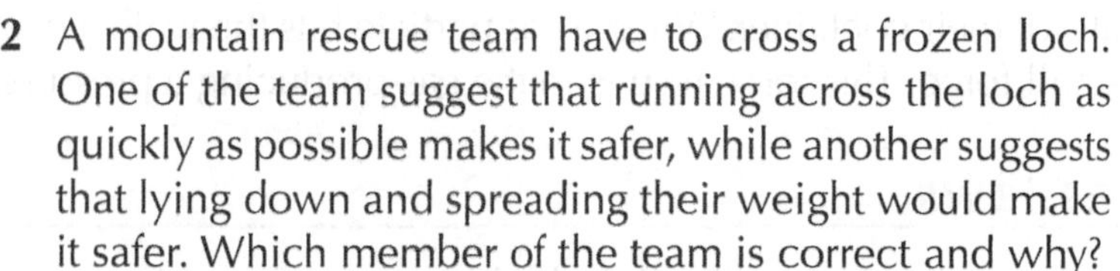

2 A mountain rescue team have to cross a frozen loch. One of the team suggest that running across the loch as quickly as possible makes it safer, while another suggests that lying down and spreading their weight would make it safer. Which member of the team is correct and why?

3 A Scorpion tank drives across a polystyrene surface and leaves an indentation from its tracks. A ballet dancer then stands on the same surface on a single toe. The imprint from the ballet dancer is deeper than the imprint

(continued)

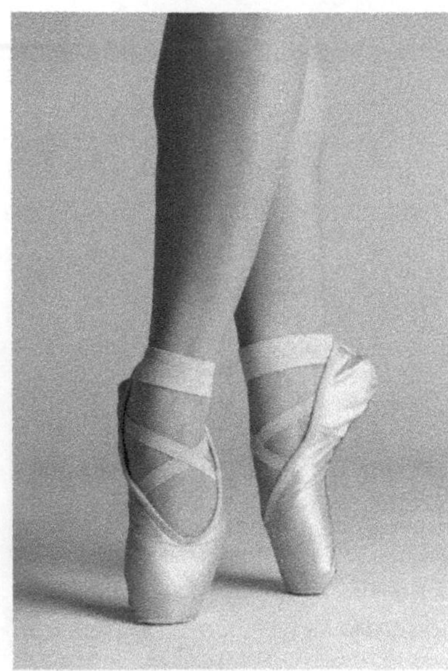

Fig 4.16.11: *Pressure exerted by the point of a ballet dancer's toes*

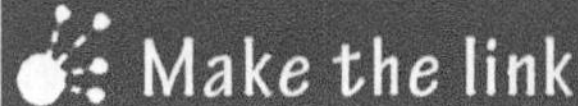

Make the link

Weather maps are considered in more detail in Geography.

from the tank. Explain why this is the case, making reference to the relative pressures applied by both the tank and the dancer.

4 As well as increasing the length of time taken for your head to come to rest in an accident, a car's airbag also increases the contact area between your head and the steering wheel. Explain why this reduces injury.

Pressure in fluids (gases)

If you watch the weather forecast, you may have seen a map similar to the one shown. It shows how the air pressure is changing – there are regions of low pressure and high pressure. Air, like all gases and liquids, is a fluid. The particles of a fluid move around and because of this they exert a pressure.

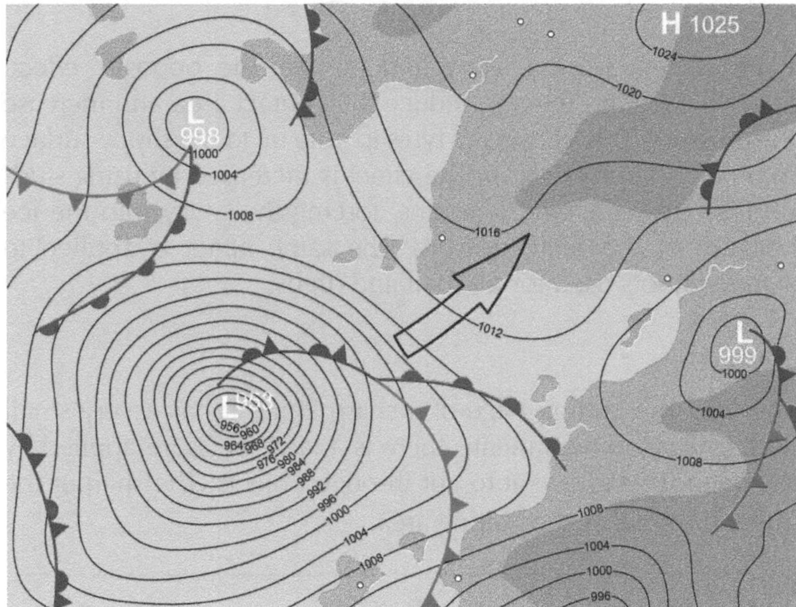

Fig 4.16.12: *High pressure and low pressure are represented by contour lines*

Gas particles are in constant motion and obey Newton's laws of motion. Particles collide with other particles and also with the walls of their container. When a gas particle hits the wall, it exerts a small force. This force results in the gas producing a pressure.

Gas laws

There are three laws that link the behaviour of gas in terms of volume *(V)*, pressure *(P)* and temperature *(T)*.

Kinetic theory of gases

The kinetic theory of gases is a model which is used to explain the behaviour of gases. Kinetic theory considers a gas to be made up of many small particles which are moving around at high speed. These particles are far apart, but they can collide with each other and also with objects or the walls of their container.

We consider three main variables for a gas:

- **Volume**: the volume of a gas is taken to be the volume of the container holding the gas. We do not consider the volume occupied by individual gas particles because this is infinitesimally small.
- **Pressure**: the pressure of a gas is caused by particles colliding with the walls of the container – each collision results in a small force over a given area which we know creates a pressure. Pressure is increased by more frequent collisions with the container walls which can arise from having more gas particles in a given volume. Pressure will also increase if the collisions are more violent or frequent, which can be caused by the gas particles moving more quickly.
- **Temperature**: the temperature of a gas depends on the kinetic energy of the gas particles – the greater the kinetic energy, the greater the temperature.

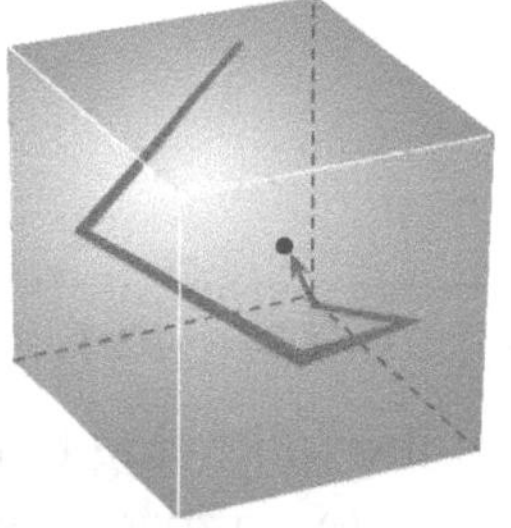

Fig 4.16.13: *Path of moving molecule*

The Kelvin temperature scale

When considering gases, we use a different temperature scale – the Kelvin temperature scale. The scale was developed by Lord Kelvin (Sir William Thomson) who discovered through experimentation that temperature did not continue to drop indefinitely but in fact had a value which it would not fall below. The lowest value that temperature can ever be is called absolute zero and is the temperature at which all motion of particles stops.

Temperature is measured in kelvin (K). Zero degrees kelvin is absolute zero which is –273°C on the Celsius scale.

It is important to note that 1°C = 1K so if a body increases its temperature by, for example, 5°C then the temperature difference in kelvin is also 5° K.

Fig 4.16.14: *Lord Kelvin*

Technique: Converting between Kelvin and degrees Celsius

To convert degrees Celsius into kelvin, add 273:

$$0°\text{C} = 273\text{K}$$

To convert kelvin into degrees Celsius, subtract 273:

$$0\text{ K} = -273°\text{C}$$

Worked example

The temperature measured on a summer's day in Aberdeen was 26°C. Convert this temperature into kelvin.

We are converting from degrees Celsius to kelvin, so we add 273:

$$T = 26 + 273$$
$$T = 299\ K$$

Exercise 4.16.3 The Kelvin temperature scale

1 Convert the following temperatures to kelvin:

a) 200°C b) –270°C c) 100°C d) –12°C e) –250°C

2 Convert the following temperatures to degrees Celsius:

a) 273 K b) 0 K c) 200 K d) 121 K e) 50 K

3 The brakes on a Formula 1 car are made from a ceramic material. During a braking manoeuvre the temperature of the brakes increases from 30 °C to 75 °C.

a) Convert the initial and final brake temperatures to kelvin.

b) Calculate the heat energy generated by the brakes during braking, assuming a specific heat capacity of 900 J/kg K and a mass of 0·9 kg.

Fig 4.16.15: *A Formula 1 car*

4 The Virgin Galactic Spaceship 1 uses a feathered air-brake system to reduce its speed on atmospheric re-entry. This is to reduce the frictional effect on the hull of the craft. The ship has a mass of 9400 kg and experiences a change in temperature from –30°C to 20°C.

a) Calculate the change in temperature in kelvin.

b) If the heat energy produced is 117 MJ, calculate the required specific heat capacity of the carbon composite hull.

Boyle's law (pressure and volume)

Boyle's law considers the relationship between the pressure and volume of a gas for a constant temperature.

Experiment: Boyle's law

This experiment will investigate how the pressure of a gas changes when the volume is changed.

You will need:

- Trapped air column (oil column inside a glass tube as shown in the diagram)
- Scale on air column to measure volume of gas. If scale measures in cm this can be converted to volume as length and volume are propotional.
- Pump
- Tap between pump and pressure gauge
- Pressure gauge (Bourdon gauge)

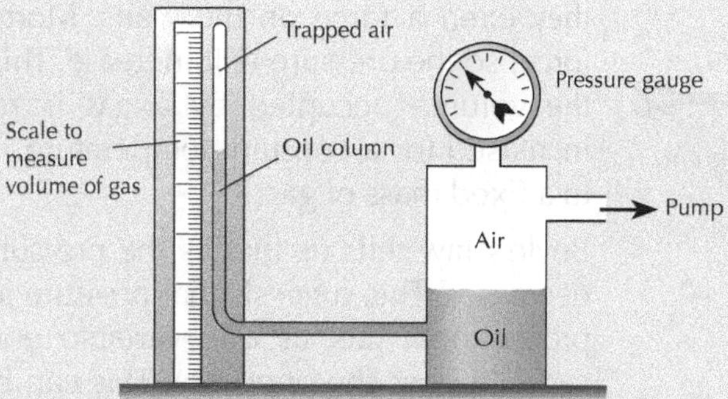

Figure 4.16.16: *Investigating Boyle's law*

Instructions

1. Connect the pump to the apparatus and use it to increase the pressure of the trapped air.
2. Once the pressure is high, use a tap to seal the apparatus and keep the pressure and volume constant.
3. Record the volume and the corresponding pressure using the Bourdon gauge.
4. Open the tap slowly and reduce the pressure by a small amount. Seal the apparatus at the new pressure using the tap.
5. Record the new volume and pressure.
6. Repeat for different values of volume and pressure and note your results in the following table.

Volume	Pressure (Pa)

Table 4.16.2

(*continued*)

7 Comment on how the pressure of a gas changes when the volume is increased or reduced – can you explain this effect based on kinetic theory?

8 Plot a graph of the volume of the gas against the pressure of the gas.

9 Plot a graph of the pressure against 1/volume for the gas.

10 Use the graphs above to determine the relationship between pressure and volume of a gas.

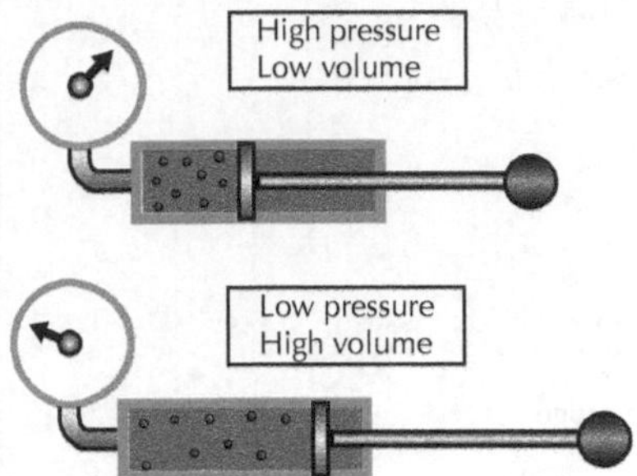

Fig 4.16.17: *Pistons increasing and decreasing in pressure*

Consider gas trapped in a cylinder by a piston as shown below. If the piston is pushed in, the gas particles will have less room to move as the volume the gas occupies has been decreased. This will result in the particles colliding more frequently with the walls of the container. Each time they collide with the walls, they exert a force on the walls. More collisions means more force, so the pressure will increase. This gives us Boyle's law: as the volume occupied by a gas is reduced, the pressure is increased for a constant temperature. Boyle's law only applies to a fixed mass of gas.

Boyle's law tells us that as the pressure increases, the volume decreases. This suggests that pressure and volume are inversely proportional (i.e. as one variable does one thing, the other variable does the opposite). This can be confirmed by plotting the graphs from the experiment above.

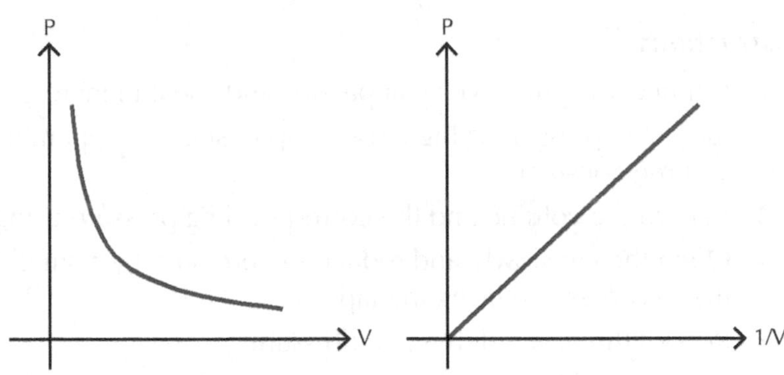

Fig 4.16.18

The straight line graph (on the right) for inverse volume proves that pressure and volume are inversely proportional. So we can write,

$$P \propto \frac{1}{V}$$

$$P = \frac{k}{V}$$

$$PV = k$$

where k is a constant. In other words, when you multiply pressure and volume together it must always come out as the same number for a given system. This allows us to quote Boyle's law in the following useful form:

$$P_1V_1 = P_2V_2$$

Technique: Using Boyle's law

1. Write down what you know from the question, leaving blank what you are trying to find:
   Pressure before, P_1 = ______ Pa
   Pressure after, P_2 = ______ Pa
   Volume before, V_1 = ______ m^3
   Volume after, V_2 = ______ m^3
2. Write down the equation linking the quantities above as you see it on the formula sheet:
   $$P_1V_1 = P_2V_2$$
3. Substitute into the equation what you know.
4. Solve for the unknown.

Worked example

A piston in a car engine moves up and down inside a cylinder. When the piston is extended the volume in the cylinder is 200 ml and the pressure is $1{\cdot}013 \times 10^5$ Pa. When the piston is compressed the volume of the cylinder is reduced to 25 ml. What is the new pressure inside the cylinder? You may assume the temperature remains constant.

Fig 4.16.19: *Piston*

$$P_1V_1 = P_2V_2$$

$$1{\cdot}013 \times 10^5 \times 200 = P \times 25$$

$$2{\cdot}026 \times 10^7 = 25P$$

$$P = \frac{2{\cdot}026 \times 10^7}{25}$$

$$P = 8{\cdot}104 \times 10^5 Pa$$

Hint

As long as the units of volume are the same on both sides of the equation the answer will be correct.

Exercise 4.16.4 Boyle's law

1 Find the missing values in the table below.

Fig 4.16.20: *Weather balloon*

Pressure 1 (Pa)	Pressure 2 (Pa)	Volume 1 (m^3)	Volume 2 (m^3)
250	500	6	(a)
145	(b)	12	9
(c)	210	100	150
25	400	(d)	245
(e)	140	200	400

Table 4.16.3

2 A weather balloon contains 25 m^3 of Helium at an atmospheric pressure of 101 000 Pa on the ground. When it reaches its working altitude, the pressure has dropped to 84 000 Pa. Find the new volume of the weather balloon.

3 A diver uses a cylinder of compressed air which holds 20 litres of air at a pressure of 12 000 kPa. At diving depth, this air occupies a volume of 100 litres. Find the pressure of the air at the depth where the diver is diving.

Fig 4.16.21: *Diver*

4 A piston in a car engine moves up and down inside a cylinder. When the volume inside the cylinder is at a maximum of 500 millilitres, the pressure is 101 000 Pa. Calculate the pressure inside the cylinder when the volume is:

a) 250 millilitres

b) 100 millilitres

c) 20 millilitres

5 Air for inflating party balloons is kept in a compressed air canister. Inside the canister, the air has a pressure of 800 kPa and occupies a volume of 0·8 m^3. Calculate the volume this air would occupy when released from the cylinder and it has a pressure of 101 kPa.

Charles' law (volume and temperature)

Charles' law considers the relationship between the volume and temperature of a gas for a constant pressure. Charles' law only applies to a fixed mass of gas.

Experiment: Charles' law

This experiment will investigate how the volume of a gas changes when the temperature is changed.

You will need:

- Water bath
- Thermometer
- Capillary tube with attached scale to measure volume
- Sulfuric acid, mecury or similar bead

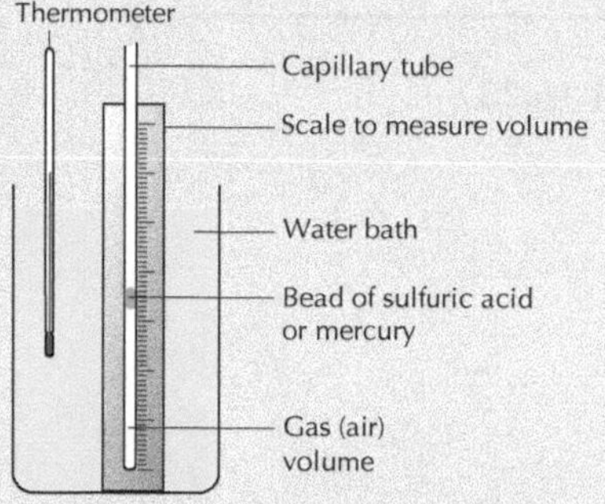

Fig 4.16.22: *Investigating Charles' law*

Instructions

1. Set up the apparatus as shown in the diagram above. A sulfuric acid bead is used to trap a column of air and keep this air dry.
2. Gently heat the water in the water bath to your starting temperature. Measure this temperature using the thermometer and note it in your results table. Remember to add 273 on to your degrees Celsius reading to record the temperature in kelvin!
3. Record the volume using the scale on the trapped air column.
4. Now record the volume for different temperatures as the water in the water bath cools. Record your results in a table similar to the following:

Temperature (K)	Volume

Table 4.16.4

5. Comment on how the volume of a gas changes when the temperature is increased or reduced – can you explain this effect based on kinetic theory?
6. Plot a graph of the volume of the gas against the temperature of the gas.
7. Use the graph to determine the relationship between volume and temperature of a gas.

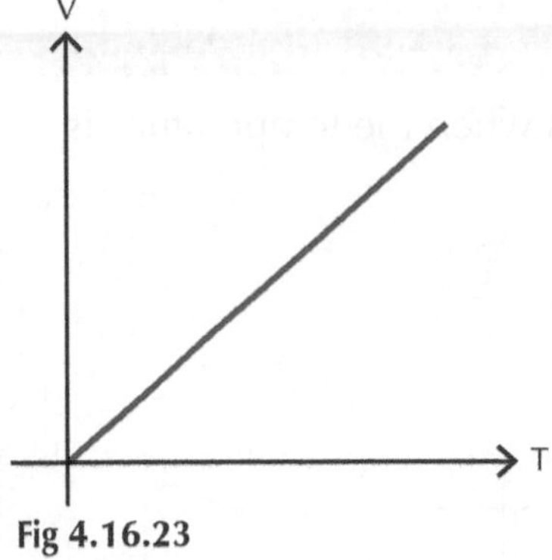

Fig 4.16.23

Consider gas particles trapped in a cylinder by a moveable piston. When heated, the particles of gas move around faster in the container because they have a greater kinetic energy. This results in them striking the walls of the container with a greater force. This will push the piston upwards, increasing the volume. In other words, as the temperature increases, the volume also increases.

If we were to measure the volume as the temperature was changed, the results would produce the graph shown (where temperature is measured in kelvin):

As the temperature increases, the volume also increases. The relationship between the two is directly proportional (i.e. when one variable does something, the other variable does the same thing) providing you have a fixed mas of gas:

$$V \propto T$$

$$V = kT$$

where k is a constant. Thus,

$$\frac{V}{T} = k$$

From this we get the following equation which can be used to solve problems involving pressure and volume where the temperature is constant:

$$\frac{V_1}{T_1} = \frac{V_2}{T_2}$$

Technique: Using Charles' law

1. Write down what you know from the question, leaving blank what you are trying to find:

   Temperature before, T_1 = ______ K *(Remember to use temperature in kelvin!)*

   Temperature after, T_2 = ______ K

   Volume before, V_1 = ______ m^3

   Volume after, V_2 = ______ m^3

2. Write down the equation linking the quantities above as you see it on the formula sheet:

   $$\frac{V_1}{T_1} = \frac{V_2}{T_2}$$

3. Substitute into the equation what you know.
4. Solve for the unknown. This can be done using a process called cross-multiplication.

Worked example

A balloon is held at constant pressure of 1 atmosphere. The temperature of the balloon when inflated is 127°C. The temperature is reduced to 90°C. Calculate the new volume of the balloon if the initial volume was 280 000 litres.

First of all, write down what we know and what we are trying to find out. Remember to convert the temperatures into kelvin:

- *Initial temperature,* T_1 *= 127°C = 400 K*
- *Final temperature,* T_2 *= 90°C = 363 K*
- *Initial volume,* V_1 *= 280 000 litres*
- *Final volume,* V_2 *= ? litres*

We are using Charles' law:

$$\frac{V_1}{T_1} = \frac{V_2}{T_2}$$

$$\frac{280\,000}{400} = \frac{V_2}{363}$$

Use cross-multiplication to solve:

$$280\,000 \times 363 = 400\,V_2$$

$$400\,V_2 = 101\,640\,000$$

$$V_2 = \frac{101\,640\,000}{400}$$

$$V_2 = 254\,100 \textit{ litres}$$

Exercise 4.16.5 Charles' law

1 Find the missing values in the table below.

Temperature 1 (K)	Temperature 2 (K)	Volume 1 (m^3)	Volume 2 (m^3)
273	500	12	(a)
(b)	320	6	9
350	270	(c)	24
290	(d)	12	30
(e)	312	11·5	11·9

Table 4.16.5

2 A volume of 27 cm^3 of air is kept at a temperature of 5°C. Assuming the pressure remains constant, what will be the temperature of the air when the volume is increased to 50 cm^3? Answer in degrees Celsius.

(*continued*)

3 Students wish to investigate Charles' law. They use the apparatus shown opposite. Air is trapped in a capillary tube by a bead of mercury. The air column is placed in hot water and the temperature of the water measured using a thermometer. At room temperature (22°C), the volume of air in the column is found to be 0·4 cm^3. The water is heated in order to heat the air. After heating, the temperature is measured to be 64°C. Find the new volume of the air which is trapped in the air column.

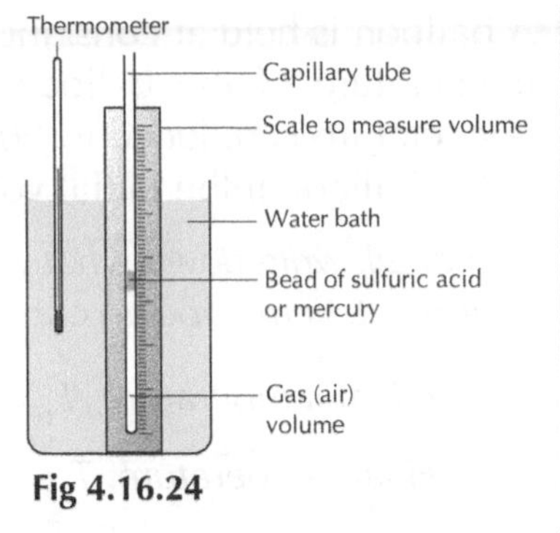

Fig 4.16.24

Guy-Lussac's law (pressure and temperature)

The third gas law investigates the relationship between the pressure and temperature of a gas for a fixed volume.

GO! Experiment: Guy-Lussac's law

This experiment will investigate how the pressure of a gas changes when the temperature is changed.

You will need:

- Flask
- Water bath
- Thermometer
- Pressure (Bourdon) gauge

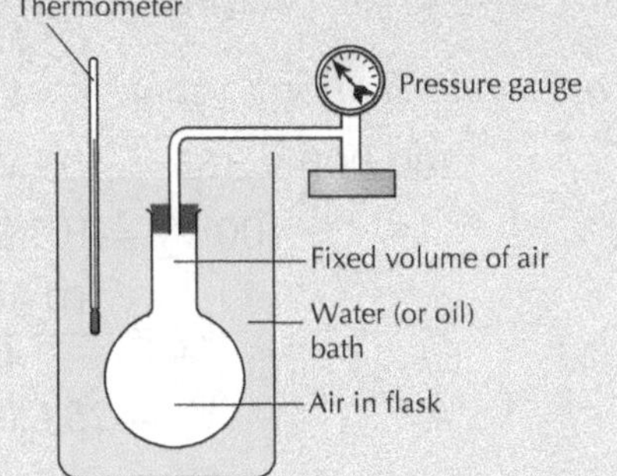

Fig 4.16.25: *Investigating Guy-Lussac's law*

Instructions

1. Set up the apparatus as shown in the diagram above. Use a rubber bung to keep the same mass of air in the flask with a constant volume.
2. Gently heat the water in the water bath to your starting temperature. Measure this temperature using the thermometer and note it in your results table. Remember to add 273 on to your degrees Celsius reading to record the temperature in kelvin!
3. Record the pressure of the air in the flask using the pressure gauge.
4. Now record the pressure at different temperatures as the water in the water bath cools. Record your results in a table similar to the following:

Temperature (K)	Pressure (Pa)

Table 4.16.6

(*continued*)

5 Comment on how the pressure of a gas changes when the temperature is increased or reduced – can you explain this effect based on kinetic theory?

6 Plot a graph of the pressure of the gas against the temperature of the gas.

7 Use the graph to determine the relationship between pressure and temperature of a gas.

Consider a gas, with a fixed mass, trapped inside a container that has a fixed volume. When the temperature of the gas is increased, the gas particles gain more kinetic energy. They move faster and therefore hit the walls of the container harder and more often – this greater force results in an increase in pressure.

In other words, as the temperature of a gas is increased, the pressure of the gas will also increase for a fixed volume. This can be represented by the graph shown here.

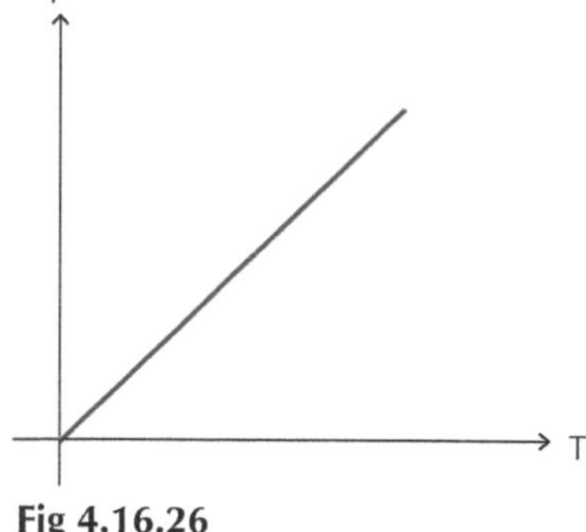

Fig 4.16.26

Temperature here is measured in kelvin. The graph shows that pressure and temperature are directly proportional. We can write this relationship mathematically:

$$P \propto T$$

$$P = kT$$

where k is a constant. We can therefore say:

$$\frac{P}{T} = k$$

This gives us the following equation for solving problems involving pressure and temperature for a constant volume:

$$\frac{P_1}{T_1} = \frac{P_2}{T_2}$$

Technique: Using Guy-Lussac's law

1 Write down what you know from the question, leaving blank what you are trying to find:

Temperature before, T_1 = ______ K

Temperature after, T_2 = ______ K

Pressure before, P_1 = ______ Pa

Pressure after, P_2 = ______ Pa

2 Write down the equation linking the quantities above as you see it on the formula sheet:

$$\frac{P_1}{T_1} = \frac{P_2}{T_2}$$

3 Substitute into the equation what you know.

4 Solve for the unknown.

Worked example

A deep-sea diver uses a rigid scuba tank to remain under water when repairing a leak on an oil rig. The temperature at the surface of the sea is 300 K and the temperature at the operating depth is 290 K. The pressure at the surface of the water is 1×10^5 Pa. Calculate the pressure at the operating depth.

First of all, write down what we know and what we are trying to find out. Remember to convert the temperatures into kelvin:

Initial temperature, $T_1 = 300$ *K*

Final temperature, $T_2 = 290$ *K*

Initial pressure, $P_1 = 1 \times 10^5$ *Pa*

Final pressure, $P_2 = ?$ *Pa*

We are using Guy-Lussac's law:

$$\frac{P_1}{T_1} = \frac{P_2}{T_2}$$

Temperatures are already in kelvin, so we can substitute into the above equation:

$$\frac{1 \times 10^5}{300} = \frac{P_2}{290}$$

Use cross-multiplication to solve for the unknown pressure:

$$300P_2 = 290 \times 1 \times 10^5$$

$$300P_2 = 2{\cdot}9 \times 10^7$$

$$P_2 = \frac{2{\cdot}9 \times 10^7}{300}$$

$$P_2 = 9{\cdot}67 \times 10^4 \, Pa$$

Exercise 4.16.6 Guy-Lussac's law

1 Find the missing values in the table below.

Temperature 1 (K)	Temperature 2 (K)	Pressure 1 (Pa)	Pressure 2 (Pa)
300	350	1250	(a)
273	300	(b)	$1{\cdot}01 \times 10^5$
(c)	350	10 500	24 000
270	(d)	1500	3500
(e)	290	1200	1250

Table 4.16.7

2 A compressed air tank which is at room temperature of 20°C normally contains air at a pressure of $2{\cdot}02 \times 10^5$ Pa. It is fitted with a safety valve which opens when the pressure reaches 10×10^5 Pa. Calculate the temperature at which the safety valve is designed to open.

3 A long and high-speed journey will cause the temperature in a car's tyres to heat up. When cold, at a temperature of 25°C, the pressure in a car tyre is 202 000 Pa.

a) Find the pressure in the car tyre after a journey where the tyre temperature has risen to 70°C.

b) Explain why car manufacturers always recommend that tyre pressures are checked when the tyre is cold.

c) A tyre which gets too hot can be dangerous – explain, using your knowledge of air pressure and temperature, why this is so.

Fig 4.16.27: *Car manufactures recommend tyre pressure is checked when tyres are cold*

4 An electric light bulb is designed so that the pressure of the inert gas inside it is 100 kPa (normal air pressure) when the temperature of the bulb is 400°C. To what pressure must the bulb be filled if it is manufactured at a temperature of 25°C?

5 The air in a hot air balloon is at a pressure of 105 000 Pa when the balloon is on the ground. The temperature of the air inside the balloon is 15°C. In order to rise into the air, the gas inside the balloon is heated to increase the pressure to 300 000 Pa. Find the temperature that the gas inside the balloon is heated to.

6 A can of compressed air can be used to clean a computer keyboard. The air is ejected out of the can, to blow dust out from between the keys. The pressure of the air inside the can is 450 000 Pa at a temperature of 25°C.

a) When the air is released from the can, its pressure drops to 101 000 Pa. Find the temperature of the air.

b) Explain, using your results above, why the can of compressed air feels cold where the air leaves the can.

The combined gas equation

So far we have considered three separate gas laws which are used to consider the three main variables: volume, temperature and pressure. The three equations can be combined to give what is known as the combined gas equation. This equation allows us to tackle realistic problems where all of volume, pressure and temperature can change rather than one of them being kept constant.

Explaining the gas laws

Kinetic theory can be used to explain the effects observed in experiments to investigate the relationship between pressure, temperature and volume of a gas.

Boyle's law: pressure-volume law

Fig 4.16.28: *Pressure-volume law*

As the volume is increased, the pressure is decreased. Pressure and volume are inversely proportional:

$$P \propto \frac{1}{V}$$

For a fixed mass of gas at a constant temperature, the particles will collide with the container walls to produce the pressure. When the volume is decreased:

- The particles will make more collisions with the walls.
- The area of the walls will be decreased.
- The average force on the walls will increase.
- The pressure on the walls will therefore increase since

$$P = \frac{F}{A}$$

Charles' law: volume-temperature law

Fig 4.16.29: *Volume-temperature law*

As the temperature is increased, the volume also increases. For a fixed mass of gas, volume and temperature are directly proportional:

$$V \propto T$$

The pressure in this experiment would have to be kept constant by using a constant mass on top of the piston. When the temperature is increased:

- The particles will move faster because they have greater thermal energy.
- The particles will therefore make more collisions with greater force with the walls of the container.
- To maintain an even pressure, the piston must rise to increase the volume.

Guy-Lussac's law: pressure-temperature law

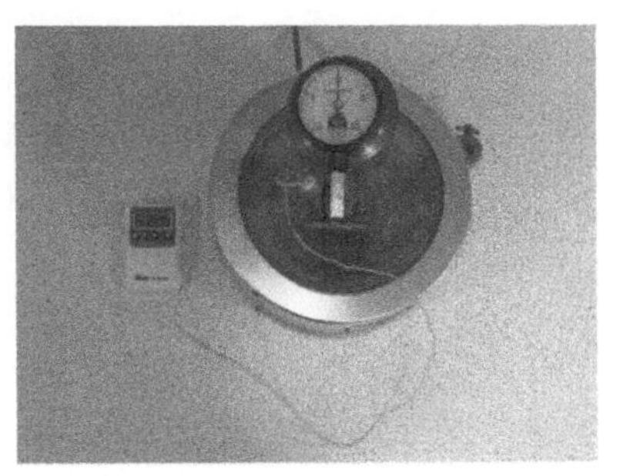

Fig 4.16.30: *Pressure-temperature law*

As the temperature is increased, the pressure also increases. Pressure and temperature are directly proportional:

$$P \propto T$$

The volume in this experiment is kept constant by ensuring we have a sealed container. When the temperature is increased:

- The particles will move faster because they have greater thermal energy.
- The particles will therefore make more collisions with greater force with the walls of the container.

- For a constant volume, this will result in a greater pressure from the greater average force because:

$$P = \frac{F}{A}$$

The combined gas equation

We have three separate gas laws shown above which give us the following three equations:

$$P \propto \frac{1}{V} \qquad V \propto T \qquad P \propto T$$

We can substitute into these equations:

$$P \propto \frac{1}{V} \qquad V \propto T \qquad P \propto T$$

$$P = \frac{k}{V} \qquad V = kT \qquad P = kT$$

$$PV = k \qquad \frac{V}{T} = k \qquad \frac{P}{T} = k$$

$$P_1V_1 = P_2V_2 \qquad \frac{V_1}{T_1} = \frac{V_2}{T_2} \qquad \frac{P_1}{T_1} = \frac{P_2}{T_2}$$

And the three equations can be combined into one single equation – the combined gas equation:

$$\frac{PV}{T} = k$$

which can be applied to problems involving a fixed mass of gas in the following useful form:

$$\frac{P_1V_1}{T_1} = \frac{P_2V_2}{T_2}$$

SPOTLIGHT ON INDUSTRY: FOUR STROKE CYCLE ENGINE

We can use the gas laws to explain how fuel can be burned in an internal combustion engine. Most car engines follow the four stroke cycle shown below.

The cycle starts by drawing a fuel-air mixture into the cylinder (intake). This is then compressed by the piston. A spark ignites the mixture, and the burning forces the piston down. Exhaust gases are then expelled from the engine.

Fig 4.16.31: *Four stroke cycle engine*

During the compression stroke, the volume that the gas occupies is reduced. From what we know about gas laws, when the volume is reduced, the temperature is increased. This increase in temperature (and pressure) makes the fuel-air mixture more flammable, so it will burn when ignited by the spark. In a diesel engine, the compression of the fuel-air mixture generates enough heat to ignite the fuel without the need for a spark! This is all down to the gas laws: reducing the volume to increase the pressure and temperature.

Technique: Using the combined gas equation

1. Write down what you know from the question, leaving blank what you are trying to find:

   Temperature before, T_1 = ______ K

   Temperature after, T_2 = ______ K

   Pressure before, P_1 = ______ Pa

   Pressure after, P_2 = ______ Pa

   Volume before, V_1 = ______ m^3

   Volume after, V_2 = ______ m^3

2. Write down the equation linking the quantities above as you see it on the formula sheet:

$$\frac{P_1V_1}{T_1} = \frac{P_2V_2}{T_2}$$

3. Substitute into the equation what you know.
4. Solve for the unknown.

Worked example

Inside a car engine, air and fuel are drawn into the cylinder at atmospheric pressure ($1{\cdot}01 \times 10^5$ Pa) at a temperature of 40°C. The volume of the cylinder is 200 cm^3. During compression, the air and fuel mixture is compressed into a volume of just 20 cm^3 at a pressure of $4{\cdot}00 \times 10^5$ Pa. Find the temperature of the fuel/air mixture after compression.

First of all, write down what we know and what we are trying to find out. Remember to convert the temperatures into kelvin:

Initial temperature, $T_1 = 273 + 40 = 313$ *K*

Final temperature, $T_2 = ?$ *K*

Initial pressure, $P_1 = 1{\cdot}01 \times 10^5$ *Pa*

Final pressure, $P_2 = 4{\cdot}00 \times 10^6$ *Pa*

Volume before, $V_1 = 200$ cm^3

Volume after, $V_2 = 20$ cm^3

Because none of the three variables are constant, we need to use the combined gas equation:

$$\frac{P_1V_1}{T_1} = \frac{P_2V_2}{T_2}$$

Substitute what we know,

$$\frac{1 \times 10^5 \times 200}{313} = \frac{4{\cdot}00 \times 10^6 \times 20}{T_2}$$

$$1 \times 10^5 \times 200 \times T_2 = 4{\cdot}00 \times 10^6 \times 20 \times 3$$

$$2 \times 10^7 T_2 = 2{\cdot}50 \times 10^{11}$$

$$T_2 = \frac{2{\cdot}50 \times 10^{10}}{2 \times 10^7}$$

$$T_2 = 1252\ K$$

Key point

If one of the three variables (temperature, pressure, volume) is constant, you can use the appropriate equation from Boyle's, Charles' or Guy-Lussac's law.

Exercise 4.16.7 The combined gas equation

1 Find the missing values in the table below:

P_1 (Pa)	V_1 (Litres)	T_1 (K)	P_2 (Pa)	V_2 (Litres)	T_2 (K)
(a)	5	50	1000	100	273
(b)	1·5	100	1×10^5	1·5	60
$1{\cdot}6 \times 10^5$	(c)	300	1×10^5	15	360
$2{\cdot}7 \times 10^6$	3	285	1×10^5	(d)	310
1×10^5	0·45	(e)	$9{\cdot}2 \times 10^4$	25	271
3×10^7	0·9	355	1×10^5	0·1	(f)

Table 4.16.8

2 Gas is initially at a pressure of 120 000 Pa, a volume of 23 litres, and a temperature of 200 K. The pressure is then raised to 140 000 Pa and the temperature increased to 300 K. What is the new volume of the gas?

3 A gas tank takes up a volume of 15 litres, at pressure of 230 000 Pa, and a temperature of 289 K. If the temperature is raised to 355 K and the pressure lowered to 150 000 Pa, what is the new volume of the gas?

4 A gas in a container has a volume of 25 litres, a temperature of 42°C and an unknown pressure. The volume is increased to 32 litres and its temperature decreased to 33°C. If the pressure after the change to be 200 000 Pa, what was the original pressure of the gas?

5 An unknown volume of gas is held at a pressure of 35 000 Pa and a temperature of 295 K. The pressure is increased to 120 000 Pa, the temperature is decreased to 302 K and the final volume is measured to be 48 litres. What was the initial volume of the gas?

6 A diesel truck engine relies on compressing the fuel air mixture in the cylinder to raise the temperature to the point where the fuel ignites. The fuel mixture is sucked into the cylinder at a pressure of 101 000 Pa and a temperature of 310 K. The volume of the cylinder is 400 cm^3. During compression, the air and fuel mixture is compressed so that its pressure is $6{\cdot}00 \times 10^6$ Pa and the temperature has been raised to 1900 K. Find the volume of the cylinder after compression.

Fig 4.16.32: *Diesel truck engine*

Learning checklist

In this section you will have learned:

- The definition of pressure in terms of force and area.
- How to use an appropriate relationship to solve problems involving pressure, force and area.
- How to describe the kinetic model for the pressure of a gas.
- The relationship between Kelvin and degrees Celsius and the absolute zero of temperature.
- How the pressure-volume, pressure-temperature and volume-temperature laws operate qualitatively in terms of the kinetic theory model.
- How to use appropriate relationships to solve problems involving the volume, pressure and temperature of a fixed mass of gas.
- How to describe experiments to verify the pressure-volume law (Boyles law), the pressure-temperature law (Gay-Lussac's law) and the volume-temperature law (Charles' law).

17 Wave parameters and behaviours

- Transverse and longitudinal waves
- Frequency, period, wavelength, amplitude and wave speed of longitudinal and transverse waves
- Diffraction

18 Electromagnetic spectrum

- Relative wavelength and frequency of bands of the electromagnetic spectrum
- Sources of electromagnetic waves
- The nature of electromagnetic waves

19 Refraction of light

- Definition of refraction
- Ray diagrams
- How light refracts when it enters a different medium.
- How refraction means a change in speed, wavelength and sometimes diffraction of light waves.
- The following terms: incident ray, refracted ray, angle of incidence, angle of refraction and the normal.
- Some applications of refraction – lenses and optical fibres.
- Dispersion of light waves.

AREA 5

Waves

17 Wave parameters and behaviours

In this chapter you will learn about:

- Waves and energy – how energy can be transferred as a wave and the difference between longitudinal and transverse waves.
- The characteristics of a wave including wavelength, amplitude, frequency, period and speed.
- The relationships between wave speed, frequency, wavelength, distance and time.
- Diffraction, including the difference between long-wave and short-wave diffraction, and the practical limitations of diffraction.

Waves and energy

Fig 5.17.1: *Water waves*

Waves are a fundamental part of physics. Waves can be seen in nature. Possibly the waves with which we are most familiar are water waves. If you go to the beach and watch the sea, you will see waves 'crashing' on the beach – but what causes these waves? You may also have heard of sound waves. Although not the same as water waves, they are related. The common factor between all types of waves is that they all transfer energy.

Think about it

Imagine throwing a stone into water. When the stone hits the water, it produces waves that move outwards. Why? What happens to these waves as they move outwards? How can you make the waves bigger?

Waves carrying energy

Waves are a means of transferring energy from one place to another. Consider the example of throwing a stone into a pool of water. The kinetic energy of the stone is transferred into the water to produce the water waves. These waves move, carrying the energy across the water. As the energy of the waves decreases, the height of the wave also decreases. We will consider this in more detail later.

Fig 5.17.2: *Outward waves produced when a stone hits the water*

Longitudinal waves and transverse waves

Many types of energy can be carried as waves. There are two main *types of wave*: **longitudinal waves** and **transverse waves**.

Key point

Waves are related because all waves transfer energy.

Experiment: The slinky

This experiment uses a slinky to examine the two main types of wave and how these waves transfer energy.

You will need:

- A slinky toy
- A board to which you can fix the end of the slinky

Instructions

1 Attach one end of the slinky to the board.

2 Hold the slinky at the other end and move it back and forth, expanding and contracting the coils.

3 Next move the slinky from side to side. You will see a pulse travelling along its length. If you look carefully you will see that each individual ring only moves by a small distance, while the pulse travels the whole way down the slinky. This pulse, or wave, is the energy being transferred down the slinky.

4 Write a short description that compares the two types of wave that you have seen.

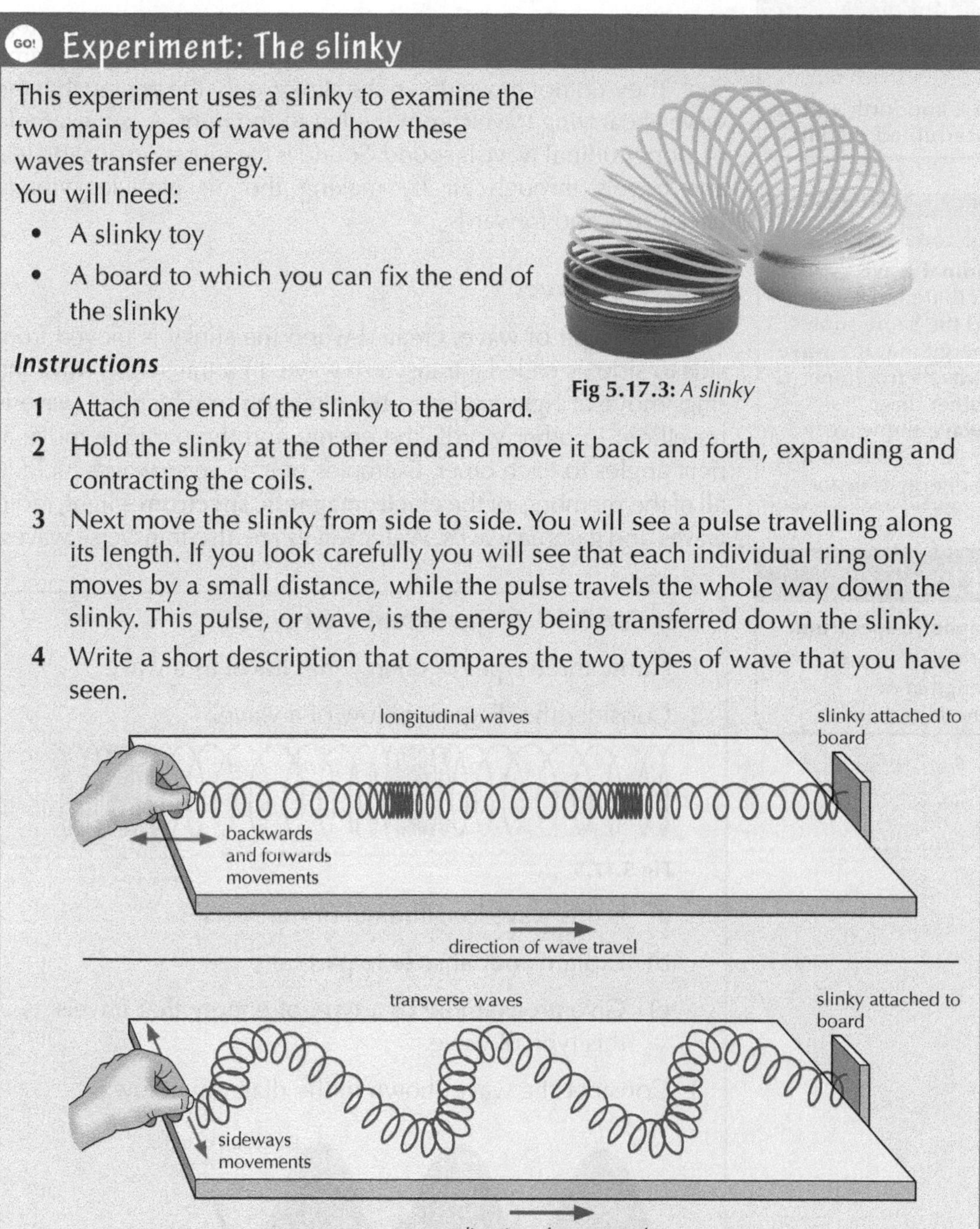

Fig 5.17.3: *A slinky*

Fig 5.17.4: *Longitudinal and transverse waves*

Longitudinal waves

The experiment showed that there are two types of wave – one type is produced when the end of the slinky is moved backwards and forwards. This is called a longitudinal wave. As the wave travels to the right, the slinky rings are moving back and forth

Word bank

- **oscillate**

to swing back and forth with a steady, uninterrupted rhythm

Key point

In a **longitudinal wave** particles oscillate back and forth around the same spot. Only the energy that the wave is carrying travels from one end to the other. In a **transverse wave** the wave oscillates at 90 degrees to the direction of energy transfer.

Word bank

- **electromagnetic spectrum**

the range of frequencies (and wavelengths) of electromagnetic radiation

from left to right – the same direction as the wave is moving. However, the particles **oscillate** back and forth around the same spot. They do not move down the slinky, only the energy that the wave is carrying travels from the left to the right. A key example of a longitudinal wave is sound. Sound is the vibration of particles, and travels through air by making the air particles vibrate backwards and forwards.

Transverse waves

The other type of wave, created when the slinky is moved from side to side, is called a transverse wave. In a transverse wave the rings move at right angles to the direction in which the wave is travelling. In other words, the energy and the particles move at right angles to each other. Examples of transverse waves include all of the members of the **electromagnetic spectrum** – light, radio waves and gamma waves. Water waves are also transverse waves.

Exercise 5.17.1 Waves and energy

1 Name three types of energy that travel as a wave.

2 Consider the diagram below of a wave.

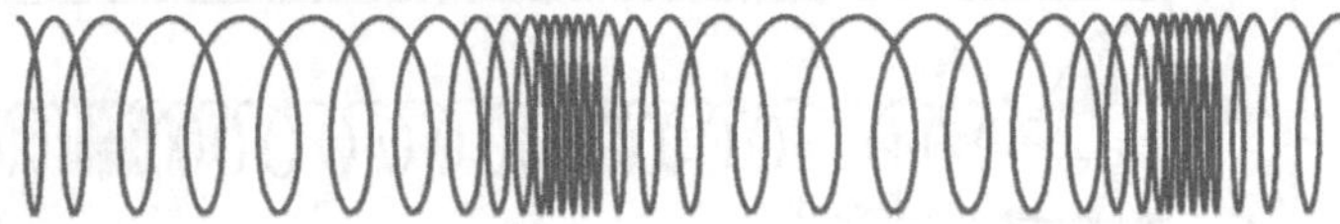

Fig 5.17.5

a) Is this wave longitudinal or transverse?

b) Explain your answer to part (a).

c) Give an example of a type of energy that travels as this type of wave.

3 Consider the wave shown in the diagram below:

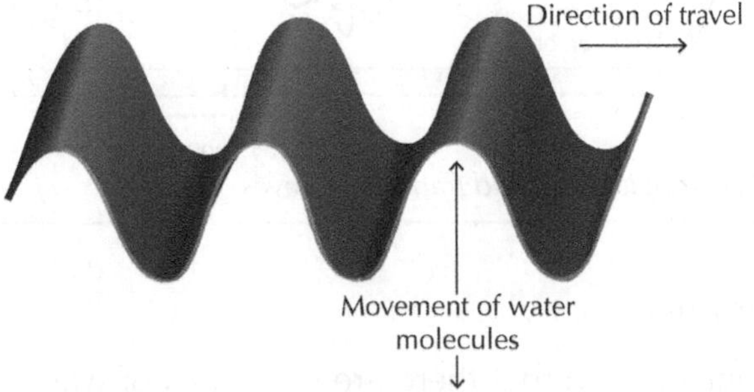

Fig 5.17.6

a) Is this wave longitudinal or transverse?

b) Explain your answer to part (a).

c) Give an example of a type of energy that travels as this type of wave.

Characterising a wave

A wave can be characterised by five main variables:

- The **wavelength**, λ, measured in metres (m). The wavelength is the distance from one part of the wave to the same part of the next wave. It does not matter where you start measuring as long as you measure to the equivalent part of the next wave.
- The **amplitude**, A, usually measured in metres (m). The amplitude of a wave is the height of the wave measured from the centre to the crest (or indeed the trough) of the wave. The greater the amplitude, the greater the energy carried by the wave.
- The **frequency**, f, of the wave which has units of hertz (Hz). The frequency of a wave is the number of waves produced every second: 1 Hz = 1 wave per second.
- The **period**, T, of the wave measured in seconds (s). The period of the wave is the length of time taken to produce one wave.
- The **speed**, v, of the wave which is measured in metres per second (ms^{-1}). The speed of the wave is the distance the wave travels every second.

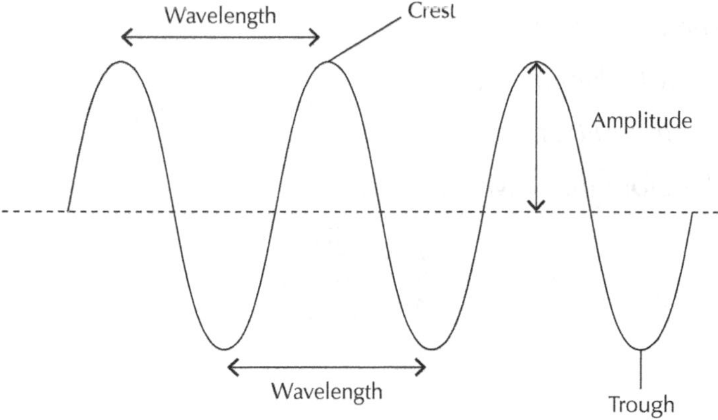

Fig 5.17.7: *Some characteristics of a wave*

Wavelength and amplitude

Two physical characteristics of a wave described above are the **wavelength** and the **amplitude.** Often we can measure these quantities for waves such a water waves in the same way as we would measure the lengths of other objects.

Wavelength

The wavelength of a wave is the length from any point of a wave to the same point on the next wave. For example, the length from one crest to the next crest or from one trough to the next trough, as shown on the diagram above.

Sometimes, the length of one wave is too short to measure on its own. In this situation, the length of several waves can be measured and divided to find the length of just one wave.

Technique: Calculating the wavelength of a wave

1. Count the number of waves in the length given – be aware that there may be part of a wave!
2. Measure the length (or read it from the question).
3. Divide the total length by the number of waves to work out the length of one wave.

Worked example

Find the wavelength of the waves shown in the diagram below.

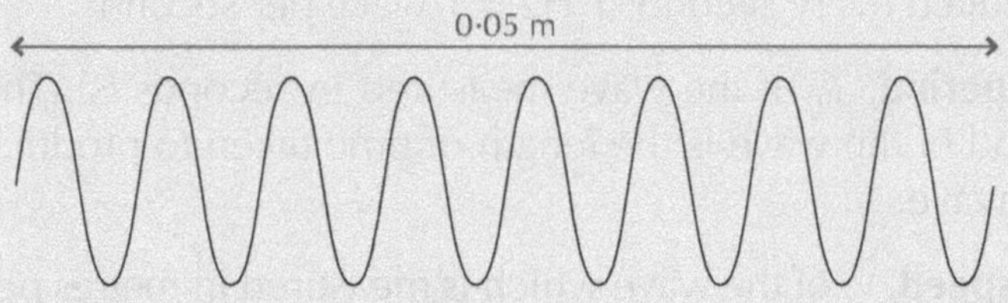

Fig 5.17.8

First of all, count the number of waves in the given length.

Number of waves = 8

The total length is given:

Length = 0·05 m

Therefore the wavelength is:

$$\lambda = \frac{0{\cdot}05}{8}$$

$$\lambda = 0{\cdot}0063\,m$$

Amplitude

The amplitude of a wave is linked to the amount of energy that the wave carries – the greater the amplitude, the greater the amount of energy. The amplitude of the wave is the **height of the wave measured from the middle** of the wave (or zero-line) as shown in the diagram below:

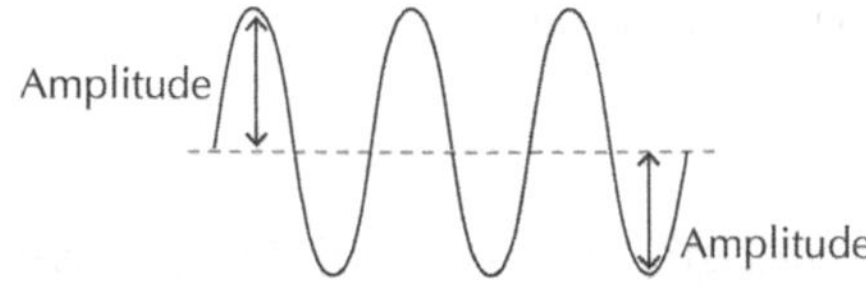

Fig 5.17.9: *Amplitude of a wave*

In some cases, for example with water waves, it is not possible to measure the height of the wave from the mid-point very easily. Instead, the lowest (trough) and highest (crest) points are measured to give the total height of the wave. The amplitude is half of the total height of the wave.

Technique: Calculating the amplitude of a wave

1. If possible, measure the height of the wave from the mid-point to the top of a crest or bottom of a trough.
2. But if this is not possible, measure the total height of the wave.
3. Divide the total height of the wave by 2 to find the amplitude.

Worked example

A pupil is measuring the amplitude of water waves hitting a pier. She measures the lowest and highest points of the wave and finds them to be separated by 4 m. Calculate the amplitude of the wave.

Here, the total height of the wave has been measured and is:

$$h = 4\ m$$

The amplitude is half of the total height, therefore the amplitude is:

$$A = \frac{4}{2}$$
$$A = 2\ m$$

Exercise 5.17.2 Wavelength and amplitude

1 Consider the diagram of a wave shown below:

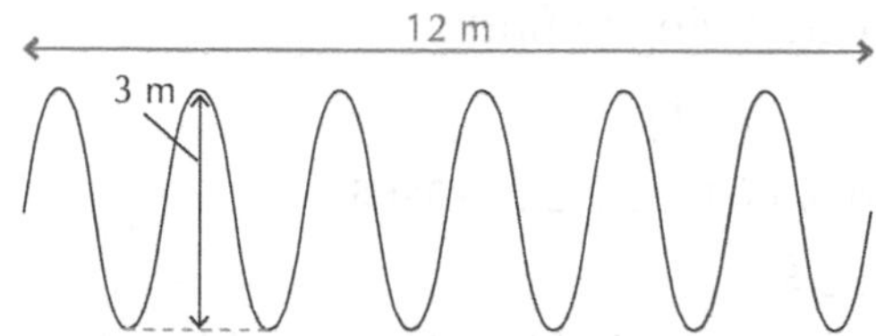

Fig 5.17.10

a) What is the amplitude of this wave?

b) What is the wavelength of the wave?

2 A student is investigating the amplitude and wavelength of water waves in a swimming pool.

(continued)

a) The swimming pool has a total length of 25 m. If there are 9 complete waves along the length of the pool, what is the wavelength of the waves?

b) It is found that the water oscillates over a total height of 1.5 m in the pool. Calculate the amplitude of the waves.

3 A sound wave carries sound energy through the air. A student uses an oscilloscope to look at a sound wave being produced by a loudspeaker. Describe how the wave would change if the volume of the sound being played was increased.

4 A musician uses a bow to make the strings of a violin vibrate. The total length of a violin string is 75 cm. It is found that there are 8 complete waves along the length of the string. Calculate the wavelength of the waves.

Period and frequency

Consider a water wave. If we focus on one point in the water as a set of waves moves past we would see the water particles oscillating up and down. The length of time it takes for one wave to pass is called the period, T, and is given by:

$$T = \frac{1}{f}$$

where f = frequency and T is measured in seconds.

The frequency of the wave is the number of waves, N, produced in a given time, t.

$$f = \frac{N}{t}$$

Frequency is measured in hertz, *Hz*. A frequency of 1 Hz means that 1 wave is produced every second.

Technique: Calculating frequency, number of waves and time

1. Write down what you know from the question, leaving blank what you are trying to find:

   Frequency, f = ______ Hz

   Number of waves, N = ______ waves

   Time, t = ______ s

2. Write down the equation linking the quantities above as you see it on the formula sheet:

   $$f = \frac{N}{t}$$

3. Substitute into the equation what you know.
4. Solve for the unknown.

Worked example

Calculate the frequency of the waves if there are 15 waves produced in 3 seconds.

$$f = \frac{N}{t}$$

$$f = \frac{15}{3}$$

$$f = 5\,Hz$$

Technique: Calculating period and frequency

1. Write down what you know from the question, leaving blank what you are trying to find:
   Frequency, $f =$ ______ Hz
   Period, $T =$ ______ s
2. Write down the equation linking the quantities above as you see it on the formula sheet:
   $$T = 1/f$$
3. Substitute into the equation what you know.
4. Solve for the unknown.

Worked example

Calculate the period of the wave if the frequency of the waves is 20 Hz.

$$T = \frac{1}{f}$$

$$T = \frac{1}{20}$$

$$T = 0{\cdot}05\,s$$

Exercise 5.17.3 Period and frequency

1. A girl stands on the end of a pier and counts the number of water waves in a given time. She counts 32 waves in a time of 100 seconds. Calculate the frequency of the waves.
2. The wave generator at a swimming pool generates waves with a frequency of 0·55 Hz. How many waves are generated by the generator in a time of 120 seconds?
3. Waves have a frequency of 1·8 Hz. A student counts 36 waves passing a point. How long is the student counting for?
4. Calculate the frequency of waves that have a period of 0·5 s.
5. A wave has a frequency of 17 Hz. Calculate the period of the wave.

Think about it

During a thunderstorm, we usually see the lightning flash before we hear the sound of the thunder even though they are produced at the same time. Both light and sound travel as waves.

Fig 5.17.11: *Lightning strikes*

What does this tell you about the speed of a light wave compared to the speed of a sound wave?

Assuming that the light wave reaches us instantly, and the speed of sound is as found in the experiment below, how could you work out how far away the thunder storm is?

Speed of a wave

We have already seen that a wave carries energy from one place to another. The speed (v) of the wave is the speed at which the energy is carried, and depends on distance (d) and time (t):

$$d = vt$$

where speed is measured in metres per second (ms^{-1}), time in seconds (s) and distance in metres (m).

The speed of a wave depends on the material the wave is travelling through. For example, sound travels faster through water than it does through air. This is because sound waves travel (or propagate) through an object by causing the atoms in that object to vibrate. These atoms cause their neighbours to vibrate and in turn those atoms cause their neighbours to vibrate. The closer together the atoms, the more quickly they can cause neighbouring atoms to vibrate, meaning the wave travels faster. The atoms in a solid or a liquid are closer together than the atoms in a gas, so sound travels faster in solids and liquids.

Experiment: The speed of sound

Sound travels as a wave. This experiment measures the speed at which the sound wave travels through the air.

You will need:

- Two microphones
- A timing device
- A hammer
- A metre stick

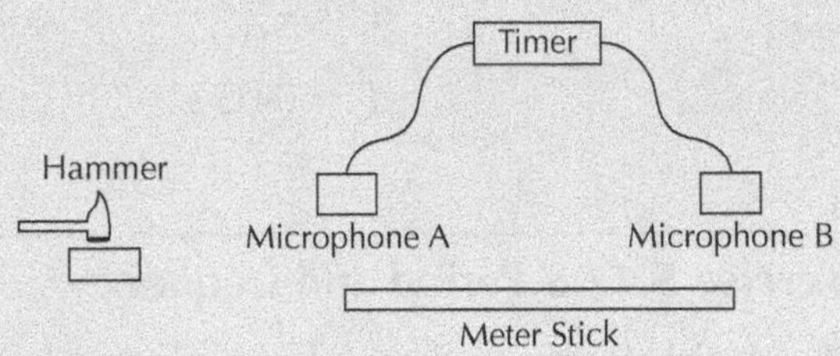

Fig 5.17.12: *Measuring the speed of sound*

Instructions

1. Connect the two microphones to the timing device so that the first microphone triggers the timer to start timing and the second microphone triggers the timer to stop timing.
2. Note the distance between the two microphones.
3. Tap the hammer on the bench in front of the first microphone.
4. Measure the time taken for the sound to travel between the two microphones using the timing device.
5. Use the equation,

$$v = \frac{d}{t}$$

to calculate the speed of sound in air.

Technique: Relating speed, distance and time for waves

1. Write down what you know from the question, leaving blank what you are trying to find:

   Distance, $d =$ _______ m

   Speed, $v =$ _______ ms^{-1}

   Time, $t =$ _______ s
2. Write down the equation linking the quantities above as you see it on the formula sheet:

$$d = vt$$

3. Substitute into the equation what you know.
4. Solve for the unknown.

Think about it

In an athletics race, it is important to ensure that all the competitors are an equal distance away from the noise source of the starting pistol. Why is this the case?

Worked example

Calculate the speed of a water wave which travels a distance of 4 m in a time of 20 s.

We are using the equation that links distance, speed and time:

$$d = vt$$

Substitute what we know and solve:

$$4 = v \times 20$$

$$v = \frac{4}{20}$$

$$v = 0{\cdot}2\ ms^{-1}$$

SONAR

Sound can be used to find how far away an object is in an application called Sound Navigation and Ranging (SONAR). A pulse of sound is sent out by a transmitter, and the time taken for the sound to reflect back to the receiver is measured. Using the equation for distance, speed and time,

$$d = vt$$

the total distance the sound has travelled can be calculated. However, it is necessary to divide the result by 2 in order to find the distance between the object and the transmitter as the sound will have travelled twice this distance in a single round trip.

This technique is also known as echo location, and is used by wildlife such as bats to hunt for prey. The bat will send out a pulse of high frequency sound and listen for the echo coming back. If a fly is near to the bat, the sound will reflect off the fly and back to the bat's ears quickly. The bat will then know that its prey is close.

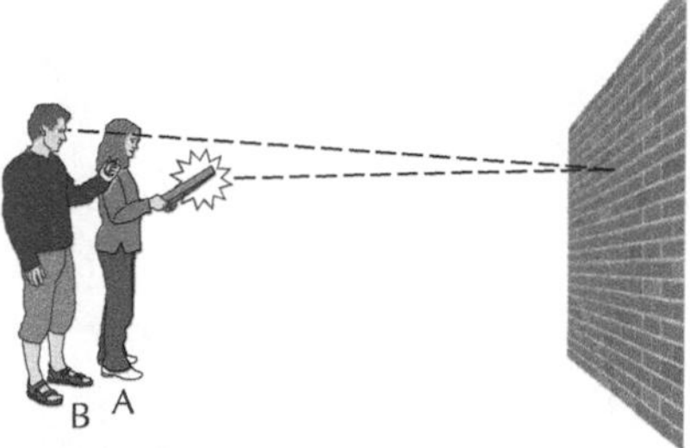

Fig 5.17.13: *Measuring the speed of sound*

Fig 5.17.14: *Bats use echo location to hunt for prey*

Technique: Echo location

1. We are using the equation for speed, distance and time so write down what you know from the question, leaving blank what you are trying to find:

   Distance, d = _______ m

   Speed, v = _______ ms^{-1}

   Time, t = _______ s

2. Write down the equation linking the quantities above as you see it on the formula sheet:

$$d = vt$$

3. Substitute into the equation what you know.
4. Solve for the unknown.
5. Divide the result by 2 in order to find the distance from the object to the receiver.

Worked example

Two students use echo location to measure the distance between them and a wall. They hit a bell and at the same time start a stop watch to measure the length of time taken for the echo to return to them. The time measured is 2·7 seconds. Assuming the speed of sound in air to be 340 ms^{-1}, calculate the distance between the students and the wall.

First of all, we can work out the total distance travelled by the sound using the equation for distance, speed and time:

$$d = vt$$

Substitute what we know and solve for the distance,

$$d = 340 \times 2{\cdot}7$$

$$d = 918\,m$$

This result must then be halved to find the actual distance between the students and the wall:

$$d = \frac{918}{2} = 459\,m$$

Exercise 5.17.4 Speed of a wave

1 A water wave travels up the beach a distance of 26 m. It takes 14 s to do this. Calculate the speed of the water wave.

2 A student is watching a water wave passing a pier. She notices that the wave is travelling with a speed of 0·8 ms^{-1}. How far will this wave travel in a time of 60 seconds?

3 Sound travels at 340 ms^{-1} in dry air. How long will it take sound to travel a distance of 4000 m?

Fig 5.17.15: *Water waves travelling up a beach*

4 In a thunder storm, the speed of sound is 500 ms^{-1} due to the moisture in the air. A thunder storm is a distance of 6 km away.

Calculate the time it takes the thunder to travel from the storm to you.

5 Thunder is heard 25 seconds after a lightning flash. The speed of sound in air is 500 ms^{-1} due to the moisture content.

a) How far away is the storm when the lightning flashed?

b) The time between seeing the next flash of lightning and hearing the thunder is reduced. What does this tell you about the distance the storm is away from you?

6 In a race the runners are at different distances away from the starter. They will hear the starting horn at different times. Assuming the speed of sound is 340 ms^{-1}, calculate the time difference in hearing the horn for two runners who are 2 m and 20 m away from the starter.

7 SONAR can be used to measure the depth of the ocean. A pulse of sound is transmitted from a boat floating on the surface. This reflects off the ocean floor and returns to the receiver on the boat.

a) If it takes 1·2 seconds for the sound pulse to return to the boat, calculate the depth of the ocean. You may assume the speed of sound is 1500 ms^{-1} in sea water.

b) Suddenly, the time taken for the pulse to return is shortened to just 0·4 seconds. Suggest a reason for this.

Speed, frequency and wavelength

Consider a water wave. If we focus on one point in the water as waves move past we would see the water oscillating up and down. If the wave is travelling quickly, we would see many oscillations in a short time. Similarly, if the wavelength is very short, we would see many oscillations in a short time.

The number of oscillations or waves that pass in one second is called the frequency. If the speed of the waves is increased, the frequency will increase. If wavelength is decreased, the frequency will increase. Clearly, there is a relationship between speed, frequency and wavelength! They are linked by the following equation:

$$v = f\lambda$$

where v = speed (ms^{-1})

f = frequency (Hz)

λ = wavelength (m)

Think about it

In terms of waves, we have two equations for speed:

$v = f\lambda$ and $v = \frac{d}{t}$

How are these equations related? They must be equivalent to each other. Can you show the equivalence?

Technique: Velocity, frequency and wavelength

1. Write down what you know from the question, leaving blank what you are trying to find:

   Speed, v = _______ ms^{-1}

   Frequency, f = _______ Hz

   Wavelength, λ = _______ m

2. Write down the equation linking the quantities above as you see it on the formula sheet:

   $$v = f\lambda$$

3. Substitute into the equation what you know.
4. Solve for the unknown.

Worked example

A sound wave has a wavelength of 0·34 metres and a frequency of 1000 Hz. At what speed is this sound wave travelling?

Here, we are using the equation that links speed, frequency and wavelength

$$v = f\lambda$$

Substitute what we know into the equation and solve for the speed:

$$v = 1000 \times 0{\cdot}34$$

$$v = 340\ ms^{-1}$$

Worked example

Calculate the frequency of a sound wave which travels at 340 ms^{-1} and has a wavelength of 10 m.

Here, we are using the equation that links speed, frequency and wavelength

$$v = f\lambda$$

Substitute what we know into the equation and solve for velocity:

$$340 = f \times 10$$

$$f = \frac{340}{10}$$

$$f = 34\ Hz$$

Exercise 5.17.5 Speed, frequency and wavelength

1 A wave has a wavelength of 0·56 m and travels at a speed of 24 ms^{-1}. Calculate the frequency of the wave.

2 A sound wave has a frequency of 850 Hz and a wavelength of 0·4 m. Calculate the speed of this wave.

3 While lying on a beach, you observe that 24 waves arrive every minute. The speed of the waves is 6 ms^{-1}.

a) What is the frequency of the waves?

b) What is the wavelength of the waves?

4 Radio 1 has a frequency of 98·1 MHz. What is its wavelength? (The speed of a radio wave is 3×10^8 ms^{-1}.)

5 A wave travels 12 km in 180 seconds and it has a wavelength of 0·25 m.

a) Calculate the speed of the wave.

b) Calculate the frequency of the wave.

c) Calculate the period of the wave.

6 A wave has a wavelength of 0·5 m and a frequency of 2 kHz.

a) Calculate the speed of the wave.

b) Calculate the time taken for this wave to travel 8 km.

7 A wave has a frequency of 50 kHz and a wavelength of 40 mm. Calculate the distance this wave will travel in 120 seconds.

8 If a wave has a frequency of 7×10^{14} Hz and a wavelength of $2{\cdot}86 \times 10^{-7}$ m, calculate the time taken for this wave to travel 100 km.

9 Blue light has a wavelength of 400×10^{-9} m and a frequency of $7{\cdot}5 \times 10^{14}$ Hz. How fast does this wave travel?

10 The apparatus shown below is used to measure the speed of sound. Two microphones are separated by a distance of 3 m. The time taken for sound to travel from one microphone to the other is measured.

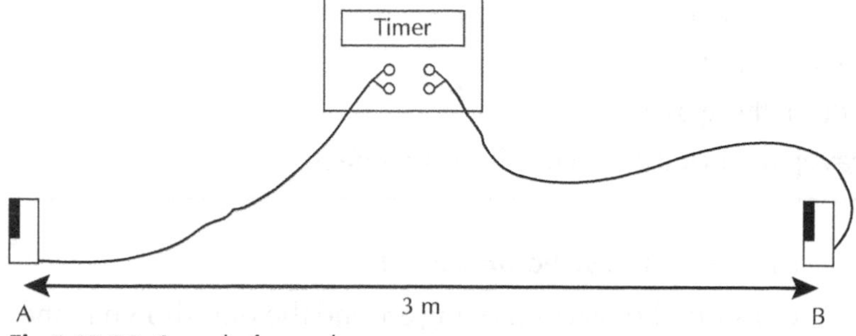

Fig 5.17.16: *Speed of sound*

a) If the timer measures a time of 0·009 s for the sound to travel from A to B, calculate the speed of sound in this example.

b) The frequency of the sound in this example is 270 Hz. Calculate the wavelength of the sound.

Fig 5.17.17: *Light refracting in a swimming pool*

Diffraction

When a wave passes an object or travels from one material into another it can experience changes. A wave can bend around obstacles in its path in a process called **diffraction**. An object can cause a wave to change its direction. The direction of a wave can also be changed by travelling from one material into another in a process called **refraction**.

Experiment: Diffraction of water waves

This experiment investigates the diffraction of a water wave as it passes different objects.

You will need:

- A ripple tank
- Wave generator
  (signal generator and oscillating paddle)
- Light
- Screen
- Clampstand

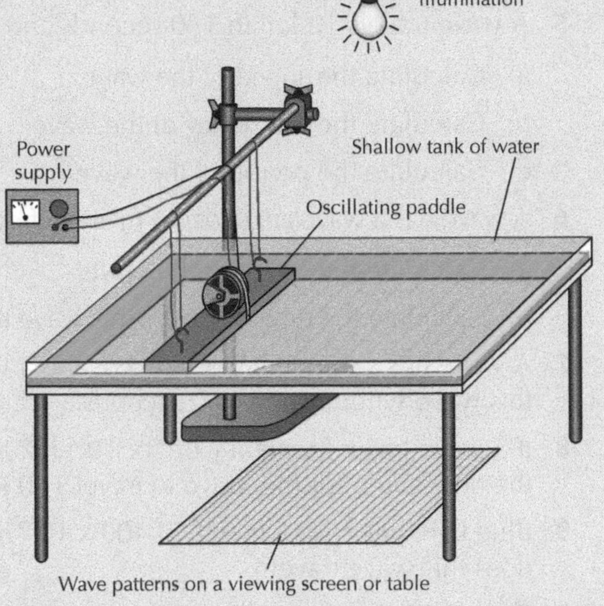

Fig 5.17.18

Instructions

1. Set up the apparatus as shown in the diagram above, using a shallow depth of water in the ripple tank. A lamp over the tank casts shadows of the wave crests onto a screen on the floor so that they can be viewed.
2. Place different objects in the path of the waves and examine the effect these have on the wave front.
3. For each object, change the frequency of the wave using the signal generator – this will change the wavelength of the wave.
4. Record your observations, noting:
   a) the effect of the object on the waves, and
   b) how changing the wavelength of the waves alters the effect.

Diffraction around an object

Consider the waves moving around the pier shown in the image.

Before reaching the pier, the full length of the waves are parallel to each other. As they pass the pier they start to bend and change direction. After the pier the waves have both a straight and a curved part. This bending of a wave is an example of diffraction.

The amount that a wave will bend around an object depends on the wavelength of the wave. A long wavelength will diffract more than a short wavelength.

Fig 5.17.19: *Diffraction of waves*

Diffraction through a narrow gap

When a wave travels through a narrow gap it will diffract around both edges of the gap. The smaller the gap, the more curved the wave will become. In other words, the smaller the gap, the greater the diffraction. If a wave is forced to pass through a very narrow gap it will appear totally curved. On the other hand, a wave will not noticeably diffract when passing through a gap which is bigger than the wavelength.

Key point

Long wavelengths diffract more than short wavelengths.

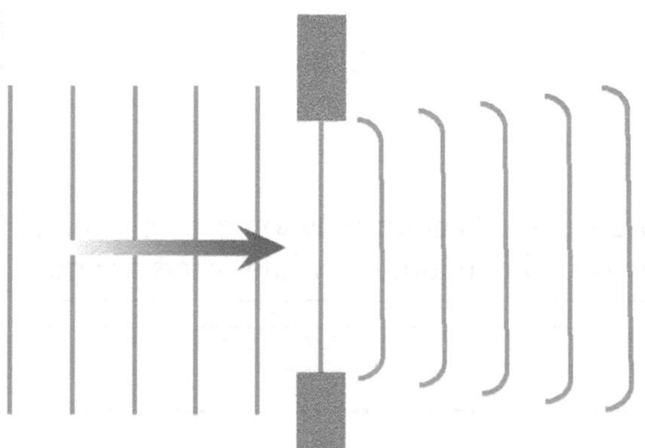

Fig 5.17.20: *Waves diffracting through a gap wider than the wavelength*

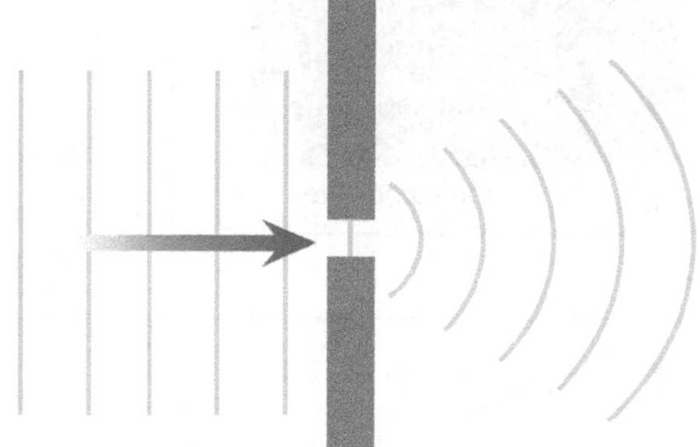

Fig 5.17.21: *Waves diffracting through a gap the same width or narrower than the wavelength*

STEP BACK IN TIME: FIRST ATLANTIC TRANSMISSION

At the turn of the twentieth century, famous physicist Guglielmo Marconi transmitted and received the first radio signal across the Atlantic ocean. Not only was the huge distance a challenge for Marconi, he also had to send the signal with no direct line of sight due to the curvature of the Earth. The signal had to have a large enough wavelength to diffract around the curvature of the Earth.

Fig 5.17.22: *Guglielmo Marconi*

Think about it

These diffraction effects are noticeable in everyday life. When listening to a radio while driving through a mountainous region, the reception of the radio may fade and a clear signal no longer be received. Why is this the case?

In mountainous terrain, radio signals transmitted using longer wavelengths are often received more clearly than those with shorter wavelengths. Can you explain this observation?

Fig 5.17.25: *Mountainous terrain*

Exercise 5.17.6 Diffraction

1. Consider the diagram below.

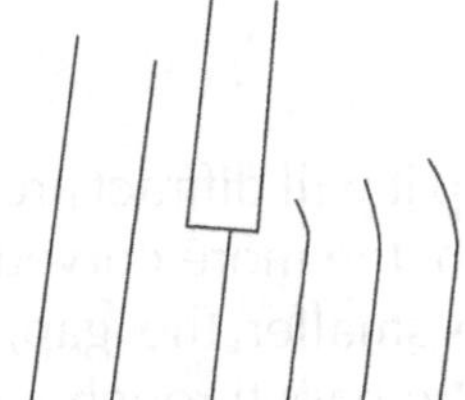

Fig 5.17.23

What is the name given to this process?

2. Copy and complete this diagram.

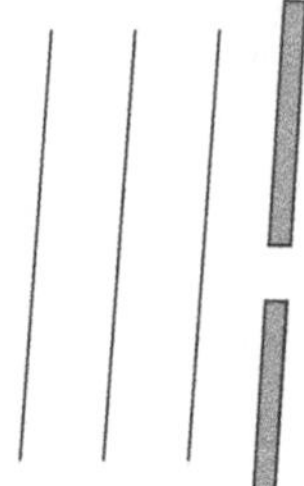

Fig 5.17.24

3. Describe, with the aid of a diagram, why some radio stations are received clearly in the mountains while others are not.

Learning checklist

In this section you will have learned:

- That waves transfer energy.
- How to define transverse and longitudinal waves.
- That sound is an example of a longitudinal wave and electromagnetic radiation and water waves are examples of transverse waves.
- How to determine the frequency, period, wavelength, amplitude and wave speed for longitudinal and transverse waves.
- How to use appropriate relationships to solve problems involving wave speed, frequency, period, wavelength, distance, number of waves and time.
- That diffraction occurs when waves pass through a gap or around an object.
- How to compare long-wave and short-wave diffraction
- How to draw diagrams using wave fronts to show diffraction when waves pass through a gap or around an object.

18 Electromagnetic spectrum

In this chapter you will learn about:

- How light waves compare to other waves.
- What the electromagnetic spectrum is.
- How all radiations in the electromagnetic spectrum are transverse and travel at the speed of light.
- Calculating frequency and wavelength of electromagnetic waves.
- The relative frequency and wavelength of bands of the electromagnetic spectrum with reference to typical sources and applications.
- The relationship between frequency and energy.

Comparing light to other waves

In the previous section, we saw that many types of energy travel as a wave. Waves can be separated into two different families. Some waves require a medium to travel and these belong to the mechanical wave family. Others can travel in a **vacuum** and these are members of the electromagnetic family of waves. All electromagnetic waves are produced in the same way but each has different characteristics, i.e. different frequencies and wave lengths. All light waves are members of the electromagnetic family.

Light can also travel through gases, liquids and some solids such as glass, but it travels fastest through a vacuum, at 300 000 000 metres per second (3×10^8 ms^{-1}). This is the fastest speed we know of – nothing travels faster.

We don't fully understand how waves travel without a medium – this remains a point of research.

Fig. 5.18.1: *Light waves*

Word bank

- **vacuum**

a space entirely devoid of all matter

Think about it

Investigate why nothing can travel faster than the speed of light.

Experiment: Comparing sound and light through a vacuum

This experiment demonstrates whether sound and light can travel through a vacuum where there are no particles to carry the wave.

You will need:

- Vacuum pump
- Bell jar
- Electric bell
- Power supply

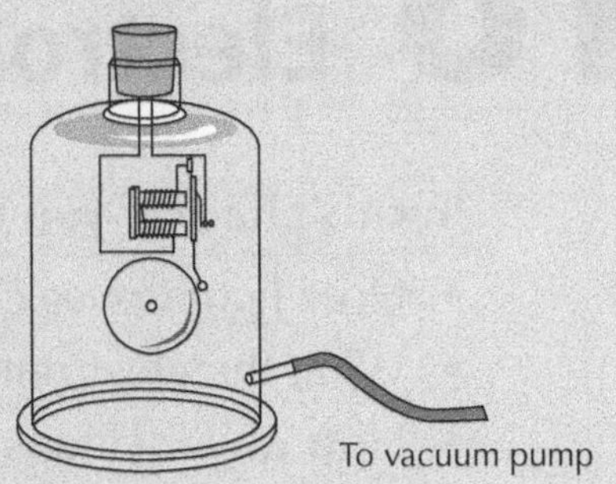

Fig. 5.18.2: *Waves through a vacuum*

Instructions

1. Connect the bell to the power supply so that the bell rings.
2. Enclose the bell in a bell jar as shown in the diagram above.
3. Use the vacuum pump to remove the air from the jar.
4. Can you hear the bell? Can you see the bell?

What is the electromagnetic spectrum?

All electromagnetic waves, including light, can travel through a vacuum.

The electromagnetic (EM) spectrum is shown in the diagram below:

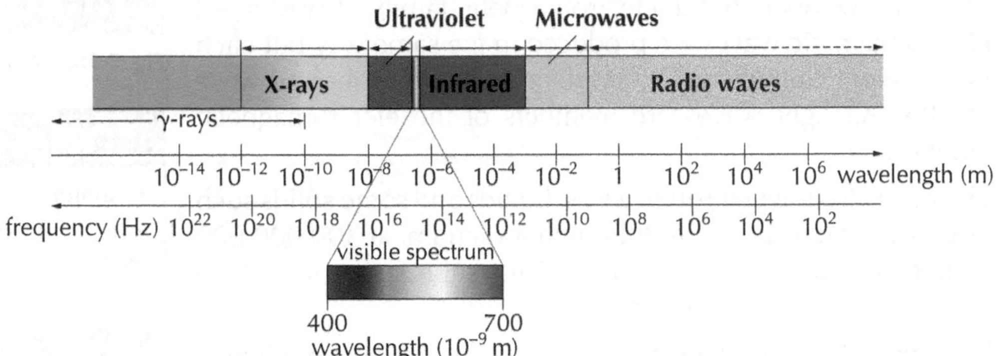

Fig. 5.18.3: *Part of the electromagnetic spectrum*

Light

Visible light makes up only a small part of the EM spectrum. Within the visible spectrum we have the colours red, orange, yellow, green, blue, indigo and violet (a useful mnemonic is ROYGBIV). Light with a frequency just too low for humans to see is called infrared radiation (IR), while light with a frequency just too high for humans to see is called ultraviolet (UV). Infrared radiation is associated with heat and is detected by thermal cameras. UV light is responsible for tanning our skin when we are exposed to the sun, but overexposure can cause skin cancer.

Radio waves, microwaves and X-rays

These waves are the same as light waves, with different frequencies and wavelengths. They have many useful applications, considered later in this chapter.

Travelling at the speed of light

All electromagnetic waves travel at the same speed, the speed of light (3×10^8 ms^{-1}). This means that if we know the frequency of an EM wave, we can calculate its wavelength using the equation:

$$v = f\lambda$$

Where v = speed, measured in metres per second (ms^{-1}), f = frequency, measured in hertz (H_z) and λ = wavelength, measured in metres (m).

Key point

All electromagnetic waves travel at the speed of light: $v = 3 \times 10^8$ ms^{-1}

Technique: Calculating the frequency and wavelength of an EM wave

1. Write down what you know from the question, including that the speed of the EM wave is the speed of light:

   Velocity, v = 3×10^8 ms^{-1}

   Frequency, f = _______ Hz

   Wavelength, λ = ______ m
2. Write down the equation linking the quantities above as you see it on the formula sheet:

   $$v = f\lambda$$
3. Substitute into the equation what you know.
4. Solve for the unknown.

Make the link

Quantum Mechanics considers the idea of bundles of light and you can learn more about this fascinating area of physics if you go on to study Higher Physics.

Worked example

A radio station broadcasts on a frequency of 102 MHz (102 000 000 Hz). Find the wavelength of the radio signal.

A radio wave is an EM wave, so it travels at the speed of light:

$$v = 3 \times 10^8 ms^{-1}$$

We are using the equation that links velocity, frequency and wavelength,

$$v = f\lambda$$

Substitute what we know and solve for the wavelength,

$$3 \times 10^8 = 102 \times 10^6 \times \lambda$$

$$\lambda = \frac{3 \times 10^8}{102 \times 10^6}$$

$$\lambda = 2{\cdot}94 m$$

Make the link

We will consider gamma rays in more detail in Chapter 20: Nuclear radiation.

Key point

The energy that electromagnetic waves carry is dependent on their frequency; increase the frequency, increase the energy.

Energy in an EM wave

High frequency EM waves have a higher energy than low frequency EM waves. This is why UV rays are dangerous to our skin – they carry more energy than infrared or visible light. As we move up the frequencies, we come to X-rays which are very high energy – so much so that we can only safely be exposed to a small amount of X-ray radiation. X-rays can pass straight through the soft tissue of our bodies – this makes them very useful but also potentially very damaging. Gamma rays are more energetic still, and can cause a lot of damage to the cells of our bodies if we are exposed to them.

STEP BACK IN TIME: JAMES CLERK MAXWELL

You may have deduced from the name that electromagnetic waves are linked to both electricity and magnetism. Scottish physicist James Clerk Maxwell is famous for his work on electromagnetism.

Find out more about his work, and what made him one of Einstein's heroes!

Regions of the EM spectrum

The following table describes the different regions of the EM spectrum and gives typical sources and applications of these waves. You will not be expected to remember all of the associated frequencies of each region of the EM spectrum but you should be able to list them, and place them in order of frequency and wavelength.

Electromagnetic wave and wavelength	Some sources	Some applications	Detector
Radio waves • ~10 cm – 1000s of metres	• Radio masts	• Transmitting signals over large distances very quickly • Radio-astronomy	• Radio receiver aerial, radar detector dishes
Microwaves • ~1 cm – 10 cm	• Stars • Magnetrons • Mobile phones	• Heating food: microwaves have just the right frequency to vibrate water and fat molecules. The quicker molecules vibrate, the hotter they are and hence the hotter the food is • Mobile phones use microwaves instead of radio waves to transmit signals as they can be produced by a smaller transmitter	• Microwave detector aerial, radar detector dishes
Infrared • ~10 μm	• Hot objects	• Remote controls for a TV: TV remotes send a pulsed IR signal to the TV • Thermal imaging cameras	• Heat sensitive paper, black bulb thermometer, thermal imaging camera, thermopile
Visible • 400 nm – 700 nm	• Stars • Light bulbs, LEDs etc	• This is what our eyes detect, allowing us to see	• Photographic film, LDR's, the retina of the eye, photodiodes, phototransistors
Ultraviolet • ~10 nm – 400 nm	• Stars • Mercury lamps, UV lamps	• Crime scene investigations • Counterfeit money detection • Art and museum inspection • Killing bacteria • Causes our skin to tan and burn	• Photographic film/fluorescent film, UV photodiode
X-rays • ~1 nm	• X-ray tube	• Viewing inside living bodies without having to cut them open • Customs at airports use X-rays to scan luggage for weapons	• Photographic film, image intensifiers, G-M tube and counter
Gamma rays • ~0.01 nm	• Radioactive decay • Interaction of cosmic rays with our atmosphere • Pulsars, Magnetars, Quasars	• Tracing – a technique used in medicine to view certain processes within our bodies • Used to alter some semi-precious stones: often used to change white topaz to blue topaz	• G-M tube & counter

Table 5.18.1: *Sources, applications and detectors of electromagnetic waves*

Exercise 5.18.1 The electromagnetic spectrum

1 What do all regions of the electromagnetic spectrum have in common?

2 Copy and complete the table showing the regions of the electromagnetic spectrum.

Radio		Infrared	Visible		X-rays	

3 Describe a practical application of X-rays.

4 What region of the electromagnetic spectrum has:

a) the highest frequency?

b) the lowest energy?

5 Green light has a wavelength of 500 nm. Calculate the frequency of this light.

6 At what speed do X-rays travel in a vacuum?

7 An X-ray machine emits X-rays with a frequency of 3×10^{17} Hz. Calculate the wavelength of these X-rays.

8 Why are UV rays more harmful than visible light?

9 After an earthquake, a rescue team searches for people trapped inside collapsed buildings. They use special cameras that can detect the heat given out by the bodies of survivors. What type of radiation is being detected by the cameras?

10 Red light has a wavelength of 700 nm. Calculate the frequency of red light.

11 Describe an application of microwaves.

12 Describe an application of UV rays.

Learning checklist

In this section you will have learned:

- About the relative frequency and wavelength of bands of the electromagnetic spectrum.
- About typical sources, detectors and applications for each band in the electromagnetic spectrum.
- That all radiations in the electromagnetic spectrum are transverse and travel at the speed of light.

19 Refraction of light

In this chapter you will learn about:

- Refraction.
- Ray diagrams.
- How light refracts when it enters a different medium.
- How refraction means a change in speed, wavelength and sometimes diffraction of light waves.
- The following terms: incident ray, refracted ray, angle of incidence, angle of refraction and the normal.
- Some applications of refraction – lenses and optical fibres.
- Dispersion of light waves.

Refraction

We have already seen that an obstacle can change the shape of a wave that passes over it. The wavelength, speed and direction of a wave can also be changed. To achieve this, the wave needs to travel from one material into another. This process is known as refraction.

Experiment: Refraction of water waves

This experiment investigates the refraction of a water wave as it passes different objects.

You will need:

- A ripple tank
- Wave generator (signal generator and oscillating paddle)
- Light
- Screen
- Clampstand
- Objects to change the depth of the ripple tank

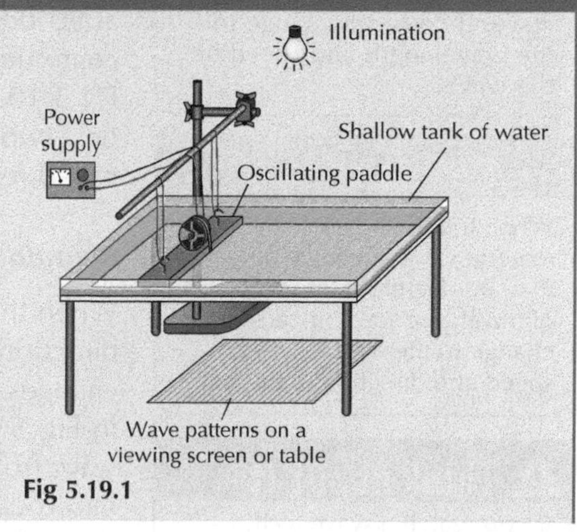

Fig 5.19.1

(*continued*)

Instructions

1 Set up the apparatus as shown in the diagram above, using a shallow depth of water in the ripple tank.
2 Place an object in the ripple tank to produce a deep section and a shallow section as shown in the diagram below (as different depths of water represent different materials in this experiment).
3 Switch on the wave generator and compare the waves in the deep section to the ones in the shallow section – how does the wavelength of the waves compare?
4 Repeat the above experiment with the shallow section turned to an angle as shown in the diagram below.

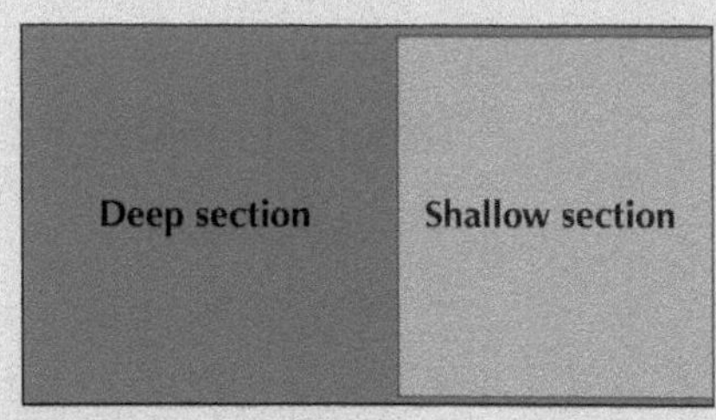

Fig 5.19.2: *A 90° boundary between materials*

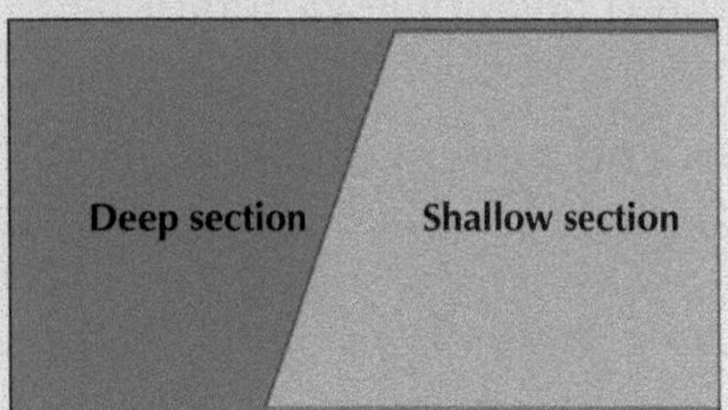

Fig 5.19.3: *A boundary between materials not at 90°*

How does the wave change as it moves from the deep section to the shallow section of the tank?

Key point

When the boundary of two materials is at 90 degrees to the direction of travel of a wave, it causes a change to the wavelength and speed of the wave.

Key point

When the boundary of two materials is at an angle other than 90 degrees to the direction of travel of a wave, it causes a change to the wavelength, speed and direction of the wave.

Think about it

If you watch a wave as it comes up a beach, you will notice that the speed of the wave changes as it gets into shallower water. Can you explain the cause of this effect?

90° boundary between materials

In the above experiment the different depths of water represent different materials. When the boundary of the two materials is at 90 degrees to the direction of travel of the wave, the wave continues moving in the same direction. This is shown in Fig 5.19.4. The wavelength and the speed of the wave are the only two things that change. The change of speed of the wave is called refraction.

Boundary other than 90°

When the boundary of two materials is not at 90 degrees to the direction of travel of the wave, the direction of the wave also changes. This is a consequence of the speed change! It is shown in Fig 5.19.5. The effect can be understood by imagining the wave front as an army of soldiers marching from smooth ground into rough ground. The soldiers hit the rough ground and slow down but they do this at different times because they are marching at an angle to the boundry so some slow before others, causing the line of soldiers, or the wave, to bend. A diagram of the soldier analogy is shown previously.

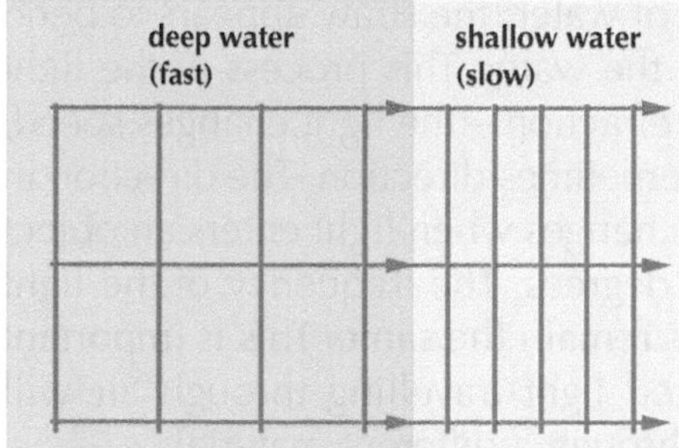

Fig 5.19.4: *Waves changing speed and wavelength when travelling from deep to shallow water*

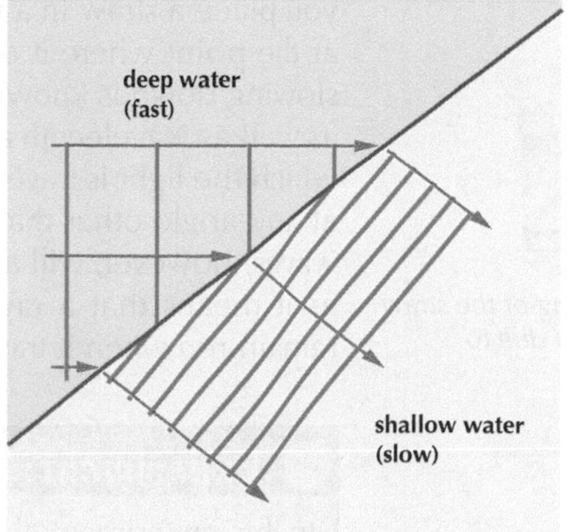

Fig 5.19.5: *Waves changing speed, wavelength and direction when travelling from deep to shallow water*

Read more about refraction in this chapter.

The behaviour of light

As we have seen, light travels as a wave. Light can travel very quickly (3×10^8 ms^{-1}) over large distances. It can travel through some liquids, gases and solids. Crucially, light can travel through a vacuum (just as well or the light from the sun wouldn't reach us and there would be no life on Earth!).

When light can travel straight through an object without being scattered (or reflected within the object) we call that object **transparent**, e.g. the windscreen of a car.

When light can pass through an object but is scattered such that a detailed image cannot be seen through the object, we say that the object is **translucent**, e.g. frosted glass or a cloud.

When light is unable to pass through an object we say that the object is **opaque**, e.g. a block of wood.

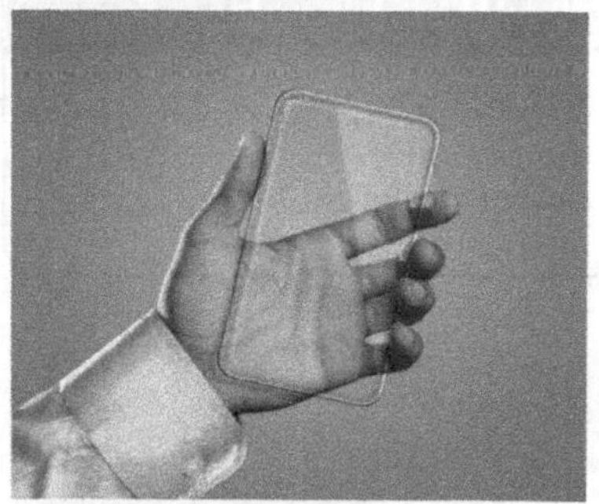

Fig 5.19.6: *A transparent object*

Fig 5.19.7: *A translucent object*

Refraction

We have already considered the process of refraction previously. As light travels as a wave, it refracts when it travels from one material to another. When light enters a transparent or translucent material, such as glass or water, it slows down. This can cause the light to change its direction. You can see this effect when

Fig 5.19.8: *The bending of the straw when it enters water is due to refraction*

you place a straw in a glass of water: the straw appears to bend at the point where it enters the water. This process of the light slowing down is known as refraction – the light changes speed, as well as wavelength and sometimes direction. The direction in which the light is travelling changes when light enters an object at any angle other than 90 degrees. The frequency of the light waves, however, will always remain the same. This is important as it means that a ray of red light travelling through air will remain red when it travels through a different material.

Key point

When a ray of light travels from one material to another material with a different density, its speed will change. The change in speed can cause a change in direction of the ray of light. This process is called **refraction**.

Experiment: Refraction through shapes

In this experiment, the basic process of refraction is investigated through different shapes.

You will need:

- A ray box and power supply
- Various shapes of glass or Perspex blocks
- Ruler

Instructions

1. Use a single slit mask to get a single ray of light from the ray box.
2. Connect the ray box to the power supply and switch on – ensure a single ray of light is emitted from the ray box.
3. Direct the single ray of light into blocks of different shapes. Make sketches of the effects that the blocks have on the rays of light (using a ruler).

Word bank

- **interface**

the boundary between two objects where their surfaces meet

Refraction: understanding the change of direction

We have seen that when light travels from one material into another it appears to change direction at the **interface** between the materials. This is because light is a wave. When the leading edge of the wave enters glass, it slows down (because glass is denser than air). This has the effect of dragging the rest of the wave round and changing its direction.

Instead of a wave, imagine lines of soldiers marching across concrete towards a belt of sand they need to cross (see diagram). The soldiers are joined, and must remain joined.

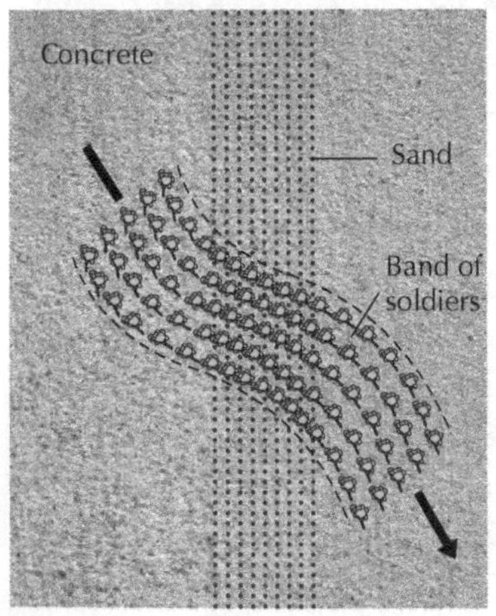

Fig 5.19.9: *A wave of soldiers*

The first soldier in the row will slow down upon entering the sand, as will the next, followed by the next. The soldiers that are moving more slowly cover less distance. The direction the soldiers march in will change because of the difference in distance that they are able to cover. Once all of the soldiers are on the sand they will continue to travel in a straight line because they are all marching with the same speed. The opposite effect will take place when the soldiers leave the sand. Once this happens they will be marching in their original direction, provided that the edges of the sand/concrete are parallel. The same thing happens with light when it is shone through a rectangular block of light.

Rays and angles

Now that we know refraction will occur when light enters a different medium, we need to know some terms so that we can accurately describe the direction in which the rays will change direction. By **convention**, the following terms are used to describe the rays and angles for light travelling between materials.

- **The incident ray:** the ray that is approaching the new medium. (*In*cident ray is the ray going *in*.)
- **The refracted ray:** the ray that is travelling through the new medium.
- **Angle of incidence:** the angle between the incident ray and the normal.
- **Angle of refraction:** the angle between the refracted ray and the normal.
- **The normal:** a line drawn perpendicular to the surface of the medium at the entry point of the incident ray.

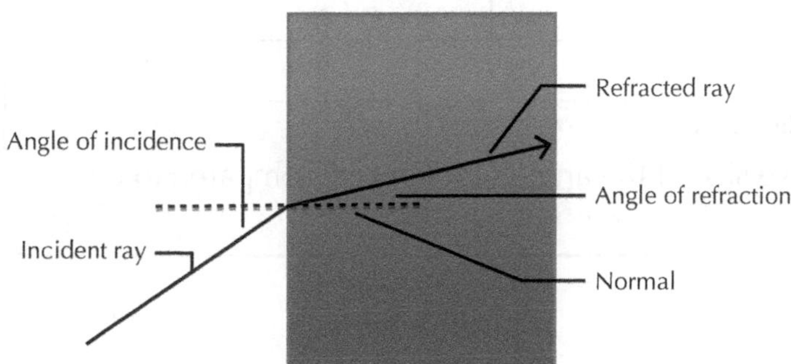

Fig 5.19.10: *The terms used in describing refraction when a ray travels from a less dense to a more dense medium*

Make the link

Different wavelengths within the EM spectrum refract by different amounts. Astronomers can use this to their advantage by choosing wavelengths that will be less affected by refraction in our atmosphere. You can read more about this in Chapter 7.

Word bank

- **incidence**

the arrival of a beam of light on a surface

Word bank

- **convention**

a generally accepted method

Experiment: Testing refraction angles

This experiment investigates the process of refraction of light going from air into glass or Perspex.

You will need:

- A ray box
- A Perspex semi-circular block
- A protractor
- A ruler

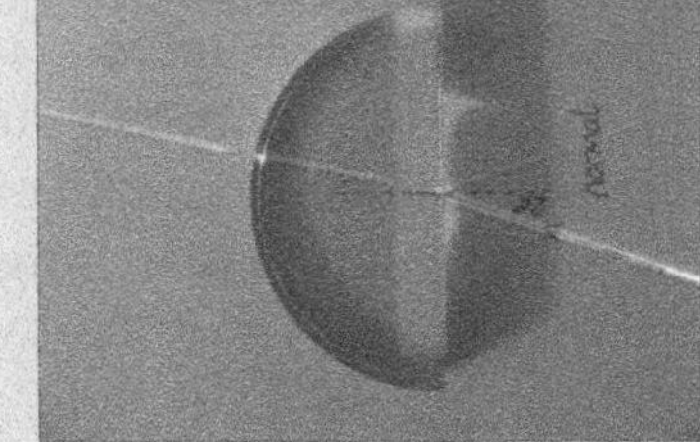

Fig 5.19.11

Instructions

1 Place the semi-circular block on a sheet of paper and draw round the block.
2 Use a protractor to draw on a normal line (at 90 degrees) to the flat surface of the semi-circle.
3 Using a protractor, measure the angle of incidence (θ_i) of 20 degrees and mark the position of the incident ray with a pencil including an arrow to show direction.
4 Using the ray box, shine a ray of monochromatic light along the incident ray into the Perspex block. Mark the position of the refracted ray.
5 Use a ruler to draw on the refracted ray including an arrow to show direction.
6 Use a protractor to measure the angle of refraction, θ_r, as shown below.
7 Record your results in the following table.

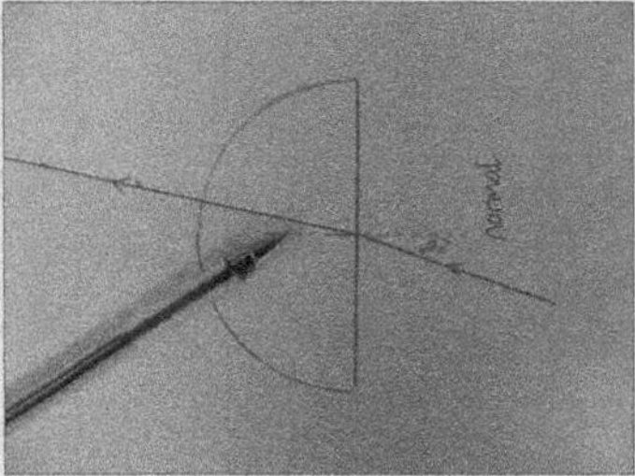

Fig 5.19.12

Angle of incidence (θ_i)	Angle of refraction (θ_r)
20	
30	
40	
50	
60	
70	
80	

Table 5.19.1

8. What do you notice about the relative size of the angle of refraction compared to the angle of incidence?

Consider the diagram below showing the path that a ray of light takes through a glass block. We can see that when the light travels from the air to the glass (a less dense to a more dense medium), the ray of light bends towards the normal. When the ray of light emerges from the block it changes direction once again, but this time the ray of light bends away from the normal. This is similar to the effect we saw when the marching soldiers moved from concrete to sand.

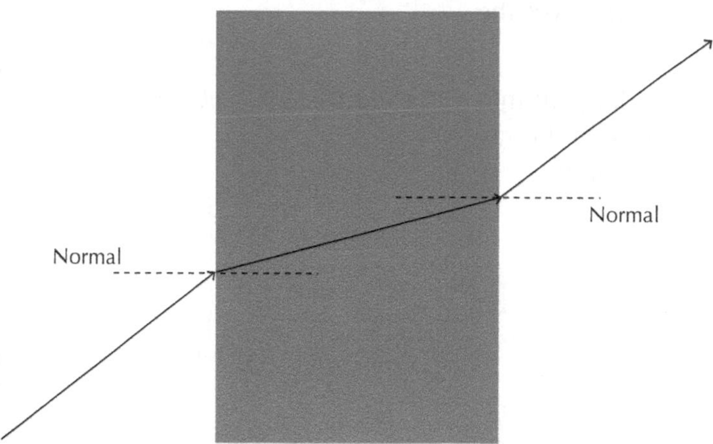

Fig 5.19.13: *A ray of light refracting when it enters and leaves a glass block*

Think about it

Why bother with the normal? Why not simply measure the angle between the incident ray and the surface of the object? Well, if the surface of the object was always flat, there would be no problem in simply measuring this angle. But if the object has a curved surface like a lens, how would you measure the angle?

It is far more accurate to measure the angle between two straight lines than measuring the angle between a straight line and a curved line.

Refraction can cause lots of interesting optical illusions in nature. Hunters had to learn to compensate for refraction when fishing with spears as the fish were not in the position they thought – as the light travelled to their eyes from the fish it changed direction on leaving the water. Our eyes cannot compensate for refraction – we automatically assume light is travelling in a straight line.

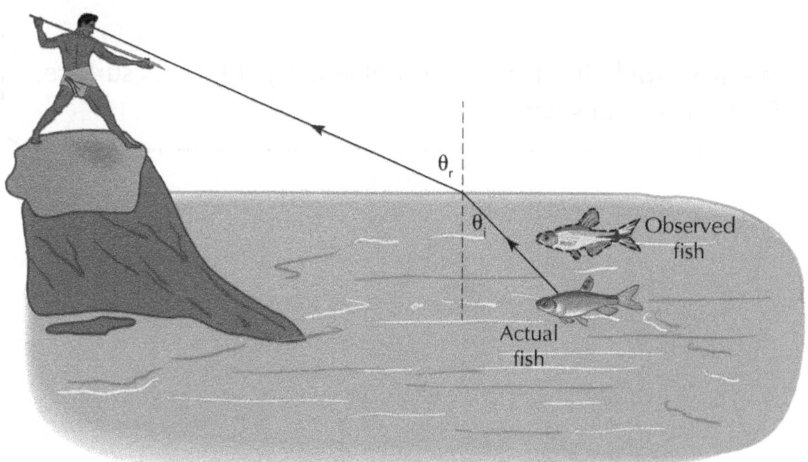

Fig 5.19.14: *Our eyes cannot compensate for refraction*

Key point

Whenever we consider a wave, such as light, travelling from one medium to another, the angles are always measured between the ray and the normal!

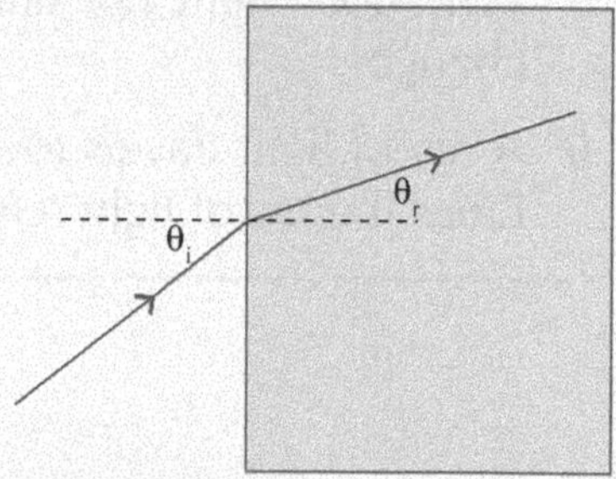

Fig 5.19.15

A common mistake is to measure between the ray and the medium – DO NOT DO THIS! Always measure between the ray and the normal.

Exercise 5.19.1 Refraction

1 Describe what is meant by the term *refraction*.

2 A ray of blue light is incident on a rectangular glass block as shown in Fig 5.19.16.

   a) State the size of the angle of incidence.

   b) Copy and complete the diagram showing the path of the ray as it travels through and then leaves the glass block.

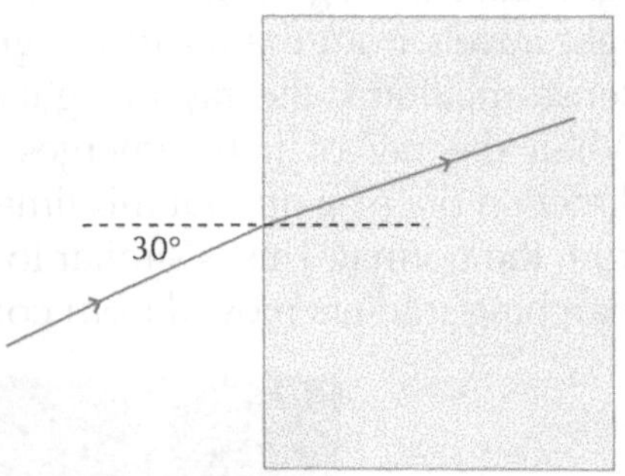

Fig 5.19.16: *Rectangular glass block*

3 For each of the shapes below, copy the diagram into your jotter and complete it to show the path the ray of light will take entering and leaving the block.

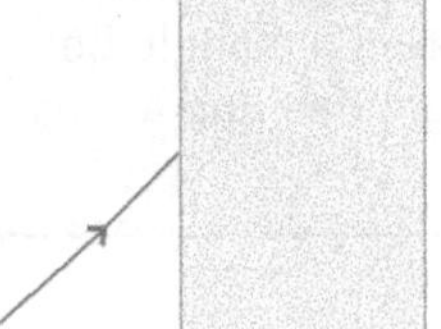
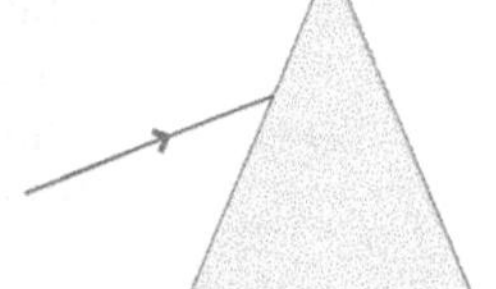

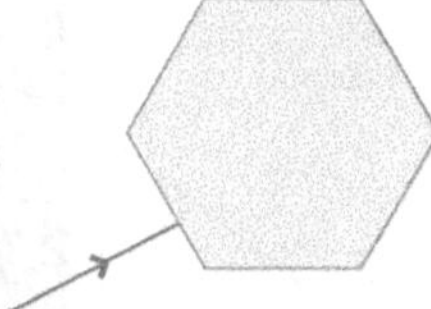

Fig 5.19.17

4 Consider the example shown of a pencil placed in water. The pencil appears to be bent as it enters the water. The ray diagram highlights why the pencil appears this way.

   a) If you look into a pool of water, is it shallower or deeper than it appears?

   b) If a bird attempts to catch a fish in the water below it, what must it take into account to succeed?

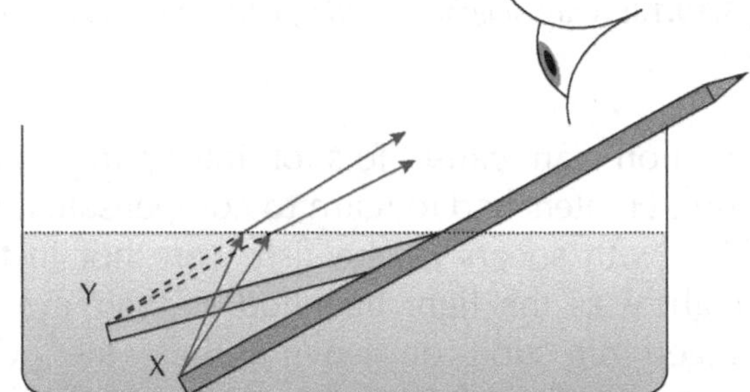

Fig 5.19.18: *Pencil in water*

5 Why does light not change colour when it is refracted even though the wavelength changes?

6 A ray of light travels towards a glass block. It hits the block at exactly 90° to its surface. Does the ray of light refract? Explain your answer.

Applications of refraction – the lens

Refraction has many useful applications. One of the most common is a lens. Since light will bend towards the normal when travelling from air to glass we can manufacture different shapes of glass to manipulate the path a ray of light will take.

The word lens comes from the Latin name for lentil, since a convex lens is lentil shaped.

Fig 5.19.19

Fig 5.19.20: *Notice how lenses and lentils are a similar shape*

Experiment: Convex and concave lenses

This experiment investigates the effects of convex and concave lenses on the path of rays of light passing through them.

You will need:

- A ray box and power supply
- A three slit mask
- Convex lenses (one thin, one thick)
- Concave lenses (one thin, one thick)
- A ruler

* Convex and concave lenses are described over the next few pages.

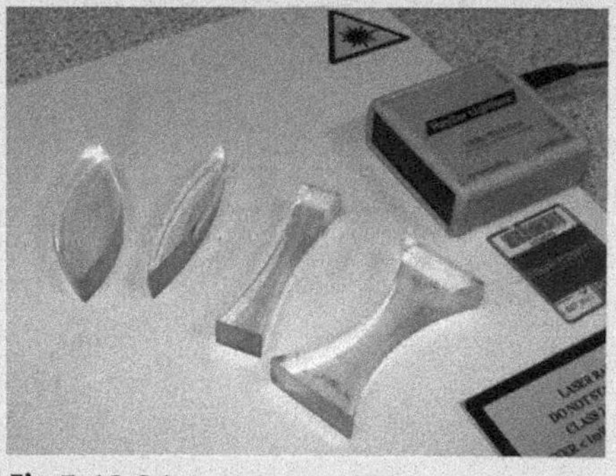

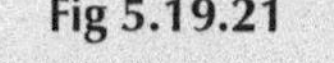

Fig 5.19.21

Instructions

1. Place the thin convex lens on a sheet of paper and draw around the lens.
2. Set the ray box to produce three parallel rays of light and direct these at the long edge of the lens as shown in this diagram.
3. Carefully mark the paths of each of the rays using two 'X's on either side of the lens.
4. Use a ruler to draw the paths the rays of light took before and after entering the lens.
5. Repeat steps 1 – 4 with the other three lenses.
6. Describe the effects of a convex lens on rays of light.
7. Describe the effects of a concave lens on rays of light.
8. How does the thickness of the lens affect the paths of the rays of light?

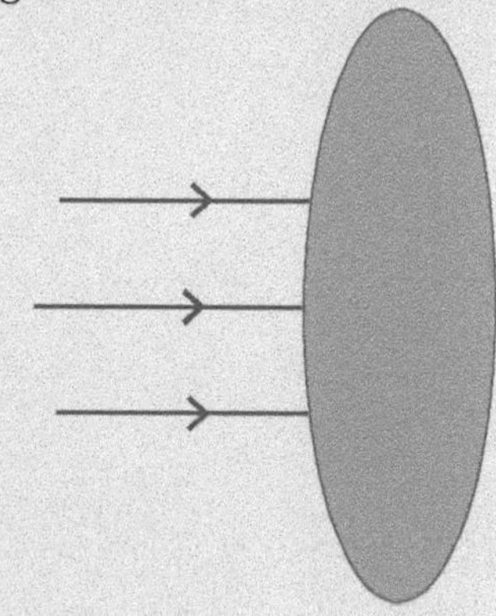

Fig 5.19.22

Think about it

Consider a convex lens represented by three glass shapes as shown below – two triangles and a rectangle. By considering how a single ray of light passes through each of the shapes, can you explain how a convex lens works to make the light rays converge?

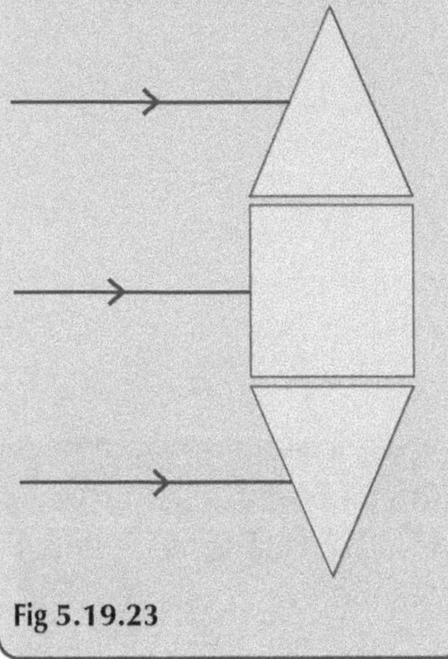

Fig 5.19.23

Convex lens

A convex lens is one that is thicker in the middle than at the edges. It is also called a converging lens.

Consider the diagram of a convex lens below.

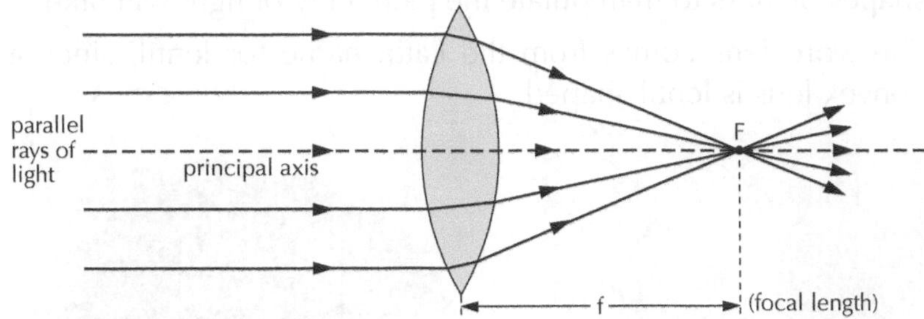

Fig 5.19.24

The rays are brought together to a single point. This point is called the **focal point** and the distance between the centre of the lens and the focal point is called the **focal length**. This focal length depends on the **strength** of the lens. The strength of the lens is determined by the difference in thickness between the centre and the edges of the lens as shown by the diagrams below.

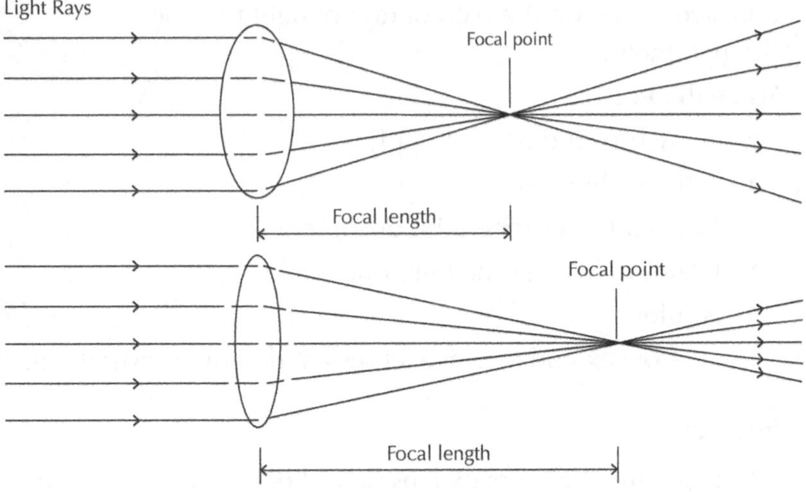

Fig 5.19.25

A thicker lens = shorter focal length = greater strength

$$\text{Strength} \propto \frac{1}{f}$$

Experiment: Focal length of a convex lens

This experiment measures the focal length of various convex lenses.

You will need:

- A distant source of light, e.g. a lab window
- A screen
- A ruler
- Convex lenses of varying thickness

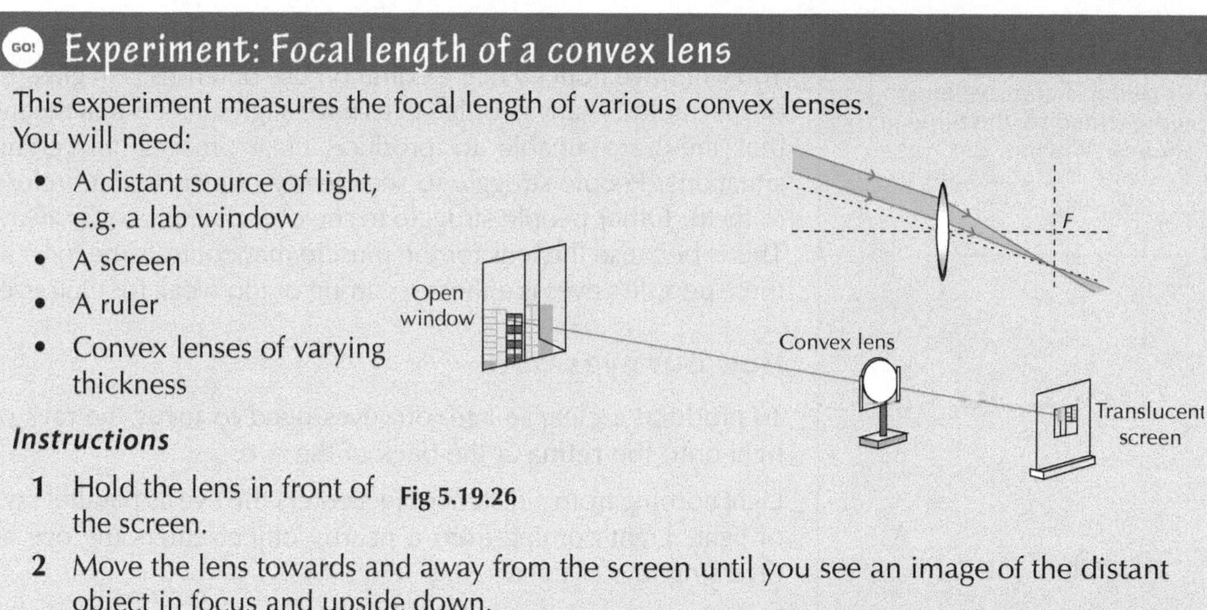

Fig 5.19.26

Instructions

1 Hold the lens in front of the screen.

2 Move the lens towards and away from the screen until you see an image of the distant object in focus and upside down.

3 Measure the distance between the lens and the screen using a ruler – this is the focal length.

4 How does the focal length of the lens depend on the thickness of the lens?

Concave lens

Another type of lens is a concave, or **diverging**, lens. This lens is thinner in the middle than at the edges. (Many people remember the name of this lens by thinking about caves – *concave* – caves in). Consider the diagram of the rays of light passing through a concave lens.

Instead of the rays coming together to form a focal point, the rays of light spread out, or **diverge**.

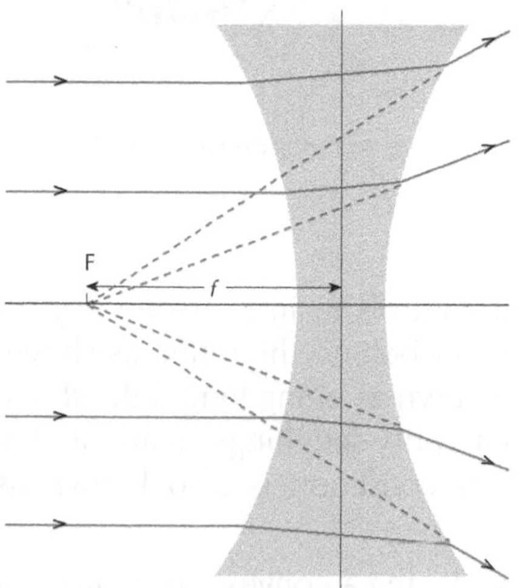

Fig 5.19.28: *Rays of light spread out when they pass through a concave lens*

Think about it

Consider the diagram of parallel rays entering a convex lens shown below.

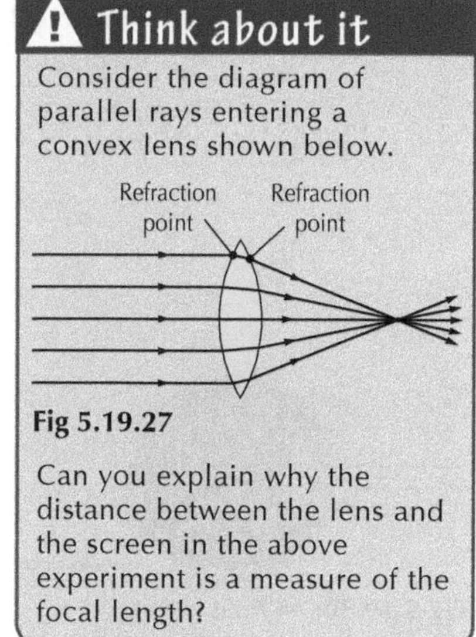

Fig 5.19.27

Can you explain why the distance between the lens and the screen in the above experiment is a measure of the focal length?

Word bank

- **diverging**

extending in different directions from a common point

> **Think about it**
>
> Consider a concave lens represented by the three glass shapes shown.
>
> Fig 5.19.29
>
> By considering how a single ray of light passes through these shapes, can you explain how a concave lens works to make the light rays diverge?

Sight defects

You will have noticed that a common use of lenses is in glasses, or spectacles. Many people's eyes have slight defects that means that they are unable to produce clear images in certain situations. People struggle to see clearly objects that are close to them. Other people struggle to see objects that are far away. This is because the lens (or the muscle that controls the lens) in these people's eyes is either too strong or too weak for their eye.

How our eyes work

To produce a clear image, our eyes need to focus the rays of light onto the **retina** at the back of the eye.

Light coming from a distant object enters the eye as parallel rays of light. Light coming from a nearby object enters the eye as diverging rays of light.

The **cornea** and a **convex lens** inside the eye act to focus the light onto the retina. The majority of the focussing is done by the cornea. The lens is the eye's fine focus control and changes its shape depending on whether you are looking at a nearby object or one that is far away (see the following diagrams). If the lens or the muscles controlling it cannot change shape as required, then light may not be focussed correctly resulting in a sight defect.

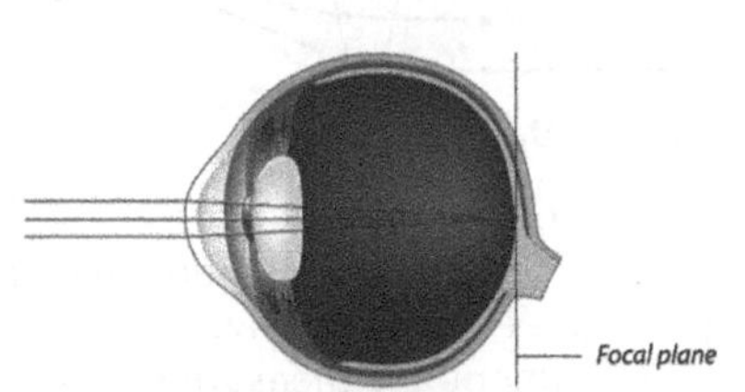

Fig 5.19.30: *Normal vision*

Fig 5.19.31: *The thickness of the lens changes to focus on close / distant objects*

Long sight (Hypermetropia)

If the lens (or the muscle that controls it) in someone's eye is too weak, the focal point will fall behind the retina as shown on the diagram below. This is known as being **long sighted** (the focal length is too long). You can only see things clearly if they are a long distance away. This condition is also known as Hypermetropia.

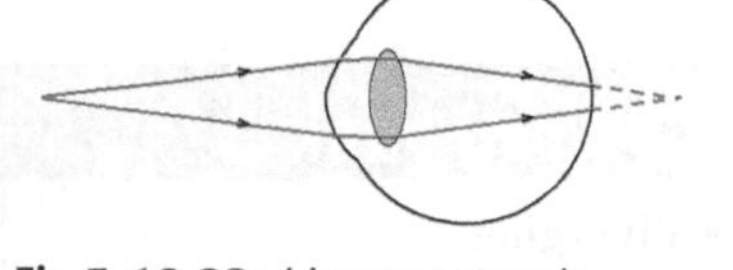

Fig 5.19.32: *Hypermetropia*

Long sight can be corrected by placing a convex lens in front of the eye as shown in the diagram below. This has the effect of reducing the focal length so that the rays of light focus on the retina.

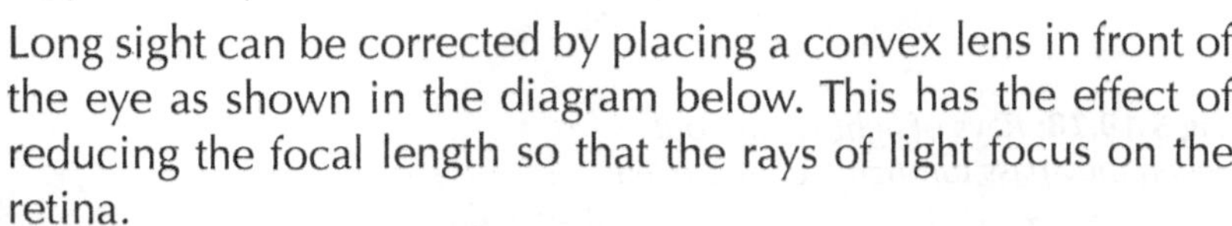

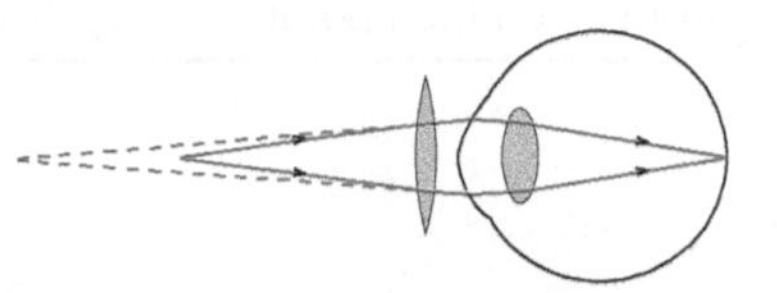

Fig 5.19.33: *Hypermetropia corrected*

Short sight (Myopia)

If the lens in someone's eye is too strong, the focal point will fall in front of the retina as shown in the diagram below. This is known as being short sighted (the focal point is short of the retina). You can only see things clearly if they are a short distance away. This condition is also known as Myopia.

Short sight can be corrected by placing a concave lens in front of the eye as shown in the diagram below. This has the effect of increasing the focal length so that the rays of light focus on the retina.

Word bank

- **retina**

a layer at the back of the eye containing cells that are sensitive to light

Word bank

- **cornea**

transparent layer forming the front of the eye

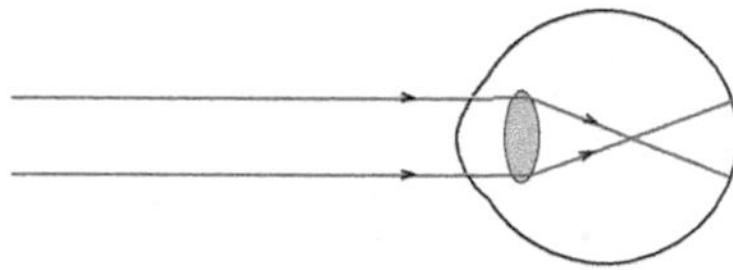

Fig 5.19.34: *Myopia*

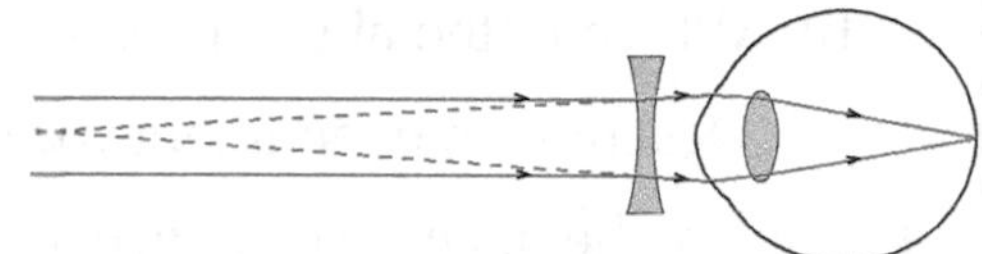

Fig 5.19.35: *Myopia corrected*

Summary of sight defects

Sight defect	Description	Correcting lens
Long sight	Light focuses **behind** retina	Convex
Short sight	Light focuses **in front** of retina	Concave

Table 5.19.2

Exercise 5.19.2 Lenses

1 Consider the diagram of a lens below.

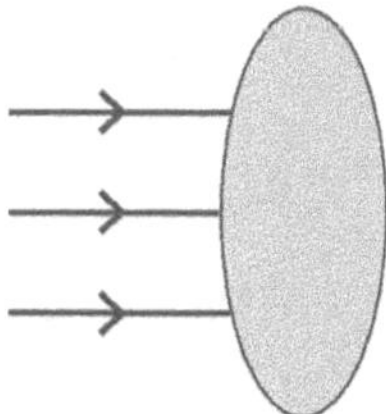

Fig 5.19.36

a) What is the name given to this type of lens?

b) Copy and complete the diagram showing what happens to the rays of light when they leave the lens.

c) What common eye defect can be corrected with this type of lens?

2 Consider the diagram of a lens below.

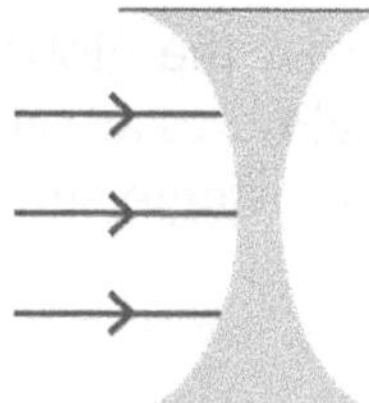

Fig 5.19.37

(continued)

a) What is the name given to this type of lens?

b) Copy and complete the diagram showing what happens to the rays of light when they leave the lens.

c) What common eye defect can be corrected with this type of lens?

3 Consider the three convex lenses shown in the diagram below.

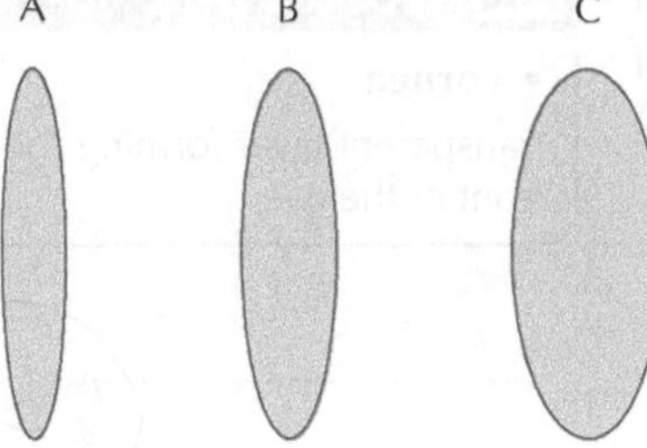

Fig 5.19.38

a) Which of the above lenses has the shortest focal length?

b) Which of the above lenses has the longest focal length?

c) Describe an experiment you could carry out to find the focal length of a convex lens.

4 Look at the diagram of an eye below:

a) What is the name given to this common eye defect?

b) What type of lens can be used to correct this problem?

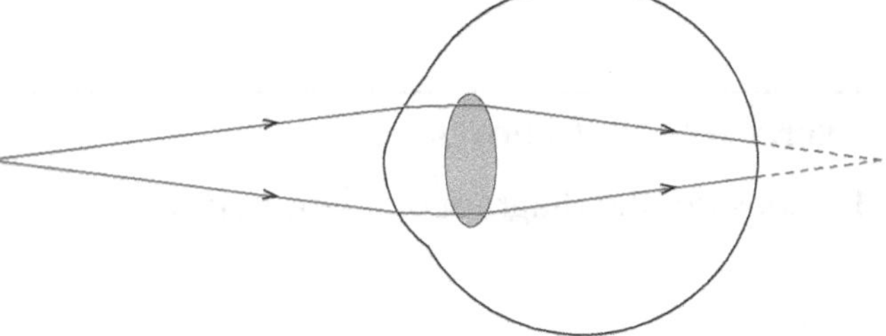

Fig 5.19.39

5 A pupil needs to wear glasses in order to read the board from the back of the laboratory. However, she can read her textbook without glasses.

a) What type of sight defect does this student have?

b) What type of lens would you use to correct this sight defect?

c) Draw a diagram of the eye to show how light rays from a distant object are focussed by the pupil's eye without additional sight correction.

Fig 5.19.40: *Optical fibres have an important role in everyday life*

Applications of refraction – optical fibres

As we become more reliant on being able to transfer large amounts of information almost instantly across the globe, we are now more reliant on optical fibres. Optical fibres are able to transmit huge amounts of information down a single strand of glass.

Experiment: Total internal reflection

This experiment investigates total internal reflection, the process by which light waves travel along optical fibres.

You will need:

- A semi-circular glass or Perspex block
- A ray box and power supply
- A protractor
- A ruler

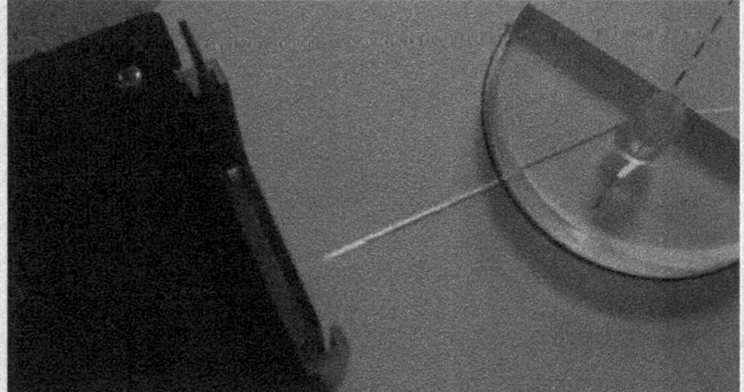

Fig 5.19.41

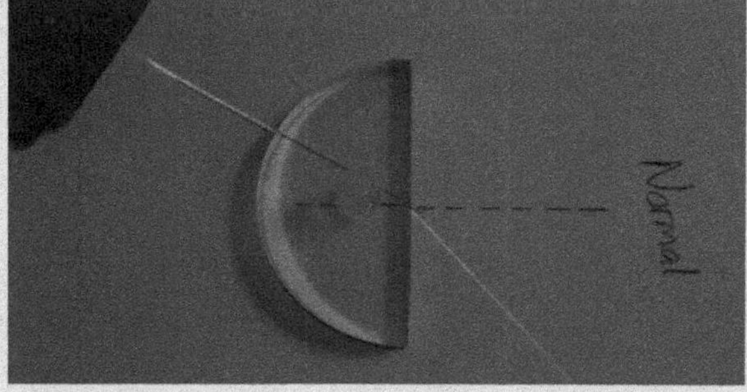

Fig 5.19.42

Instructions

1. Place the semi-circular block on the paper and draw round its outline.
2. Use a protractor to measure a 90 degree angle to the flat surface of the block and mark on the normal through the semi-circular block.
3. Use the ray box to shine a single ray of light into the curved side of the semi-circular block so that it meets the straight edge at the normal – see diagram above right.
4. Mark the positions of the ray with X. Switch off the ray box and use a ruler to draw on the ray of light, remembering to include arrows showing the direction of the ray.
5. Move the ray box round, increasing the size of the angle of incidence, until the ray is reflected by the straight edge of the block – see diagram below. The light here is said to be totally internally reflected.

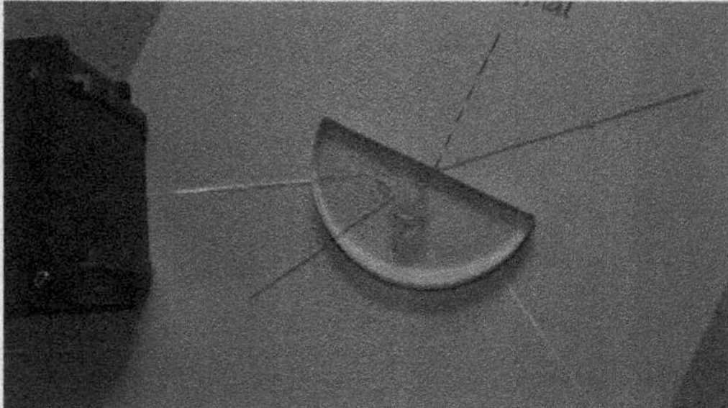

Fig 5.19.43

6. Mark on the rays as described in part 4.
7. Use a protractor to measure the angle of incidence where all of the light is reflected by the straight edge – this is the **critical angle**. The critical angle is the smallest angle for which all of the light is reflected.

Total internal reflection

Optical fibres rely on a result of refraction called total internal reflection. When a ray of light travelling through glass is incident on the glass-air boundary, it can be refracted or reflected as shown by the diagram below.

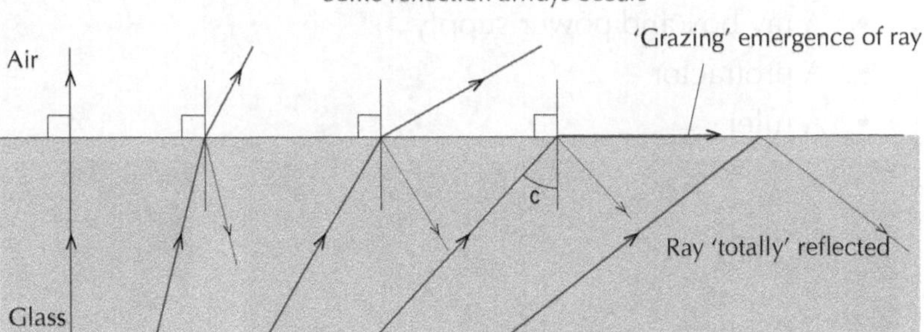

Fig 5.19.44: *Light rays at increasing angles of incidence*

When the refracted ray travels along the surface of the glass, we call the angle of incidence the critical angle(θ_c). If the angle of incidence is greater than this critical angle then we will see total internal reflection. All of the light is reflected without the use of mirrors. Some light will be reflected for angles below the critical angle.

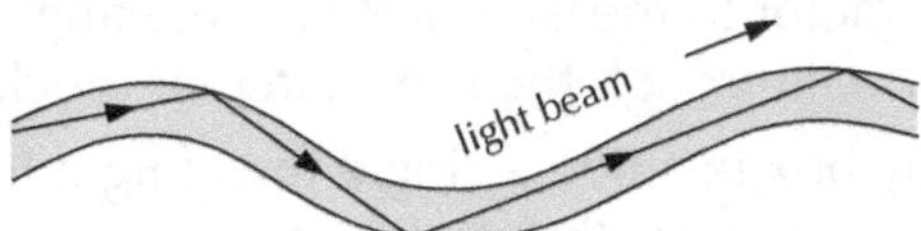

Fig 5.19.45: *Light travelling down an optical fibre by total internal reflection*

Key point

The critical angle is the angle of incidence for which the angle of refraction is 90°. Beyond the critical angle, all light is totally internally reflected.

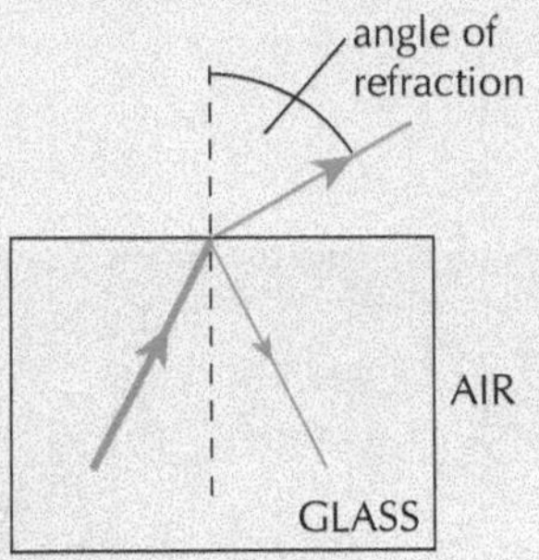

Fig 5.19.46: *There is a weak reflection*

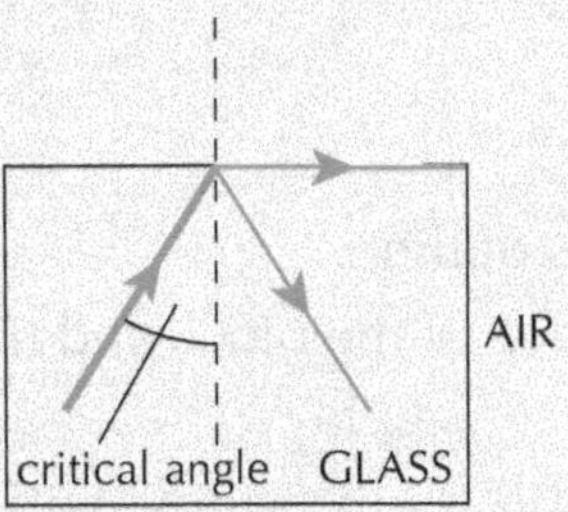

Fig 5.19.47: *There is a stronger reflection*

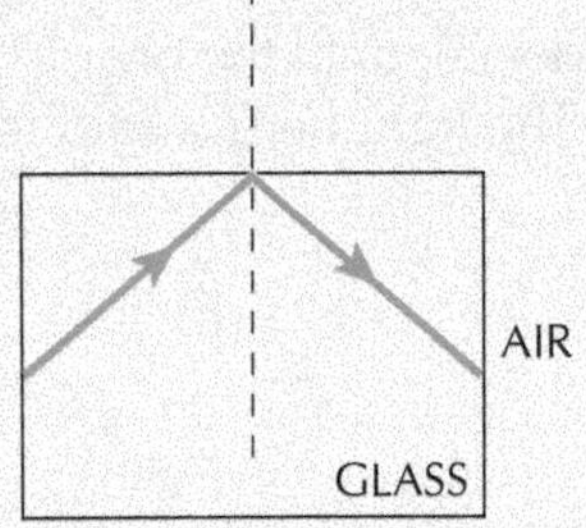

Fig 5.19.48: *All light is reflected*

Exercise 5.19.3 Optical fibres

1. Why does light not escape from the sides of a fibre optic fibre?
2. With the aid of a diagram, explain how light travels along an optical fibre.
3. Light travels at 2×10^8 ms^{-1} in glass. Calculate how long it will take a signal to travel along an optical fibre between Scotland and the USA, a distance of 5000 km.
4. Describe, with the use of a labelled diagram, an experiment that can be used to find the critical angle of a glass block.

Refraction: a renewable source of light

An interesting application of refraction has been identified by Brazilian mechanic Alfredo Moser. He was trying to find a cheap method of lighting his home during the day; many of the poorest people across the world cannot afford to pay for electric lighting. His simple but clever solution was to fill a plastic bottle with water (and a touch of bleach to prevent algae from growing in it) and seal it in a hole cut into the roof of his house. Sunlight refracts through the water-filled bottle and lights the room. Over the last 2 years, his invention has spread across the world, and it is expected that 1 million people will be using his simple water bottle light bulb by early 2014.

Fig. 5.19.49: *Child reading in the light of a water bottle light bulb*

Dispersion

Just as different wavelengths *diffract* by different amounts (as we saw in Chapter 17), different wavelengths also *refract* by different amounts. This difference in the amount of refraction caused by various wavelengths leads to an effect called dispersion. In refraction, shorter wavelengths change direction by a greater amount – they refract more.

Experiment: Dispersion by a prism

This experiment investigates the effect of dispersion of white light through a prism.

Fig 5.19.50

You will need:

- A ray box and power supply
- A prism
- A ruler
- Coloured pencils

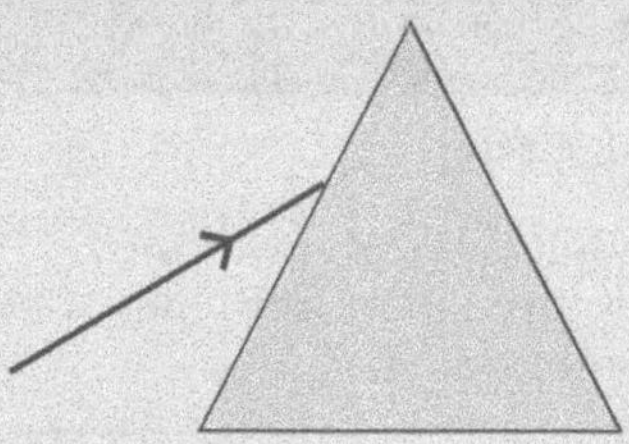

Instructions

1. Set up the ray box to produce a single ray of white light.
2. Place the prism on to a sheet of paper and draw around its outline.
3. Switch the ray box on and direct the ray of white light at the glass or Perspex prism as shown in the diagram above.
4. Use two 'X's to mark the position of the incident ray.
5. What do you notice about the light leaving the prism? Note the colours you see on the diagram.
6. Switch off the ray box and use a ruler to mark the position of the incident ray using the 'X's to help you.

Fig 5.19.51: *Dispersion of white light*

Dispersion of white light

Consider this photograph of white light incident on a prism.

Notice how the white light splits into different colours. White is not actually a colour – it is the effect of adding all of the colours of visible light together. You can see in the next picture that the longer wavelength red light does not refract as much as the shorter wavelength blue light. "Blue bends best" is one way to remember this.

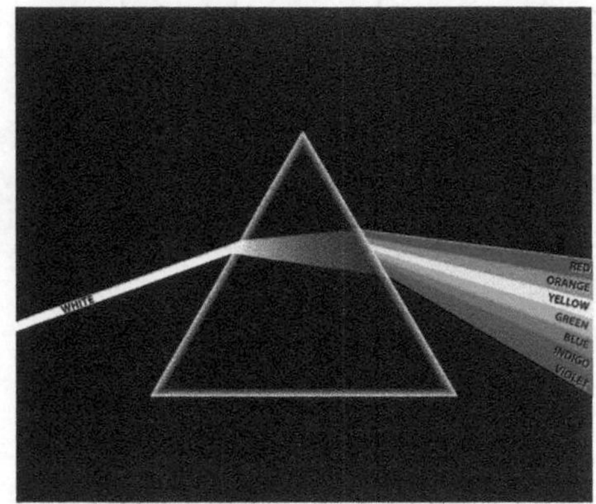

Fig 5.19.52: *Refraction of visible light*

Think about it

How are rainbows formed? Hint: think about a ray of white light from the sun travelling through a water bead (rain drop).

Fig 5.19.53

What is wrong with this picture (apart from the fact it isn't raining!)?

Exercise 5.19.4 Dispersion

1 A glass prism is used to split white light into the colours of the rainbow. Which colour of light is refracted by the greatest angle?

2 Why does white light that passes through a glass prism get split into different colours?

Learning checklist

In this section you will have learned:

- That refraction occurs when waves pass from one medium to another.

- How to describe refraction in terms of change of wavespeed, change in wavelength and change of direction (where the angle of incidence is greater than 0°), for waves passing into both a more dense and a less dense medium.

- How to identify the normal, angle of incidence and angle of refraction in ray diagrams showing refraction.

20 Nuclear radiation

- The nature of alpha (α), beta (β) and gamma (γ) radiation
- Ionistaion
- Activity
- Background radiation
- Dangers associated with ionising radiations
- Equivalent dose (rate)
- Applications of nuclear radiation
- Half-life
- Experiment used to determine the half-life of a radioisotope
- Fission and fusion

AREA 6
Nuclear radiation

20 Nuclear radiation

In this chapter you will learn about:

- The nature of alpha, beta and gamma radiation, and how to detect and identify radiation.
- The effects of radiation on living cells.
- Dosimetry, including activity in Becquerels, absorbed dose, equivalent dose and equivalent dose rate.
- Half-life, including the use of graphs or numerical data to determine half-life.
- Sources and applications of radiation, including sources of background radiation.
- Nuclear fission, the generation of nuclear power and nuclear fusion.
- Half-life.
- An experiment used to determine the half-life of a radioactive material.

Ionising radiation

We are exposed to radiation all the time. Radiation has a wide variety of uses, especially in medicine where it is relied on for some diagnostic and treatment purposes. Radiation can be powerful and can be used to produce electricity. It has also been used to devastating effect in nuclear weapons. In this section we will look at the different types of ionising radiation, their properties, effects on living things, and practical uses.

Fig 6.20.1: *International radioactive symbol*

The structure of the atom

Nuclear radiation is the result of a sub-atomic process, so in order to understand it we need to have an understanding of the structure of an atom.

Since the beginning of the 20th century, scientists have concluded that the atom consists of a positively charged nucleus which contains almost all of the mass of the atom, surrounded by a cloud of negatively charged electrons. Atoms are almost completely taken up with empty space. You can see a diagrammatic model of this structure opposite.

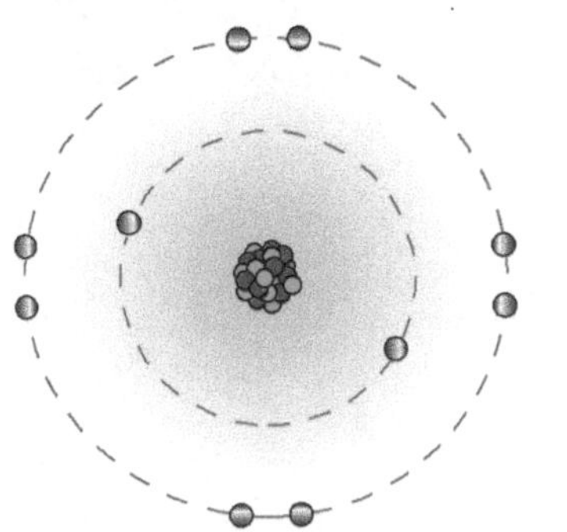

Fig 6.20.2: *The structure of an atom with protons and neutrons in the nucleus surrounded by orbiting electrons*

The nucleus of an atom is made up of protons and neutrons. Protons are positively charged particles; neutrons have no charge.

While all of the matter in the universe is made up of atoms, clearly not all matter is the same. This must mean that there are different types of atom. A substance that is made up of only one type of atom is called an element. These elements were first organised by Dmitri Mendeleev into the periodic table of the elements. What makes one element different from another is the number of protons, neutrons and electrons that make up its atoms.

Hydrogen is the simplest, lightest and most abundant of all the atoms. It consists of one proton and one electron. Helium is the next most simple atom, consisting of two protons, two neutrons and two electrons. Lithium has three protons, three electrons and three neutrons.

Fig 6.20.3: *Dmitri Mendeleev, 1804–1907*

Hydrogen	Helium	Lithium
Fig 6.20.4 1 proton, 1 electron	**Fig 6.20.5** 2 protons, 2 neutrons, 2 electrons	**Fig 6.20.6** 3 protons, 3 neutrons, 3 electrons

Table 6.20.1

Isotopes and ions

It is possible to have different versions of the same element. For example, Deuterium (also called Heavy Hydrogen) is a Hydrogen atom that contains one neutron as well as one proton and one electron. Helium-3 is like Helium but with only one neutron rather than two. This variation of the same type of atom is called an **isotope.**

Key point

The number of protons in the atom determines the type of atom it is:

- 1 proton = Hydrogen
- 2 protons = Helium
- 3 protons = Lithium

and so on...

Word bank

- **isotope**

a variation of an atom that has a different number of neutrons

Word bank

- **ion**

an atom that does not have the same number of protons as electrons

An atom will always have the same number of electrons as protons. This gives the atom an overall neutral charge. If there is an imbalance of protons and electrons we call that atom an **ion**. If there are more electrons than protons, the ion is a negative ion (because there are more negative charges than positive ones). Conversely, if there are more protons than electrons the ion is a positive ion.

Types of radiation

Some isotopes are stable – therefore they will remain unchanged for long periods of time. Other isotopes are unstable due to an imbalance between protons and neutrons in the nucleus. This is shown in the diagram on the left (Fig 6.20.7).

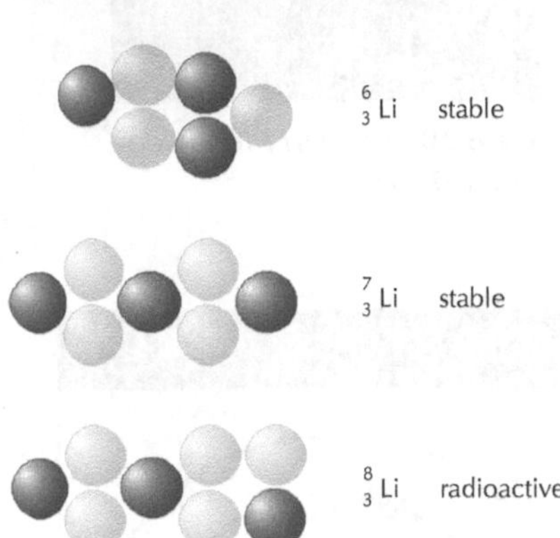

Fig 6.20.7: *Different isotopes of Lithium*

Some unstable isotopes will break down into two smaller, stable atoms. This process is known as radioactive decay. When an unstable isotope goes through radioactive decay it releases energy in the form of particles and/or rays. These particles and rays are known as nuclear radiation. There are three main types of nuclear radiation: alpha (α), beta (β) and gamma (γ). Alpha and beta radiation are particles, and gamma radiation is a high energy wave from the electromagnetic spectrum.

Alpha radiation

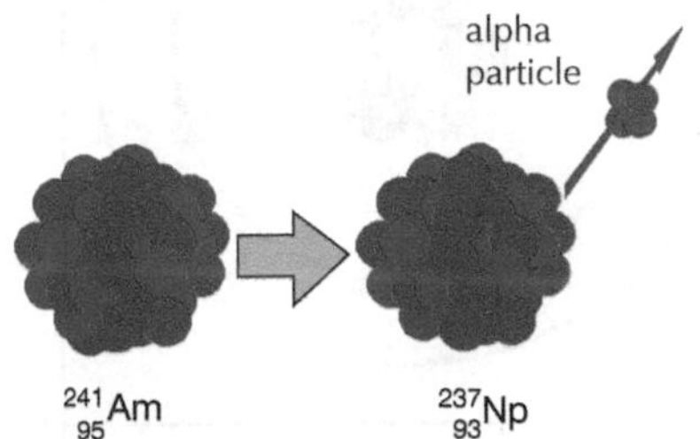

Fig 6.20.8: *An example of alpha radiation*

Alpha particles are the nuclei of helium atoms. They have two protons and two neutrons and are positively charged. Alpha particles are highly damaging radiation, but they do not travel very far – a couple of sheets of paper will stop them.

Beta radiation

Beta particles are fast moving electrons. They are produced within the nucleus of an atom when a neutron breaks up causing the emission of an electron from the atom (to conserve the atomic charge).

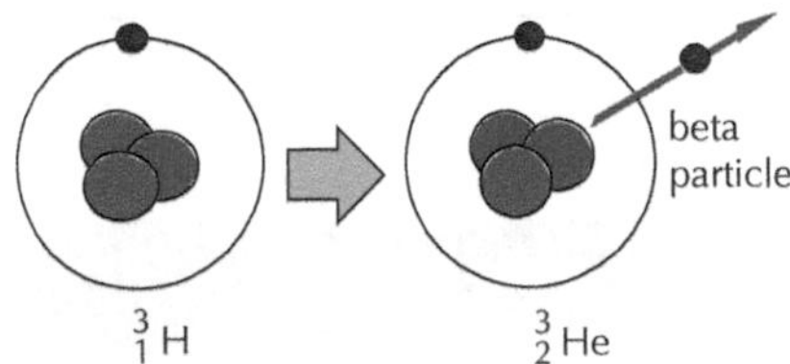

Fig 6.20.9: *An example of beta radiation*

Beta particles are typically less ionising than alpha particles but they can travel further through materials, requiring a couple of centimetres of aluminium to stop them.

Gamma radiation

Gamma radiation is a high energy wave emitted by the nucleus of the atom – no actual particle is emitted. Gamma rays have no mass or charge and carry energy from the nucleus leaving it in a more stable state.

Gamma particles are typically less ionising than alpha particles but can travel great distances through materials requiring several centimetres of lead to stop them!

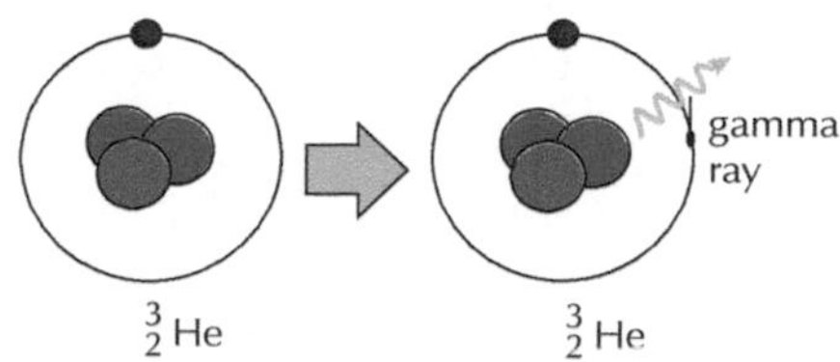

Fig 6.20.10: *An example of gamma radiation*

Ionisation

So far we have considered the three main types of radiation but not yet considered in detail the effects of radiation. The biological harm that radiation can cause is looked at in more detail in the Dosimetry section. However, on an atomic level, radiation has the effect of ionising other atoms.

Consider a stable atom which is electrically neutral – it has the same number of electrons orbiting the nucleus as protons in the nucleus. When a radioactive particle collides with this atom, it is ionised – it turns into an ion. This means it becomes electrically charged. An alpha particle can have the effect of knocking one of the electrons out of the atom – this imbalance results in the atom becoming positively charged as it has one less electron.

Make the link

Mass number and atomic number are considered in more detail in Chemistry.

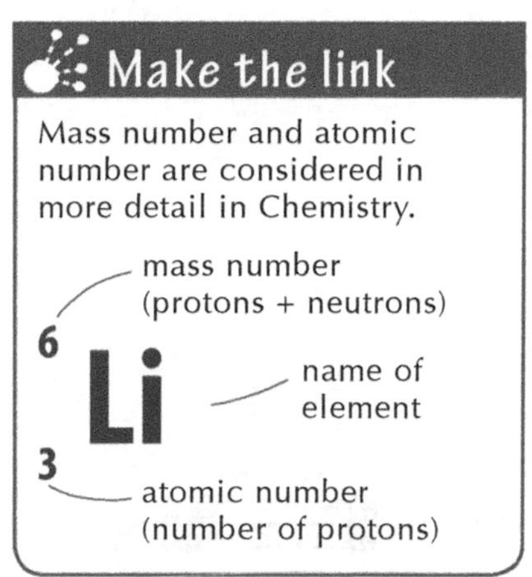

The effects of radiation

This variation between the different types of particle has some important consequences. For example, absorbing an alpha particle into your body could do far more damage than a beta particle. However, because alpha particles are much bigger than beta particles they are much easier to stop.

Consider the atoms in your skin to be like trees in a forest and alpha particles to be like a bowing ball thrown at the trees. It might break off some branches but it would not penetrate too deep into the forest. However, the trees it did hit would be badly damaged. The alpha particle would damage the atoms it hit and ionise them, but it would not be able to penetrate through your skin.

The beta particle is like a marble catapulted at the trees – more penetrating than an alpha particle but not as damaging and will not ionise as many atoms.

A grain of sand launched at very high speed is like a gamma ray – extremely penetrating but it may only ionise a few atoms. It would, however, pass through your skin and therefore towards your vital organs.

It is not clear which radiation is most damaging. Alpha particles are the most ionising but least penetrating. Gamma rays are the most penetrating but cause the least amount of ionisation, and beta particles are somewhere in the middle. Of course, the number of alpha/beta particles or gamma rays is important too – more particles/rays lead to more ionisation and therefore more damage to your cells.

Type of radiation	Electrical charge	Ionisation	Can be stopped by
Alpha (α)	Positive	High	A few cm of air, a sheet of paper, your skin
Beta (β)	Negative	Medium	A few cm of aluminium
Gamma (γ)	No charge	Low	Many cm of lead, a few metres of concrete

Table 6.20.2: *Types of radiation*

Detecting radiation

Geiger–Müller tube

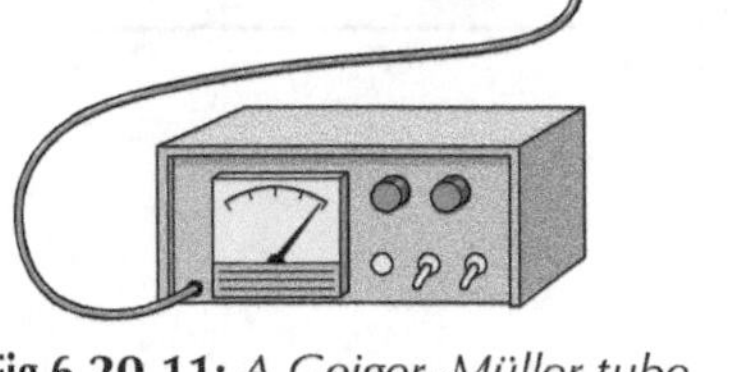

Fig 6.20.11: *A Geiger–Müller tube*

A Geiger-Müller (GM) tube can be used to detect ionising radiation. When radiation enters the GM tube, ions are produced. The ejected electrons flow to form a current. The number of times a current is produced in a given period of time gives an indication of how many radioactive particles are being emitted. This is a measure of the activity of the radioactive source, and is considered in more detail in the section on Dosimetry.

Photographic film

Radioactivity darkens photographic film, and the extent to which this happens can be used to determine the level of radiation involved. This technique of measuring radioactivity is used to ensure that workers dealing with nuclear radiation do not receive too large a dose of radiation.

The spark counter

The spark counter was an early method of detecting nuclear radiation. A metal gauze was placed near a wire and a voltage applied across them. The voltage was reduced until sparks were no longer produced. When a source of radiation was brought near, it caused the air between the gauze and the wire to ionise, resulting in sparks being produced again.

Scintillation detector

Scintillation detectors can be used to monitor radiation because when the radiation strikes a suitable material such as sodium iodide, it produces a small amount of light energy. Amplified, this results in a burst of electrons large enough to be detected. Scintillation detectors are used in monitoring contamination in nuclear power stations.

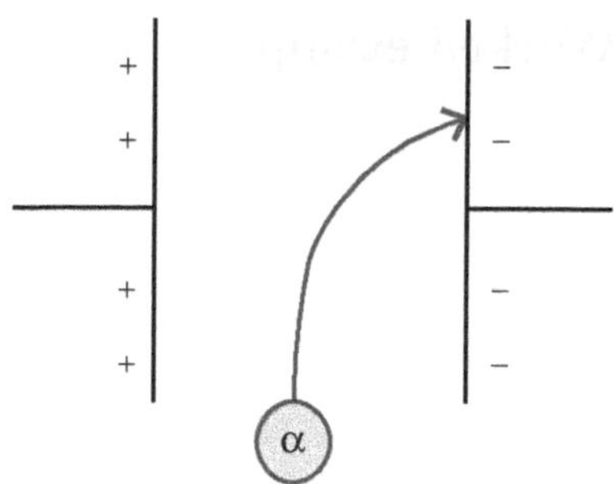

Fig 6.20.12: *Detecting an alpha particle: the positive alpha particle is attracted to the negative plate*

Identifying radiation

As well as the activity of the radiation (the number of counts in a given time), the type of radiation can also be identified. We have seen above that the different types of radiation have different charges. They are also stopped by different materials. We can use these facts to identify a specific type of radiation.

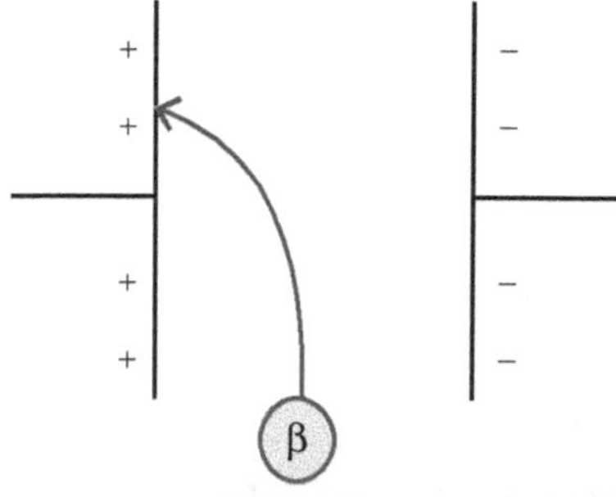

Fig 6.20.13: *Beta particles are negatively charged so are attracted to the positive plate*

Identifying alpha radiation

If the count rate is reduced to zero by a sheet of paper, then the radioactive source must be emitting alpha radiation. This can also be confirmed by firing the radiation between two charged plates as shown on the diagram. As alpha particles are positively charged, they will be attracted to the negatively charged plate.

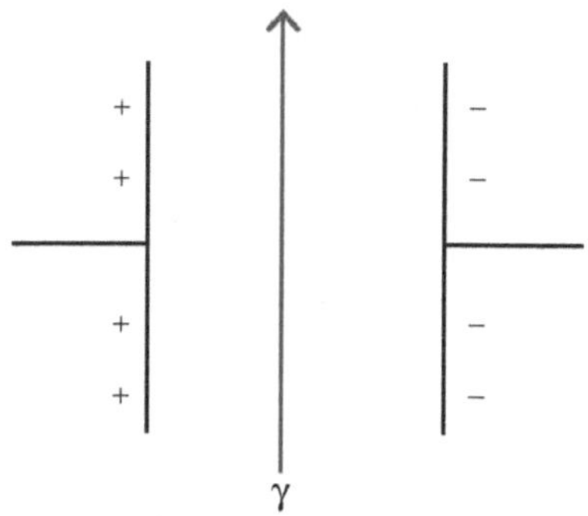

Fig 6.20.14: *Gamma radiation has no charge so is unaffected by the charged plates*

Identifying beta radiation

If the count rate is not reduced by paper, but is reduced to zero by a couple of centimetres of aluminium then the radioactive source must be emitting beta radiation. This can also be confirmed by firing the radiation between two charged plates as shown on the second diagram. As beta particles are negatively charged, they will be attracted to the positively charged plate.

Identifying gamma radiation

If the count rate is only reduced by several centimetres of lead then the radioactive source must be emitting gamma radiation. This can also be confirmed by firing the radiation between two charged plates as shown on the third diagram. As gamma rays are a part of the EM spectrum, they have no charge thus they will not be deviated by the charged plates.

Key point

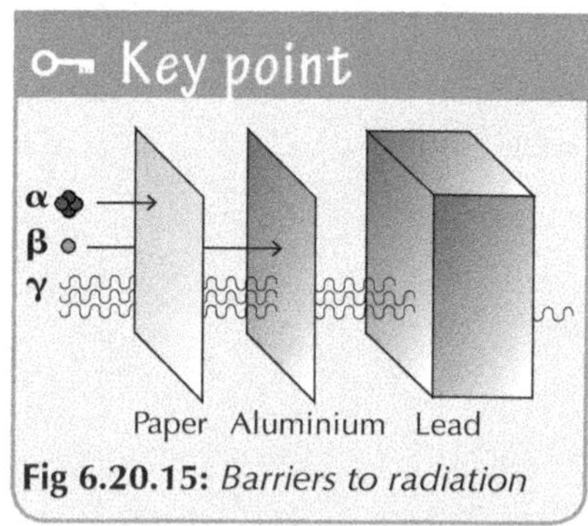

Fig 6.20.15: *Barriers to radiation*

Worked example

A pupil uses charged plates to determine the type of radiation being emitted from a source. She finds that the radiation is attracted to the positive plate as shown in the diagram below. What type of radiation is being emitted from the source?

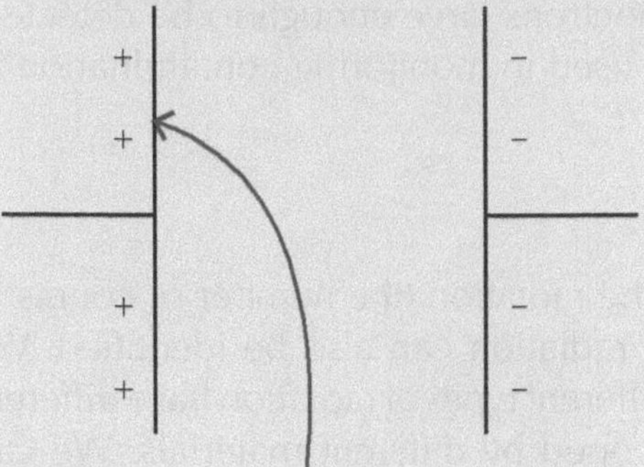

Fig 6.20.16

As the radiation is attracted to the positive plate, it must be negatively charged. The type of radiation which is negatively charged is beta radiation.

Worked example

A student is investigating the type of radiation being emitted from a radioactive source. He measures the count rate with different materials placed in front of the detector and records his results in the table shown below. Using the table above, find which type(s) of radiation are emitted by this source.

Material in front of detector	Count rate
Air	4250
Sheet of paper	2200
2 cm aluminium	2200
10 cm lead	90

Table 6.20.3: Count rate

The sheet of paper in the above example significantly reduces the count rate, so the source must be emitting alpha radiation.

Aluminium has no additional effect, so there is no beta radiation being emitted.

The lead significantly reduces the count rate, so the source must also be emitting gamma radiation.

Exercise 6.20.1 Ionisation and types of radiation

1 What is meant by the term ionisation?

2 Describe a simple model of the atom. Include in your description the following terms:
- electron
- proton
- neutron
- nucleus

You may find it useful to include a diagram in your answer.

3 What are the three main types of radiation that can be emitted from a radioactive source?

4 In an experiment to determine the type of radiation being emitted by a source, a technician sets up an experiment involving two charged plates. The radiation passes between the plates and the deflection is measured. The results are shown in the diagram below.

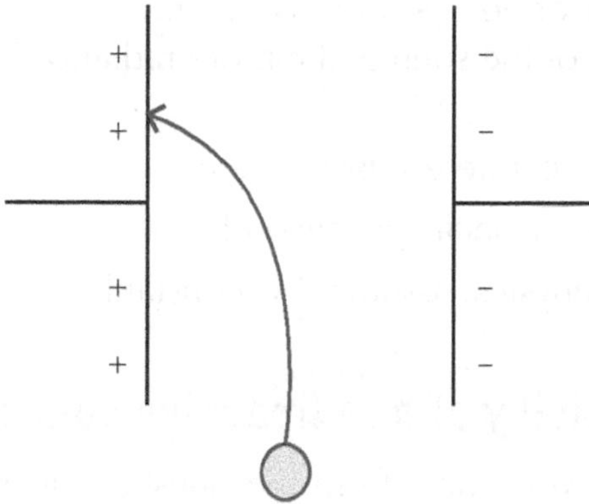

Fig 6.20.17

a) What type of radiation is being emitted? Explain your answer.

b) The apparatus used has been shielded in a lead container. Why do you think a lead container is used to shield the apparatus?

5 A pupil is examining the effect of different absorbers on the count rate of radiation received from a radioactive source.

a) What device can be used to detect radioactive particles?

b) The pupil records the following results in the table.

Material in front of detector	Count rate
Air	976
Sheet of paper	974
2 cm aluminium	433
10 cm lead	12

Table 6.20.4

What types of radiation are being emitted by the source? Explain your answer.

6 List the three types of radiation in order of their penetration through different materials, starting with the most penetrating radiation.

Word bank

- **dosimetry**

the measurement and calculation of radiation dose

Dosimetry

In the previous section we investigated the properties of the three main types of radiation. In this section we consider the actual amount of radiation emitted and the damage this radiation can do.

Activity of a source

Imagine we have a sample of an unstable isotope – this is a radioactive source. Every time one of the isotopes in the source decays into a stable isotope we will have one burst of radiation. If we measure how many of these decays happens in one second we have a measure of how radioactive our source is. We call this the activity, A, of the source:

$$A = \frac{N}{t}$$

where N is the number of decays and t is the time in seconds. The greater the activity of the source, the more radiation it will emit in a given time.

Activity is measured in becquerels (Bq), where:

1 Bq = 1 decay per second

For the purpose of N5 physics, assume A is constant.

Technique: Activity of a radioactive source

1. Write down what you know from the question, leaving blank what you are trying to find:
   - Activity = _____ Bq
   - Time = ____ s (time must be converted to seconds!)
   - Number of decays = _____ decays
2. Write down the equation linking the quantities above as you see it on the formula sheet:

$$A = \frac{N}{t}$$

3. Substitute into the equation what you know.
4. Solve for the unknown.

Worked example

Calculate the activity of a radioactive source which has 450 decays every minute.

First of all, we must convert the time into seconds:

t = 1 minute = 60 s

(*continued*)

Write down the equation that links activity, number of decays and time:

$$A = \frac{N}{t}$$

Substitute into the equation and solve for the activity:

$$A = \frac{450}{60}$$

$$A = 7{\cdot}5\ Bq$$

Worked example

A radioactive source has an activity of 50 Bq. How many decays will there be in a time of 30 seconds?

Write down what we know from the question, ensuring time is in seconds.

- $A = 50$ Bq
- $t = 30$ s
- $N = ?$

Write down the equation that links activity, number of decays and time:

$$A = \frac{N}{t}$$

Substitute what we know and solve for the number of decays:

$$50 = \frac{N}{30}$$

$$N = 50 \times 30$$

$$N = 1500\ decays$$

Exercise 6.20.2 Activity of a source

1 What do we mean by the activity of a radioactive source? What is the unit of activity?

2 Calculate the activity of the following radioactive sources:

a) 50 decays in 8 seconds

b) 25 decays in 1·5 seconds

c) 240 decays in 1 minute

d) 1500 decays in 3 hours

e) 42 decays in 11 milliseconds

f) 930 decays in 2 hours

3 A source of radiation is found to emit 9 million radioactive particles in 15 minutes. Calculate the activity of this source.

4 The activity of a source is 2 MBq. How many unstable isotopes decay in 4 minutes?

Think about it

The equation D = E/m tells us that a large amount of energy absorbed by a small amount of tissue will lead to a large absorbed dose and therefore a large risk of damage.

Based on this, why do you think doctors do not use X-rays (radiation) to examine unborn babies in the womb? What do doctors use instead for this?

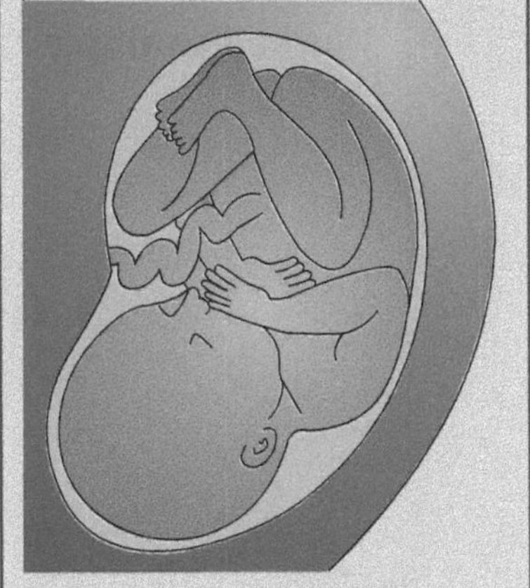

Fig 6.20.18: *Unborn child*

Absorbed dose

Physicists have developed methods to assess how much damage *may* be done by exposure to radiation. The first of these is known as absorbed dose and assumes radiation has been absorbed. Once the radiation is absorbed two things need to be considered when assessing potential damage: the amount of energy that has been absorbed from the radiation, and the amount of tissue absorbing this energy.

The absorbed dose, D, is the energy absorbed per unit mass of material,

$$D = \frac{E}{m}$$

where E is the energy absorbed (J) and m is the mass (kg). The greater the absorbed dose, the more energy that is transferred to a given mass of tissue by the radiation.

Absorbed dose is measured in Grays (Gy):

$$1\ \text{Gy} = 1\ \text{J/kg}$$

Worked example

1 mJ of nuclear radiation is absorbed by 0.5 kg of tissue. Calculate the absorbed dose that this tissue receives.

Write down what we know from the question, remembering to convert milli-joules into joules – notice the use of scientific notation:

- $E = 1mJ = 1 \times 10^{-3} J$
- $m = 0{\cdot}5\ kg$
- $D = ?$

Write down the equation that links absorbed dose to energy absorbed and mass:

$$D = \frac{E}{m}$$

Substitute what we know into the equation and solve for absorbed dose:

$$D = \frac{1 \times 10^{-3}}{0{\cdot}5}$$

$$D = 2 \times 10^{-3}\ Gy$$

$$D = 2\ mGy$$

Exercise 6.20.3 Absorbed dose

1 A technician of mass 60 kg is handling a source of radiation. She receives an absorbed dose of 5×10^{-5} Gy. Calculate the total energy absorbed by the technician.

2 While being treated for cancer a patient absorbs of 0.1 mJ of energy from the radiation being used to kill the cancer cells. This radiation is absorbed by 0·5 kg of tissue. What is the absorbed dose that this patient receives?

3 A scientist is working with a radioactive sample. She finds that her hand, which has a mass of 0·7 kg, absorbs 4 mJ of energy from the sample. Calculate the absorbed dose received by the scientist.

4 While working with radioactive tracers in a hospital laboratory a medical technician receives an absorbed dose of $1{\cdot}7 \times 10^{-4}$ Gy. The total energy absorbed by the technician was 0·011 J. Calculate the mass of the technician.

Equivalent dose

Absorbed dose does not take into account the effects of the type of radiation absorbed. To do this physicists use what is called a weighting factor. Each type of radiation has a weighting factor, w_R. The weighting factor is an indication of how damaging that type of radiation is – the greater the weighting factor, the more damaging the radiation. Alpha radiation has the highest weighting factor and gamma radiation the lowest. Combining the weighting factor with the absorbed dose gives us the equivalent dose, H, which is a measure of the **biological harm** caused by the radiation. Equivalent dose is given by the formula,

$$H = Dw_R$$

where D is the absorbed dose (Gy) and w_R is the weighting factor (unitless). The equivalent dose is measured in sieverts (Sv). From this equation, a large absorbed dose matched with a large weighting factor will lead to a large equivalent dose and therefore a high risk of biological harm.

Typical weighting factors are shown in the table below:

Type of radiation	Weighting factor (w_R)
Alpha (α)	20
Beta (β)	1
Gamma (γ)	1

Table 6.20.5

Word bank

- **biological harm**
harm done to living things by, for example, damaging or killing cells

Key point

The risk of biological harm from radiation depends on three main factors:

- The **absorbed dose** – how much energy is transferred.
- The **type of radiation** – how damaging the radiation is.
- The **part of the body** – how easily damaged the organ or tissue is.

The measure of this damage is the equivalent dose,

$$H = Dw_R$$

Technique: Equivalent dose

1. Write down what you know from the question, leaving blank what you are trying to find:
   - Equivalent dose, H = _____ Sv
   - Weighting factor, w_R = ____ (unitless)
   - Absorbed dose, D = ____ Gy
2. Write down the equation linking the quantities above as you see it on the formula sheet:
$$H = Dw_R$$
3. Substitute into the equation what you know.
4. Solve for the unknown.

Worked example

A worker in a nuclear power station is exposed to 2·0 mGy of alpha particles. The weighting factor for alpha particles is 20. Calculate the equivalent dose received by the worker.

We are using the equation that links equivalent dose, absorbed dose and weighting factor:

$$H = Dw_R$$

The absorbed dose, converted into Gy using scientific notation, and the weighting factor taken from the question:

$$D = 2{\cdot}0\ mGy = 2{\cdot}0 \times 10^{-3}\ Gy$$

$$w_R = 20 \text{ for } \alpha \text{ particles}$$

Substitute into the above equation and solve for the equivalent dose:

$$H = 2{\cdot}0 \times 10^{-3} \times 20$$

$$H = 0{\cdot}04\ Sv$$

Worked example

An engineer working with nuclear waste is exposed to radiation during his work. The absorbed dose per year for two types of radiation exposure are shown below:

- 20 mGy of gamma radiation, $w_R = 1$
- 400 mGy of fast neutrons, $w_R = 10$

Calculate the equivalent dose for the engineer.

We have two different types of radiation here, so we need to calculate the equivalent dose separately for each type.

Using the equation for equivalent dose:

$$H = Dw_R$$

For gamma radiation:

$$H = Dw_R$$

$$H = \left(20 \times 10^{-3}\right) \times 1$$

$$H = 20 \times 10^{-3}\,Sv$$

For fast neutrons:

$$H = Dw_R$$

$$H = \left(400 \times 10^{-3}\right) \times 10$$

$$H = 4\,Sv$$

Then find the total equivalent dose by adding together the individual equivalent doses:

$$H = \left(20 \times 10^{-3}\,Sv\right) + 4\,Sv = 4{\cdot}02\,Sv$$

Fig 6.20.19

Exercise 6.20.4 Equivalent dose

1 What factors affect the biological harm from radiation?

2 State the units of measurement for:

a) activity

b) absorbed dose

c) equivalent dose

3 A scientist measuring the amount of nuclear radiation in a nuclear power station receives an absorbed dose of 10 μGy. The source of the radiation is an alpha emitter with a weighting factor of 20. Calculate the equivalent dose received by the scientist.

4 A technician working with a radioactive source receives an absorbed dose of 1·3 μGy of alpha radiation and 15 μGy of beta radiation. Her mass is 58 kg. Calculate the equivalent dose received by the technician. (The weighting factor of alpha radiation is 20 and the weighting factor of beta radiation is 1.)

(*continued*)

5 A rugby player breaks his leg in a bad tackle. He receives an X-ray with an equivalent dose of 1·5 mSv to examine the injury. If the weighting factor for X-rays is 1, calculate the absorbed dose of the patient.

6 Alpha particles produce an equivalent dose of 60 mSv from an absorbed dose of 3 mGy. X-rays produce a dose equivalent of 4 mSv for an absorbed dose of 4 mGy.

a) Calculate the radiation weighting factor for alpha particles.

b) Calculate the radiation weighting factor for X-rays.

c) Explain why exposure to alpha particles causes more biological harm than exposure to X-rays.

Equivalent dose rate

The amount of time that an individual is exposed to ionising radiations is very important since more biological harm can be caused by over-exposure.

To reflect this a quantity called the 'equivalent dose **rate**' ($\dot{H}$) is used. The relationship between this quantity and equivalent dose is given by the formula below,

$$\dot{H} = \frac{H}{t}$$

$\dot{H}$ is measured in sieverts per unit time which could be in seconds, minutes, hours etc..., i.e. Svh^{-1}.

Worked example

A radiologist in a hospital is exposed to 1·0 μGy of X-rays for a period of 5 seconds. The weighting factor for X-rays is 1. Calculate the equivalent dose rate received by the radiologist.

As before, we use the equation that links equivalent dose, absorbed dose and radiation weighting factor:

$$H = Dw_R$$

$$H = 1{\cdot}0 \times 10^{-6} \times 1$$

$$H = 1{\cdot}0 \times 10^{-6}\ Sv$$

Substitute the above including the time of 5 seconds into the relationship for equivalent dose ***rate:***

$$\dot{H} = \frac{H}{t}$$

$$\dot{H} = \frac{1{\cdot}0 \times 10^{-6}}{5}$$

$$H = 0{\cdot}2 \times 10^{-6}\ Sv\ s^{-1}$$

Exercise 6.20.5 Equivalent dose rate

1 A technician is working with a radioactive source. The technician's hands receive an absorbed dose rate of 4.0 μGy h^{-1} for 2 hours. The radiation from the source has a radiation weighting factor of 3. Calculate the equivalent dose received by the technician's hands.

2 An x-ray scanner is used at an airport security terminal to examine the contents of passenger's baggage prior to boarding an aircraft. Occasionally a member of the security team requires to insert their hands into the x-ray scanner to clear blockages. In doing so, the security member receives an absorbed dose of 0.030 μGy of x-ray radiation.

According to the regulations, the occupational exposure limit for a member of the security team is 60 μSv h^{-1}. Calculate how many times the security member would have to put their hands into the scanner each hour in order to reach this limit.

3 A researcher's hands are exposed to alpha radiation for a total time of 4 hours during the course of a week. The absorbed dose received during this time period is 0.50 μGy. Calculate the total equivalent dose rate in Sv h^{-1}.

The equivalent dose rate and exposure safety limits for the public and for workers in the radiation industry are as follows:

- Average annual background radiation in the UK: 2.2 mSv
- Annual effective dose limit for member of the public: 1 mSv
- Annual effective dose limit for radiation worker: 20 mSv

A knowledge of these values is expected for examination purposes.

Sources and applications of radiation

Background radiation

Radiation is around us all of the time. This is known as background radiation. Life on Earth has evolved with this so natural background radiation is nothing to be concerned about. On average, people in the UK receive around 2.7 mSv of radiation each year from background radiation. This dose will not adversely affect your health.

- In Scotland the main source of background radiation is from granite. Granite is a rock which produces a radioactive gas called radon. It produces this in low levels and is nothing to worry about – our cells have self-repair mechanisms that can cope with exposure to background radiation like this.

Fig 6.20.20: *Radon gas from granite is responsible for most of Scotland's background radiation*

Fig 6.20.21: *Cosmic rays provide some background radiation*

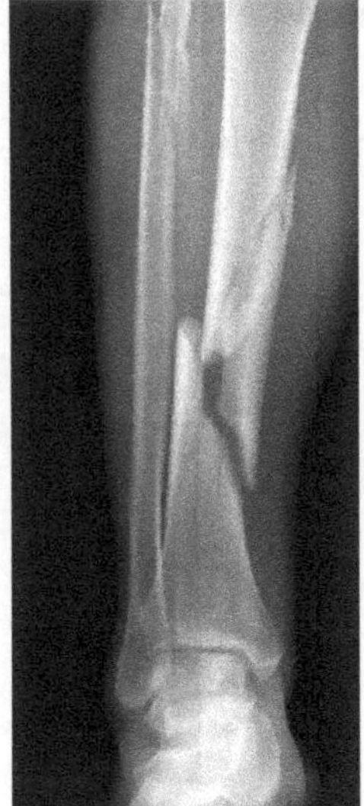
Fig 6.20.22: *X-ray of a leg bone*

Fig 6.20.23: *Even some foods can contribute to background radiation*

- Cosmic rays are another significant source of background radiation. These are released from the sun and collide with the nuclei of molecules in our atmosphere. This collision produces high speed particles which can cause ionisation – some of these reach the surface of Earth, contributing to the background radiation. However, our atmosphere protects us from much of the harm from cosmic rays!
- Each time you get an X-ray, perhaps to see if you have a broken bone, you receive a small dose of nuclear radiation.
- Even some of the food we eat contains low doses of nuclear radiation. For example, a banana contains low doses of nuclear radiation. However, you would have to eat around 200 bananas to receive the same dose as a chest X-ray! All of these factors contribute to the yearly dose of radiation that we receive.

The pie chart shows the proportions of different sources of background radiation that we are exposed to every year.

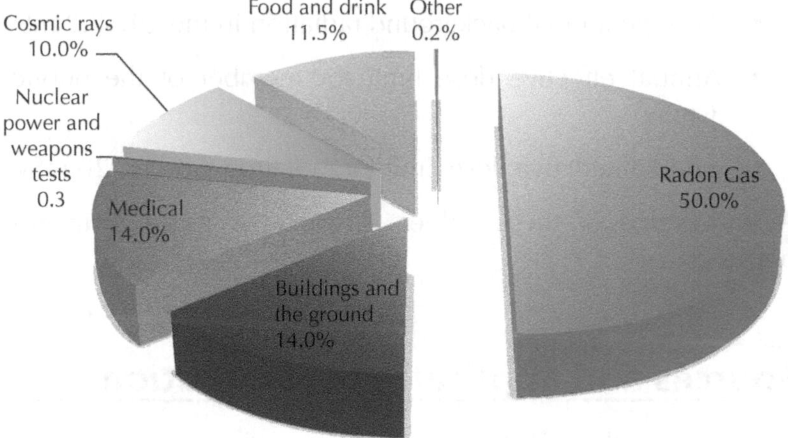

Fig 6.20.24: *The various sources of background radiation*

Whenever we carry out an experiment involving radiation we must account for background radiation. For example, when measuring the activity from a radioactive source, the measured activity will in fact be a combination of activity from the source plus background radiation! To get around this, we measure the background radiation first and then subtract this value from overall activity to obtain the activity of just the source.

Key point

Whenever the activity of a radioactive source is measured, the background radiation must first be subtracted from the reading to obtain the true activity of the source. The value obtained once background radiation has been subtracted is called the corrected count rate.

Exercise 6.20.6 Background radiation

1 Name two sources of background radiation.

2 Explain why background radiation must be measured before making any measurements of the activity of a radioactive source. How do you obtain the corrected count rate?

3 The activity of a source is measured to be 1280 Bq on a day where the background radiation is measured at 80 Bq. Six hours later the source is found to have an activity of 230 Bq. How long is the half-life of this source, assuming the background radiation is constant?

4 The background count rate is measured at 48 counts per second. When a source of nuclear radiation is present the count rate drops from 8240 to 50 counts per second over 4 days. Calculate the half-life of the source, assuming background radiation remains constant.

Applications of radiation

The fact that nuclear radiation can damage or kill cells can be used to our advantage in medical applications. Radiation can be usefully applied elsewhere too – for example, in smoke alarms.

Sterilisation

It is essential that hospital equipment is very clean – so clean it can be said to be sterile (free from bacteria or other living organisms). Equipment such as scalpels and needles are cleaned with chemicals and then packaged. To ensure no germs get into the packaging and the equipment remains clean, the sealed package is subjected to nuclear radiation. This radiation kills any germs and ensures that the scalpel or needle remains sterile until the package is opened.

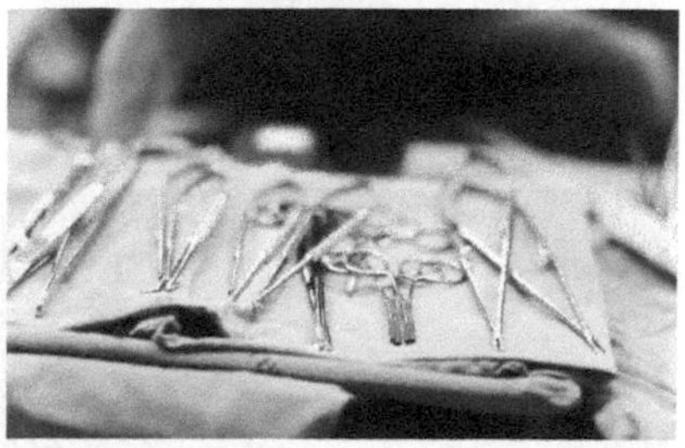

Fig 6.20.25: *Nuclear radiation can be used to sterilise hospital equipment*

Cancer treatment

Another use of nuclear radiation is to treat some forms of cancer. Cancerous cells are cells that have mutated and are growing out of control. These cancer cells must be killed before the cancer spreads around the body. The cancer cells can be bombarded with nuclear radiation to kill them. Of course, healthy cells around the cancer cells may also be damaged so doctors must be very careful to use just enough radiation to kill the cancer without causing too much damage to the healthy cells.

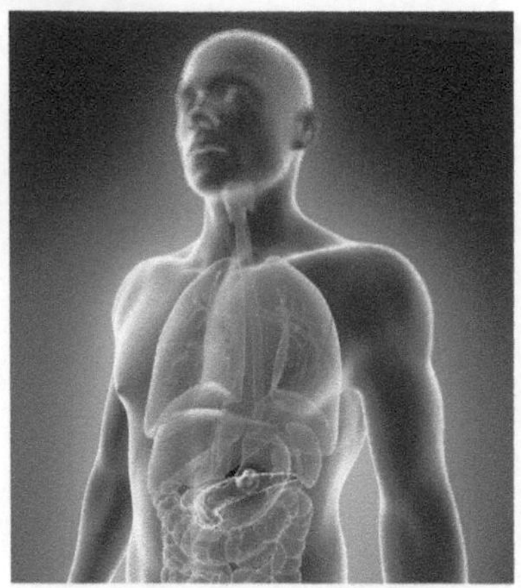

Fig 6.20.26: *Nuclear radiation can be used as a tracer*

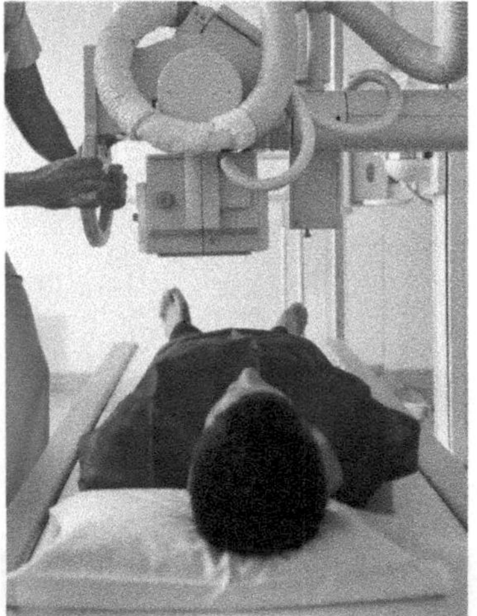

Fig 6.20.27: *An X-ray machine*

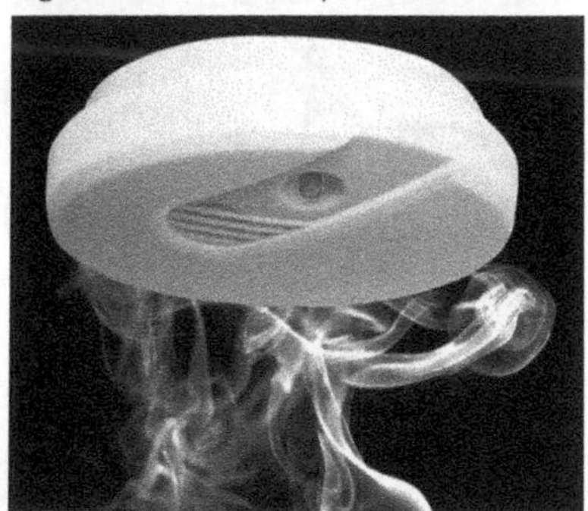

Fig 6.20.28: *Smoke alarms contain a small amount of alpha radiation that can be used to detect smoke*

⚠ Think about it

When taking an X-ray of a patient, the radiographer leaves the room during the exposure. Why is it important for the radiographer to protect himself when it is ok for the patient to be exposed?

Tracers

Nuclear radiation can be used as a tracer within the body to study whether things are moving correctly around the body. For example, if a doctor suspects there is a blockage in a kidney such as a kidney stone, a radioactive source can be injected into the body which will be detected from outside the body using a Geiger counter. A blockage will cause the radioactive source to collect there and a larger than expected reading will show at that point. This technique is called tracing.

Another use for tracing is in searching for a cancerous tumour. A tracer is injected into the blood and, as a tumour has a higher than usual supply of blood, the tracer will accumulate at the location of a tumour.

The source of radiation used in such medical procedures must be carefully selected.

- The source must emit gamma radiation – alpha radiation can't penetrate far enough to leave the body so it won't be detected. Alpha radiation also causes harmful ionisation.
- The source must have a short half-life – doctors only need to detect the radiation for long enough to aid their diagnosis. Once this is done there is no need to further expose the patient to dangerous nuclear radiation.

X-rays

X-rays are a form of radiation that can be used to detect broken bones in the body. X-rays are absorbed by bone but pass freely through muscle and flesh. By placing a piece of photographic film on the other side of the body, doctors can view images of bones as these will show as white sections where the radiation has been absorbed. As the X-rays pass through any breaks in the bone, these will show up as black on the photographic film.

Smoke alarms

As well as medical applications, radiation is useful in other ways, including in smoke alarms. Inside a smoke alarm there is a source of alpha radiation (americium-241). Alpha particles from this ionise the air and allow a small current to flow. When smoke enters the alarm, the alpha particles are absorbed by the larger smoke particles. This causes the amount of ionisation to fall, and therefore the current flowing drops. The drop in current triggers the alarm.

Exercise 6.20.7 Applications of radiation

1 Explain how nuclear radiation is used to ensure that medical instruments such as scalpels are sterile.

2 Nuclear radiation is used to locate illness in the human body. To do this doctors use gamma radiation as a tracer.

a) Suggest a suitable half-life for the source of the gamma radiation. Justify your answer.

b) Give 2 reasons why alpha radiation is unsuitable for use as a tracer.

3 Describe how nuclear radiation is used to detect cancer tumours in organs which are deep within the body.

4 What medium is used to detect X-ray radiation? Explain how an X-ray can be used to find a broken bone – your answer should include an explanation of how radiation passes through different tissues in the human body.

5 A smoke alarm uses a source of alpha radiation to detect the presence of smoke.

a) Why is a source of alpha radiation used for a smoke alarm?

b) Suggest a suitable half-life for the source of alpha radiation used in the smoke alarm.

c) Explain why a source of gamma radiation would be unsuitable for use in a smoke alarm.

Half-life

Once one particular unstable isotope has decayed into stable isotopes, it will not decay any further. This means that if we have a radioactive source, it will become less active over time as each unstable isotope decays into a stable one. The activity of a radioactive source decreases with time. The rate of this decrease depends on the radioactive source. Some sources will be active for a matter of minutes while others will be active for millions of years.

The length of time that a source will remain active can be predicted by its half-life. As the name suggests, it is the time taken for the activity to fall to half of its initial value. For example, the half-life of Uranium-235 (used in some nuclear power stations) is 703·8 million years. After 703·8 million years the activity of lump of Uranium-235 would have halved from its initial value.

The emission of a radioactive particle is a random process – we cannot say exactly when a particular nucleus will decay. However, we can make predictions as to how many nuclei will decay in a given time. Consider having 200 coins. You throw each coin separately and if it lands on 'heads' you keep the coin, 'tails' you lose the coin. Each lost coin represents an isotope that decays and becomes stable. After throwing all the coins once, you will have lost roughly 100 of them because there is a 50/50 chance of it landing on 'tails'. If you repeat the experiment, you will lose around half the coins again, leaving you with roughly 50. Plotting a graph of the number of coins lost (radioactive decays) each time you throw all the coins will give something like the following:

Fig 6.20.29: *The decay of a nucleus is a random process a bit like heads or tails*

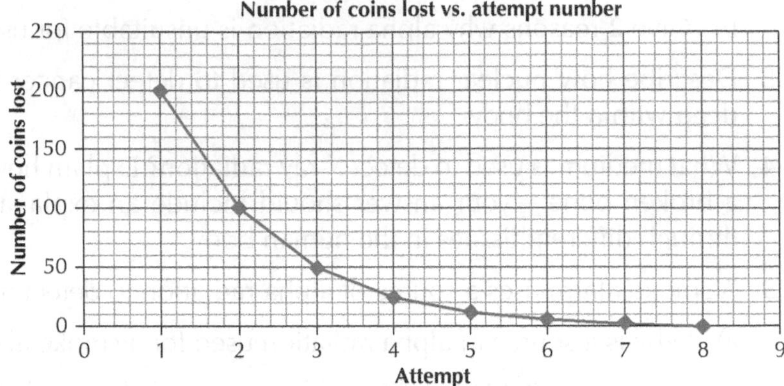

Fig 6.20.30

As the number of throws increases, the number of coins lost decreases, quickly at first and then more slowly.

A graph for the activity of a radioactive sample against time will look the same as that for the coins experiment described above.

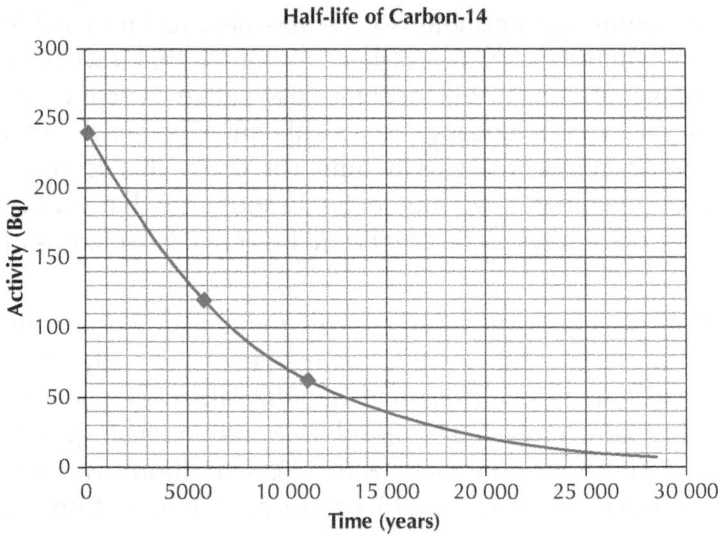

Fig 6.20.31

Consider Carbon-14, which is present in trees and decays over a period of thousands of years. If the activity of a block of wood is measured to be 240 Bq, half of this value is 120 Bq. You can see from the graph that it takes around 5700 years to reach this value. A further 5700 years after this (11 400 years after the initial reading) the activity has dropped to 60 Bq – it has fallen by half again. This

Key point

The half-life of a radioactive source, $t_{\frac{1}{2}}$, is the length of time taken for the activity to fall to half of its original value.

The method you choose to calculate the half-life of a source (by graph or numerically) will depend on the information that you are given. If it seems that you are unable to calculate the half-life using one method then you will likely have more success using the other.

means that the half-life of Carbon-14 is 5700 years. Scientists have used this knowledge of the half-life of Carbon-14 to determine the age of certain objects. This is known as carbon dating.

Technique: Finding the half-life by graph

1. Plot a graph of the activity of the source against time.
2. Divide the initial value of the activity, A_0, by 2 to find the activity after one half-life.
3. Use this to find the half-life – the corresponding time for which the activity is half of the initial value as shown on the graph below.

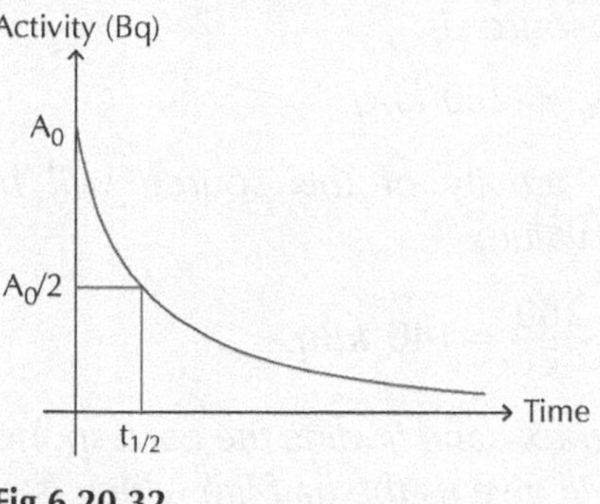

Fig 6.20.32

Worked example

The activity of a source was measured every 60 s. The results are shown in the table below. By plotting a graph of activity against time, find the half-life of the radioactive source.

Time (s)	Activity (kBq)
0	280
60	200
120	140
180	100
240	70
300	50
360	35

Table 6.20.6

First of all, plot a graph of the activity against time to show how the activity of the source has decayed. Activity is always plotted on the y-axis (vertical axis).

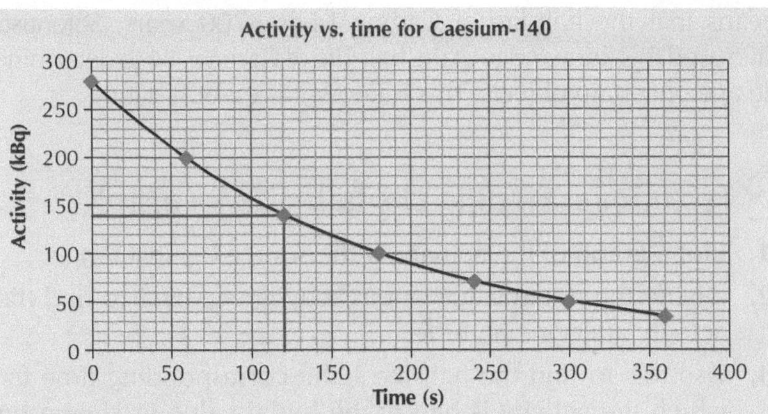

Fig 6.20.33

The initial activity of the source is:

$$A_0 = 280\ kBq$$

After one half-life, the activity of this source will have decreased to half of this value:

$$A = \frac{280}{2} = 140\ kBq$$

Finding 140 kBq on the graph, and finding the corresponding time (red lines above) will give us the half-life which is:

$$t_{\frac{1}{2}} = 120\ s$$

Technique: Finding the half-life numerically

1. Work out how many half-lives there are from the initial activity to the final activity by dividing the initial activity by 2 until the final activity is reached.
2. Divide the length of time by the number of half-lives that have occurred to work out the time for one half-life.

Worked example

A sample of Technetium-99m has an initial activity of 256 Bq. 24 hours later this sample's activity is measured at just 16 Bq. Calculate the half-life of Technetium-99m.

Firstly, work out how many half-lives have occurred in the time mentioned.

$$256\ Bq \xrightarrow{1} 128\ Bq \xrightarrow{2} 64\ Bq \xrightarrow{3} 32\ Bq \xrightarrow{4} 16\ Bq$$

Since this happened over a period of 24 hours, we can calculate the half-life of Technetium-99m to be:

$$t_{\frac{1}{2}} = \frac{24}{4}$$

$$t_{\frac{1}{2}} = 6\ \text{hours}$$

Exercise 6.20.8 Half-life

1 Describe what the term 'half-life' means in the context of radioactivity.

2 The graph below shows how the activity of a radioactive source decays over time. What is the half-life of this source?

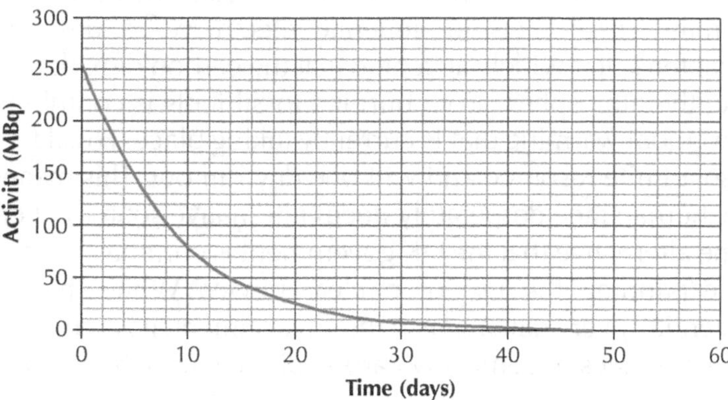

Fig 6.20.34

3 In an experiment to measure the half-life of a radioactive source, the activity is recorded every minute. The results are shown in the table below. Plot a graph of activity versus time for this source and determine the half-life of the source.

Time (in minutes)	Activity (Bq)
0	200
1	116
2	64
3	36
4	20
5	11·2

Table 6.20.6

4 A radioactive material has a half-life of 19 hours. Its original activity is 240 Bq. What will the activity be after 76 hours?

5 The initial activity of a radioactive source is 700 Bq. After 14 days the activity has reduced to 175 Bq. Calculate the half-life of this source.

6 A radioactive material has a half-life of 20 years. Its initial activity is 240 kBq. How long will it take for the activity of this source to decay to 15 kBq?

7 Cobalt-60 has a half-life of approximately 5 years. A student measures the activity of a cobalt-60 sample to be 450 Bq. She finds that the source was made 35 years ago. What was the activity of the source when it was made?

8 Scientists can use carbon dating to estimate the age of objects by measuring the activity of radioactive Carbon-14 in the object. A fossil of a lizard was recovered from an archaeological dig. The activity of the fossil was 75 μBq.

When the lizard died, its activity was 4800 μBq due to Carbon-14 atoms. If the half-life of Carbon-14 is 5700 years, calculate the age of the fossil.

Fig 6.20.35: *Scientists use carbon dating to estimate the age of things like fossils*

Nuclear fission

As humans, we have come to rely on the availability of large amounts of energy in our everyday lives. Try to imagine how different your life would be without electricity. This energy has to come from somewhere. For many years we have relied on burning fossil fuels to generate electricity as described in Chapter 1. It is widely accepted that the carbon dioxide released into the atmosphere from burning fossil fuels is contributing to the global warming that is currently taking place. Fossil fuels are also a finite resource and will one day run out. This means that the human race desperately needs alternative sources of energy. One alternative that the UK government is currently considering is expanding the use of nuclear power. While this is also potentially very damaging to the environment, it is very attractive in that it does not emit any carbon dioxide into the atmosphere.

In the nuclear reactor

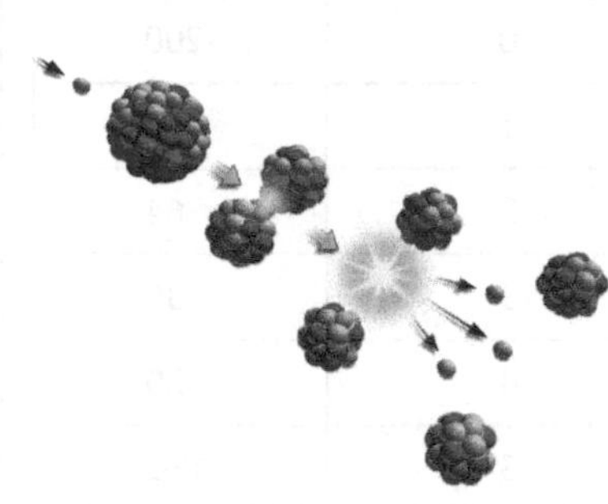

Fig 6.20.36: *An unstable isotope decays into smaller, more stable isotopes releasing nuclear radiation and energy*

At the heart of a nuclear power station is the nuclear reactor. The nuclear reactor produces energy by the process of nuclear fission. Nuclear fission is where a heavy nucleus is split into two lighter daughter nuclei. Crucially, when the nucleus is split, there is also a release of energy which can be used in the generation of electricity. The mass of the two daughter nuclei do not add up to that of the parent nucleus; it is this missing mass that is converted to energy.

Nuclear fission can take place spontaneously – that is, an unstable nucleus can split into two lighter nuclei and emit energy. In a power station, however, the process is stimulated. A neutron or similar particle is fired at an unstable nucleus and this causes it to split and release energy as shown in the diagram. Not all atoms are fissionable.

Chain reaction

Other neutrons are released when a nucleus is split. These neutrons go on to split other nuclei, in turn releasing more neutrons and energy. The reaction continues, getting bigger and bigger, releasing more and more energy! This is known as a chain reaction.

This chain reaction means that we can obtain great amounts of energy from a radioactive source. This energy can in turn be used to heat water to turn it into steam. The steam drives huge turbines to generate electricity.

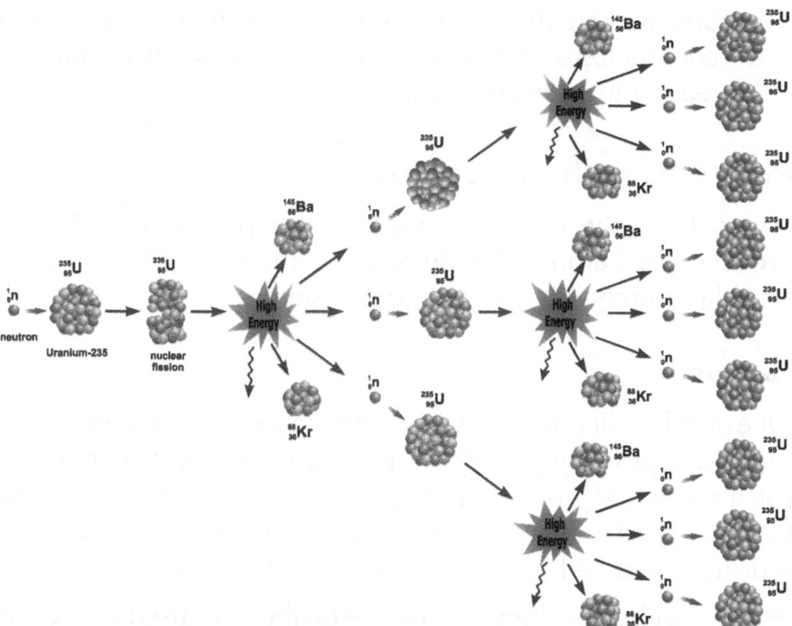

Fig 6.20.37: *Each nuclear decay goes on to cause further decays creating a chain reaction and releasing energy*

Controlling the reaction

In a nuclear power station, the chain reaction needs to be controlled. If the chain reaction continues uncontrolled, too much energy will be released in a short space of time – this is the basis for a nuclear bomb. The technology used to control a nuclear chain reaction in a power station is described below.

The crucial parts of a nuclear power station are:

- ***Fuel rods:*** These rods contain the radioactive source – e.g. Plutonium or Uranium – the unstable isotopes that decay, releasing nuclear radiation and energy.
- ***Boron control rods:*** These rods absorb the neutrons that go on to stimulate further nuclear decays. They are used to control the rate of the reactions by being raised or lowered to absorb more (or fewer) neutrons, and thus the amount of energy the power station is capable of producing.
- ***Moderator:*** The moderator slows down the neutrons, making them more likely to stimulate further nuclear decays.
- ***Containment vessel:*** This thick concrete shield houses the core of the reactor and prevents nuclear radiation escaping from the nuclear power plant.
- ***Coolant:*** This liquid or gas is pumped around the fuel rods and becomes very hot. The hot coolant is pumped to the

Fig 6.20.38: *The mushroom cloud created by a nuclear bomb*

boiler where the heat is used to boil water into steam. The coolant loses heat here and is then pumped back into the reactor to be heated again.

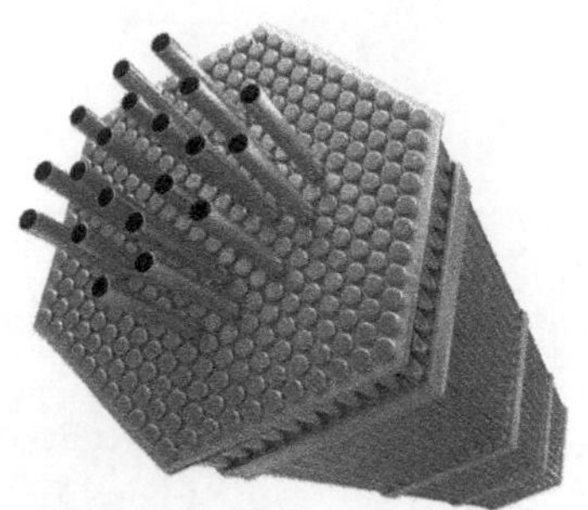

Fig 6.20.39: *Fuel rods containing the radioactive source*

Problems with nuclear power

So what are the problems with nuclear power? If it doesn't produce any Carbon Dioxide and won't run out for a very long time then surely it is an ideal power source?

Fuel rods

One problem lies in what to do with the fuel rods once they are no longer producing a useful amount of energy. When a fuel rod has stopped producing enough energy to generate sufficient heat, it is said to be spent. Spent fuel rods still produce a lot of radiation, and will continue to do so for a very long time.

Isotopes with a very long half-life are deliberately used as nuclear fuel because it is important that they produce a steady flow of radiation. The half-life of Plutonium-239 used in some reactors has a half-life of 24 100 years. The half-life of Uranium-235 that is used in others has a half-life of 703·8 million years! This means that since the earth formed around 4·5 billion years ago, the Uranium-235 has only gone through 6 half-lives!

These fuel rods are very harmful to all types of life, so it is extremely important that they are shielded and isolated from all life forms for many thousands of years. This, obviously, presents major engineering challenges.

Nuclear accidents

Another issue is what happens when something goes wrong in a nuclear reactor and radiation is leaked into the surrounding environment. On 11 March 2011, a tidal wave severely damaged the Fukushima 1 nuclear power plant in Japan. This led to a leak of radiation into the environment.

In 1986, an explosion at a nuclear power plant in the Ukraine had devastating and lasting effects.

Fig 6.20.40: *Chernobyl Nuclear Power Plant*

STEP BACK IN TIME: THE CHERNOBYL DISASTER

On 26 April 1986, the Chernobyl nuclear power plant in the Ukraine (then part of the Soviet Union) exploded, releasing a cloud of radioactive dust and gas into the atmosphere. This cloud was blown across Europe and was carried to the ground by rain. Even in Scotland, more than 2000 km away, sheep farming in some areas was restricted because of a risk that nuclear radiation in the soil could contaminate sheep and therefore affect the health of people who ate them. These restrictions were only lifted in July 2010, more than 24 years after the disaster. The photographs show Chernobyl before and after the explosion.

Fig 6.20.41: *Chernobyl after the explosion*

Exercise 6.20.9 Nuclear fission

1 Explain how energy is released during a nuclear fission reaction.

2 Describe what is meant by a chain reaction.

3 In a nuclear power station, why is it important to control the chain reaction?

4 In a nuclear reactor, describe the purpose of the following items:

a) fuel rods

b) control rods

c) moderator

Fig 6.20.42: *Cooling towers at a nuclear power station*

5 Some students are discussing the use of nuclear power to generate electricity. They make the following statements:

'The risk of environmental damage due to nuclear power stations is just too high. They should not be used at all.'

'The amount of carbon dioxide emitted into the atmosphere through the use of fossil fuels must be reduced and quickly. The only feasible way to do this is by making use of nuclear power.'

'We should invest all of our efforts in renewable sources of power such as wind farms, wave energy and hydro-electric power stations.'

'Humans should worry less about how to generate clean electrical energy and concentrate more on saving energy and being more efficient with the energy we already produce.'

What do you think and why?

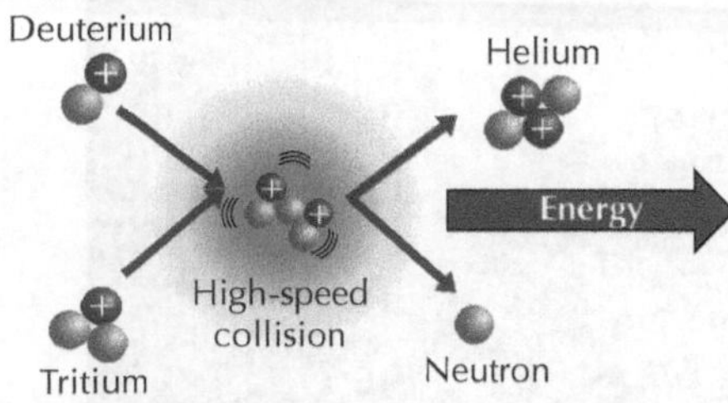

Fig 6.20.43: *The release of energy during nuclear fusion*

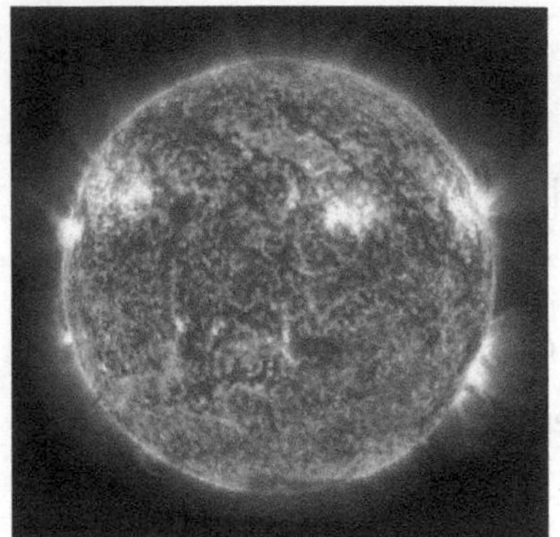
Fig 6.20.44: *The sun is powered by nuclear fusion*

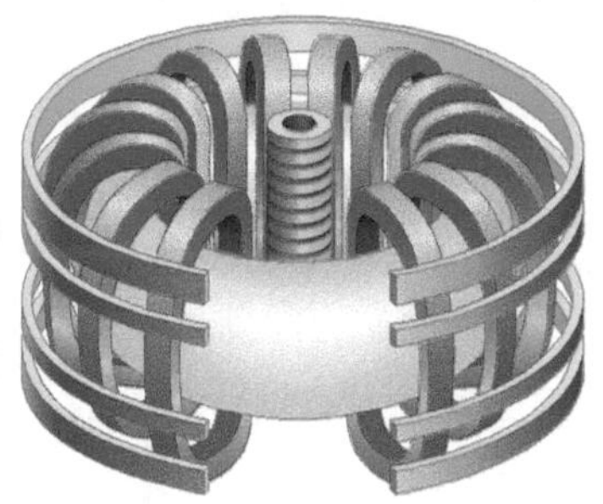
Fig 6.20.45: *Tokamak nuclear fusion reactor. Hyper warm plasma (yellow colour) is contained by magnetic forces inside a vessel with a toroidal shape formed by magnets (blue and red).*

Nuclear fusion

Nuclear fusion is another type of nuclear reaction. Instead of the atom splitting into smaller atoms, two atoms are smashed together under such immense pressure that they fuse together forming a new heavier atom. Just as fission reactions release large amounts of energy, so do fusion reactions. Two light nuclei combine to form one heavy nucleus, however, the mass of the two light nuclei do not add up to that of the single nucleus, and it is this missing mass that is converted to energy. Fusion reactions take place in the core of our sun, releasing the huge amounts of heat and light energy that we see on Earth.

Stars are made up of Hydrogen. In the centre of a star, this Hydrogen is being crushed at high pressure by the star's large gravitational field. At the centre of the star, types of Hydrogen called Deuterium (1 proton and 1 neutron) and Tritium (1 proton and 2 neutrons) are fused together creating Helium. This process is shown in the diagram.

This reaction releases vast amounts of energy. Our sun releases about 385 billion billion mega joules of energy every second which can travel huge distances through space!

With regards to generating electricity from nuclear materials, nuclear fusion is a much better option than nuclear fission for a variety of reasons. Firstly, it does not produce anything like the highly radioactive by-products that fission does. Secondly, it uses hydrogen which is extremely abundant in sea water, and most of all, it produces a far greater amount of energy per kilogram of reactants than nuclear fission.

There are however major problems with producing energy from nuclear fusion. In order to fuse hydrogen, extremely large pressures and very high temperatures are required (in excess of 150,000,000 °C!). At these temperatures the material being fused is in the form of a 'plasma' where the electrons and protons are separated from each other. The problem with plasma is that it is so hot, how do you contain a material at millions of degrees Celsius? Any man- made container would simply melt. In practice the plasma is confined using magnetic fields which keeps it away from the walls of the container. The container itself is a torus (doughnut shaped) vessel which has no open ends (which would allow plasma to spill out). However, in practice it does sometimes touch the walls. Fortunately, there is only a very small amount of plasma in there (~0.1 g), so although it is exceptionally hot it is counteracted by the wall of the torus, which has a mass of many hundreds of tonnes. So far, **sustained** energy production by nuclear fusion has eluded scientists for a variety of reasons but it is hoped that one day in the future, it may be possible.

Learning checklist

In this section you will have learned:

- The nature of alpha (α), beta (β) and gamma radiation (γ).
- About the term 'ionisation' and the effect of ionisation on neutral atoms.
- About the relative ionising effect and penetration of alpha, beta and gamma radiation.
- How to define activity in terms of the number of nuclear disintegrations and time.
- How to use an appropriate relationship to solve problems involving activity, number of nuclear disintegrations and time.
- About sources of background radiation.
- About the dangers of ionising radiation to living cells and of the need to measure the exposure to radiation.
- How to use appropriate relationships to solve problems involving absorbed dose, equivalent dose, energy, mass and weighting factor.
- How to use an appropriate relationship to solve problems involving equivalent dose rate, equivalent dose and time.
- How to compare the equivalent dose due to a variety of natural and artificial sources.
- About the equivalent dose rate and safety limits for the public and for workers in the radiation industry in terms of annual effective equivalent dose.
- About applications of nuclear radiation including electricity generation, cancer treatment and other industrial and medical uses.
- How to define half-life.
- How to use graphical or numerical data to determine the half-life of a radioactive material.
- How to describe an experiment to measure the half-life of a radioactive material.
- How to qualitatively describe fission, chain reactions and their role in the generation of energy.
- How to qualitatively describe fusion, plasma containment and their role in the generation of energy.

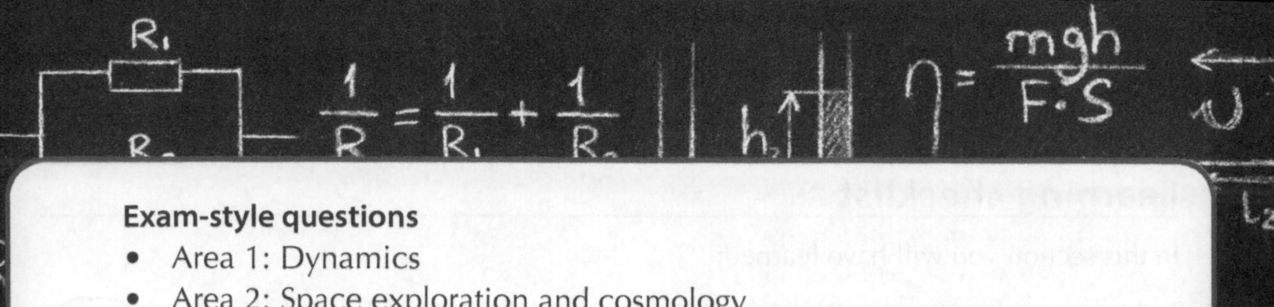

Exam-style questions

- Area 1: Dynamics
- Area 2: Space exploration and cosmology
- Area 3: Electricity
- Area 4: Properties of matter
- Area 5: Waves
- Area 6: Radiation

Exam-style questions

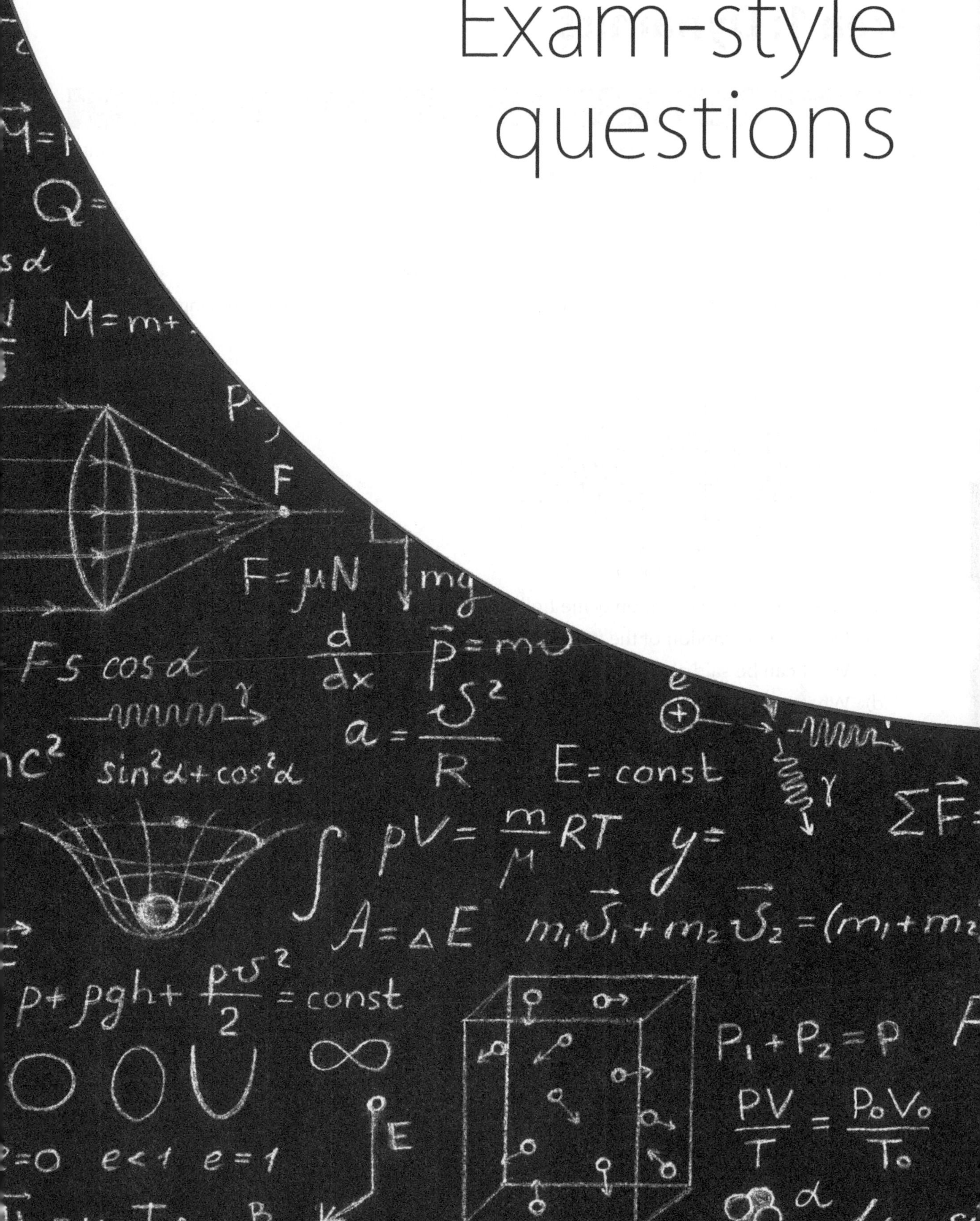

Area 1: Dynamics

QUESTIONS

1 Explain the difference between vector and scalar quantities. Sort the following quantities into a table to show whether they are vector or scalar:

- Speed
- Distance
- Velocity
- Displacement
- Time
- Force

2 The speed-time graph shown below represents the motion of a trolley moving along a lab bench.

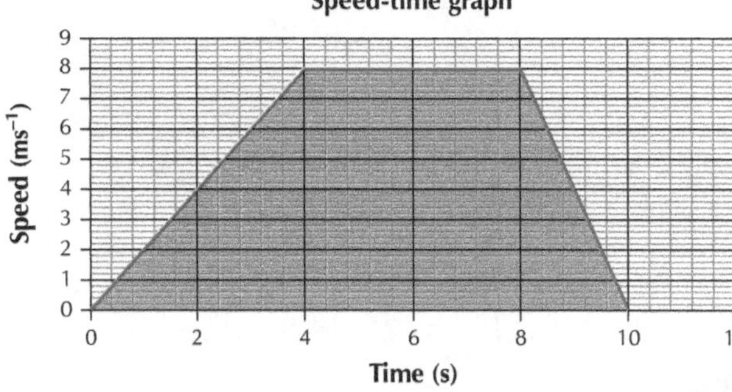

a) Calculate the acceleration of the trolley in the first 4 seconds.

b) Describe the motion of the trolley between 4 and 8 seconds.

c) What can be said about the forces acting on the trolley between 4 and 8 seconds?

d) What quantity is given by finding the shaded area?

3 A lorry accelerates from rest to its cruising speed of 20 ms^{-1} in a time of 30 s. Calculate the acceleration of the lorry.

4 Explain the difference between the mass of an object and the weight of an object.

5 The gravitational field strength on Earth is 9·8 N/kg. On the moon the gravitational field strength is 1·6 N/kg. If an astronaut has a mass of 80 kg, find:

(i) his weight on Earth

(ii) his mass on the moon

(iii) his weight on the moon

6 When a darts player throws a dart at the board, the dart will follow a projectile path.

a) Describe the horizontal motion of the dart during its flight.

b) Describe the vertical motion of the dart during its flight.

c) If the dart left the player's hand with an initial horizontal velocity of 12 ms^{-1}, how long would it take to reach the board which is 1·5 m away?

7 By means of a diagram, explain how a curved reflector can be used to receive signals from space.

8 A satellite in orbit around the Earth can be used as part of a Global Positioning System for a cycle computer.

a) State the speed of radio waves in air.

b) If the distance between the cycle computer and satellite is 25 000 km, find the time taken for a signal to travel from the satellite to the computer.

9 A Cairngorm cable car travels from sea level to the first ski drop-off point at 549 m. It gains potential energy of 6.99 MJ. The motor that pulls the winch is required to apply an electrical energy of 10.5 MJ so that the cable car can reach the top. Using your knowledge of conservation of energy, explain why there is a difference in these two values.

10 A cyclist is cycling along a straight stretch of road.

a) Describe the method you would use to measure her average speed, including what measurements you would make, how you would make the measurements and how you would process this information to calculate the average speed.

b) Imagine the cyclist is then travelling with an average speed of 12 ms^{-1} when a boy crosses the road in front of her. After 1 second, the cyclist applies the brakes and comes to rest 2 seconds later.

(i) Calculate the deceleration of the cyclist when she applies the brakes.

(ii) Sketch a velocity-time graph for the cyclist's motion for the 3 seconds described above.

(iii) Use the graph to work out the distance travelled by the cyclist during braking.

11 A camera mounted on a car is used to film the Grand National horse race. The velocity-time graph shows the velocity of the camera car over the course of the race.

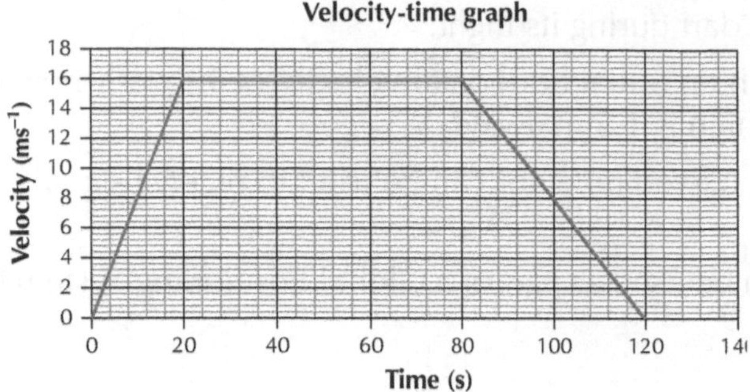

a) Describe the motion of the camera car over the 120 seconds of the race.

b) Calculate the acceleration of the camera car during the first 20 seconds of the race.

c) Find the total distance travelled by the camera car.

12 A swimmer is aiming to cross a river which is flowing from west to east at a velocity of 10 ms^{-1}. The swimmer is capable of swimming from north to south with a velocity of 5 ms^{-1}.

a) Calculate the resultant velocity of the swimmer when her velocity is combined with that of the river.

b) If the swimmer takes 30 minutes to cross the river, calculate the resultant displacement of the swimmer.

c) The swimmer would like to reach a point directly opposite her on the other side of the river, i.e. she would like to swim in a direction north to south. Explain, using your knowledge of vectors, the direction the swimmer should swim to ensure she reaches a point directly opposite her starting point.

13 Engineers are studying the forces applied by a seatbelt in a typical collision. They are considering a collision where a car is brought to rest from a speed of 25 ms^{-1} in a time of 0·4 s.

a) Explain, in terms of Newton's first law, how a seatbelt works to keep you in your seat in the event of a collision.

b) Calculate the deceleration of the car.

c) Calculate the force the seatbelt would need to apply to a body of mass 70 kg. You may assume that friction between the body and the seat is negligible.

14 In July 2011, the space shuttle Atlantis lifted off from Cape Canaveral heading for the International Space Station. Atlantis has a mass of 80 000 kg with its three rocket engines. For the purpose of the following calculations, air resistance is negligible.

a) Calculate the weight of Atlantis.

b) Calculate the thrust required to give Atlantis an upward acceleration of 20 ms^{-2}.

c) Assuming the thrust of the engines remains constant, give two reasons why the acceleration of the rocket increases as its height above the surface of the Earth increases.

15 Explain, in terms of Newton's second law, why a crumple zone in a car will help to reduce the average force exerted on its occupants during a collision.

16 Stunt man Felix Baumgartner set the world record for the world's highest skydive in 2012, skydiving an estimated 39 km. During his fall he reached a speed of 1357·64 km/hr. (Mach 1·25; that is, 1·25 times the speed of sound!)

a) Convert Baumgartner's dive speed from km/hr to ms^{-1}.

b) Explain, in terms of Newton's first law, why Baumgartner reached a terminal velocity during his dive.

c) Explain, with reference to Newton's laws of motion, why Baumgartner needed to dive from such a great height to reach as high a terminal velocity as possible.

d) Explain, in terms of Newton's second law, why Baumgartner decelerated once his parachute was opened.

17 Boeing recently launched the *Dreamliner (787)*, a medium to long haul wide-body passenger air craft. The 787-8 plane has a mass of 110 000 kg and a maximum take-off mass of 228 000 kg. It is powered by two Rolls Royce Trent 1000 engines which provide a total thrust of 560 kN.

a) Explain, in terms of Newton's third law, how a jet engine produces a forward thrust to move the aircraft forwards.

b) Calculate the acceleration of the Dreamliner when both engines are operating at maximum thrust and the plane is at its maximum take-off weight.

c) Assume that the take-off speed for a Dreamliner is 110 ms^{-1}.

(i) Calculate the time taken for the plane to reach its take-off speed.

(ii) Plot a velocity-time graph for the plane as it starts from rest and accelerates to its take-off speed. You may assume it has a constant acceleration. Use the graph to find the distance taken by the plane to reach its take-off speed.

d) The wings of an aircraft force air downwards. Explain why this results in the wings producing lift, making reference to the relevant Newton's laws.

18 A satellite in orbit around the Earth can be used to relay signals across the globe. A sports event in the USA is being broadcast via satellite to Scotland.

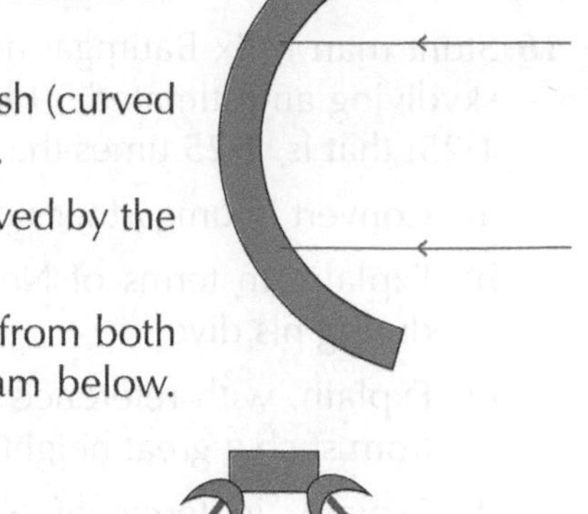

a) Copy and complete the diagram to show how a satellite dish (curved reflector) focusses the light which is incident on its surface.

b) Explain how a curved reflector enhances the signal received by the receiver.

c) The satellite used for the broadcast is 20 000 km away from both the transmitter and the receiver as shown on the diagram below. Assuming the signal is re-transmitted instantaneously calculate the time taken for the signal to travel from the transmitter to the receiver.

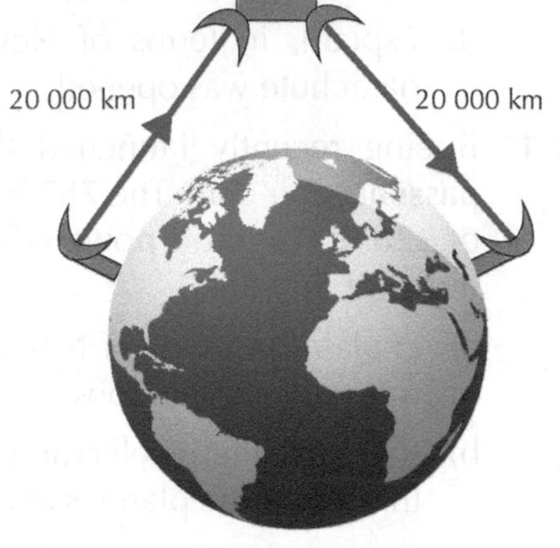

d) The satellite used for the transmission is said to be geostationary. Explain what a geostationary satellite is and what advantage it has for global communication.

e) Use your knowledge of curved reflectors and the principle of reversibility of a ray path to explain how a curved reflector can be used as part of a car headlight to produce a beam of light directed out in front of the vehicle.

19 In an experiment to investigate projectile motion, a ball of mass 2 kg is placed on a ramp on top of a desk as shown in the diagram below.

The ball starts at a height, *h*, of 0·6 m as shown in the diagram below.

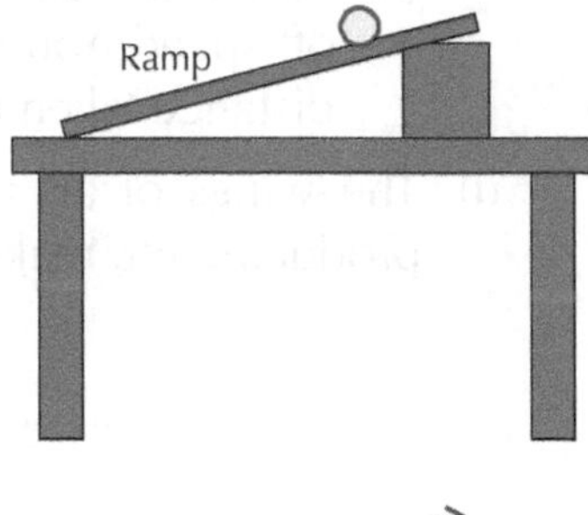

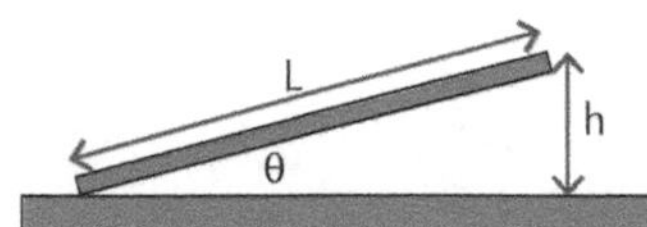

a) Calculate the potential energy of the ball at a height *h* above the bench.

b) Assuming no energy losses, find the horizontal velocity of the ball at the bottom of the ramp.

c) State the initial horizontal and vertical velocities of the ball as it leaves the table.

d) If the ball is in the air for a time of 0·4 s, calculate the horizontal distance travelled by the ball.

e) Calculate the vertical velocity the ball hits the ground with.

Area 2: Space exploration and cosmology

QUESTIONS

1 When studying space, astronomers do not use metres to measure distance. Instead they use the light year (Ly) where 1 Ly = $9{\cdot}5 \times 10^{15}$ m

a) Explain why metres are not used to describe distance in space.

b) A star is 240 Ly away from Earth. Calculate this distance in metres.

Area 3: Electricity

QUESTIONS

1 Using your knowledge of charged particles, describe the types of materials that make good conductors, making reference to the type and charge of carriers that aid conductivity.

2 An escalator engineer notices that when he has his arm against the rubber handrail of an escalator the hairs on his arm sometimes 'stand on end'. Using your knowledge of charge, describe why this happens.

3 An electric motor is attached in parallel to a bulb as shown. The motor is used to turn a DVD inside a games console. The motor and bulb are rated at 6 V with an operating current of 250 mA.

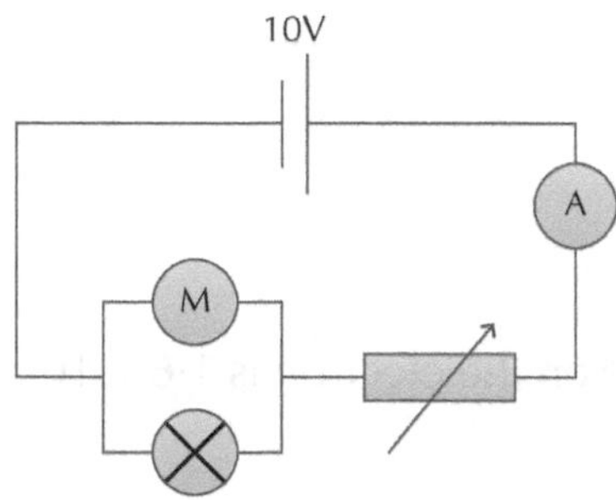

a) Calculate the reading on the ammeter.

b) Calculate the resistance of the motor.

c) What is the total resistance of the circuit?

d) What value of resistance is the variable resistor set to?

e) How much charge will flow through the motor in 2 minutes?

4 Three identical resistors are connected in series with a 9 V supply and an ammeter. A voltmeter is connected across one of the resistors.

a) Draw this circuit.

b) If the resistors have a resistance of 100 Ω, what will be the reading on the voltmeter?

c) What is the current in the circuit?

5 a) Using your knowledge of conservation of energy, and ignoring the effects of friction, describe how the energy and speed of a stunt bike would vary if it starts from rest at the top of a slope and accelerates due to gravity to the bottom of the hill.

b) If the speed was measured at the bottom of the hill, why would it be less than predicted by the total potential energy at the top of the hill?

c) Where does this 'lost' energy go?

6 a) Define what is meant by electric current.

b) Draw the trace that would be seen on an oscilloscope with a setting of 5 V/cm for a DC voltage of 10 V.

c) How would you describe the motion of the charge carriers in the circuit as described on the oscilloscope screen?

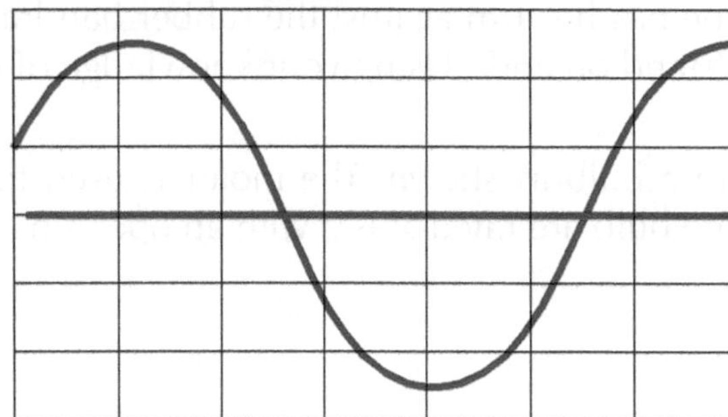

d) A USB charger has a current of 0·5A.

(i) Calculate the charge flowing per second.

(ii) How many electrons are flowing in this time? (Charge on the electron is $1{\cdot}6 \times 10^{-19}$ C)

7 a) State what is meant by a potential difference of 6 V.

b) Certain particle accelerators use an electron gun to create a beam of charged particles. Often these charge particles are accelerated to near light speed. Describe how these charged particles can be controlled to ensure that they are in the correct position.

8 Why do filament light bulbs almost always break when they are switched on from cold?

9 a) Using a combination of 10 Ω resistors, find a combination that draws a current of 0·5 A with a supply voltage of 7·5 V.

b) What is the device labelled A in the circuit diagram shown?

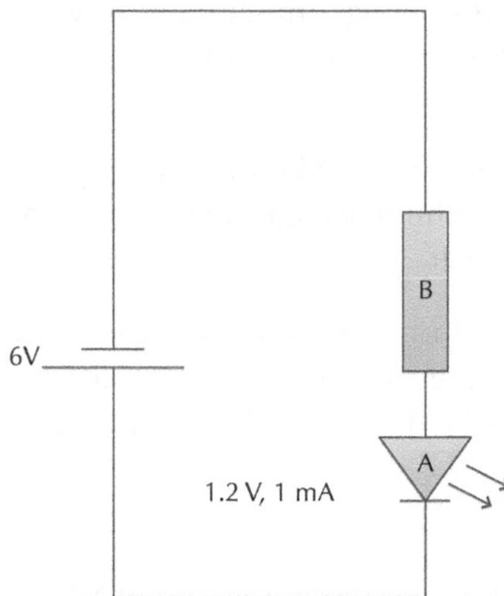

c) Review the circuit shown and describe the functions of the resistor.

d) A customer reports a fault with this circuit. What changes, if any, would you make to this circuit?

e) If the circuit is working correctly, what value of resister R would be required?

10 a) What is the reason for putting fuses into appliances?

b) Certain devices use more energy than others and therefore cost more to run. Give examples of devices which cost more to run than others. Explain your answer.

c) An electric kettle uses 2 kW to boil 100 ml of water. If the kettle takes 5 minutes to boil, what is the energy requirement?

Area 4: Properties of matter

QUESTIONS

1 When a space-craft re-enters the Earth's atmosphere following a mission, it slows down due to air resistance. Explain the effect this has on the space-craft and why the shape and material used for its construction are important.

2 A student is investigating the melting points of different solids. She places samples of equal mass of aluminium and lead into a heat bath. After 6 minutes she notices that the lead is warmer than the aluminium. Explain why this is the case.

3 A tea bag manufacturer wants to design a tea bag that only requires water to be heated to 88°C and not 100°C and therefore requires less energy to make. A standard cup of tea contains 0·2 kg of water which requires 66 880 J of heat energy. What is the reduction in energy in making a cup of tea at 88°C rather than 100°C?

(The specific heat capacity of water is 4180 J/kg°C)

4 Gas is compressed in order to fit into a fire extinguisher and has a pressure of 18 Pa at 20°C.

a) Calculate the temperature of the gas in degrees kelvin.

b) During a very cold winter, the temperature of the extinguisher reduces to 15°C. Calculate the new pressure of the gas, assuming a constant volume and mass of gas.

5 a) A sealed chamber contains a mixture of gases of constant mass and fixed volume. Use the kinetic theory of gases to describe how the pressure changes as the temperature increases.

b) The piston of a Formula 1 car contains a mixture of gases. The diameter (known as the bore) of the piston chamber is 98 mm.

(i) Calculate the initial pressure in the chamber if the engine generates a force of 3800 N.

(ii) Calculate the pressure of the piston if the initial volume is 100 ml and the final volume is 10 ml, assuming a constant temperature of 88°C. Describe and explain the law that allows you to calculate this.

6 a) When using a sensor to determine where heat energy is lost from your school, what kinds of waves are detected?

b) It is discovered that the school canteen is the hottest part of the school. Explain why this might be the case and describe the energy changes that occur in the school kitchen.

c) Pizza is served every Wednesday. When the pizza is fresh out of the oven you check that it is not too hot to eat by touching the crust. However, when you bite into the pizza you find that the sauce feels much hotter than the crust and almost burn yourself. Using your knowledge of specific heat capacity, explain why this happens.

d) The school invests in a new coffee machine. The machine can hold 2 kg of water which must be heated from 15°C to 90°C. Find the heat energy required to ensure the water is hot enough to make the coffee. Assume that the specific heat capacity of water is 4180 J/kg°C.

7 In October 2012 Felix Baumgartner jumped from high altitude to break the speed of sound. On the ground his oxygen tank had a pressure of 1×10^5 Pa and a volume of 20 litres at an ambient temperature 22°C. At his jump altitude, the pressure is less than that at sea level and the temperature is –40°C. Assuming a constant volume, find the pressure of the oxygen tank at the new altitude.

Area 5: Waves

QUESTIONS

1 Draw a diagram of a transverse wave labelling the wavelength and amplitude.

2 X-rays are part of the electromagnetic spectrum.

a) State the speed of X-rays in a vacuum.

b) Name the other 6 bands of the electromagnetic spectrum.

c) Which of these bands has the lowest energy?

3 A student is investigating the refraction of light as it travels from glass to air. She draws what she observes, shown below.

Name the parts labelled A, B and C

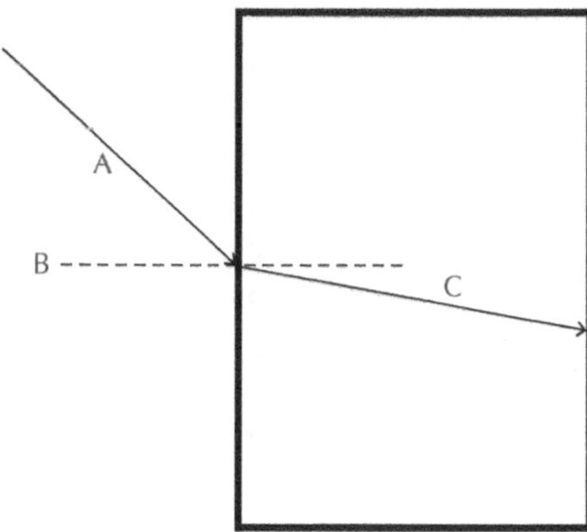

4 In a hospital an ultrasound signal is being used to investigate a pregnancy. The ultrasound signal has a frequency of 50 kHz. The speed of sound in human tissue is 1500 ms^{-1}.

a) Calculate the wavelength of the sound signal in the tissue.

b) If the reflected sound is sent out, then picked up by the receiver 0.1 milliseconds after it is transmitted, what is the depth of the foetus?

5 Complete the diagram showing how waves will diffract as they pass the object:

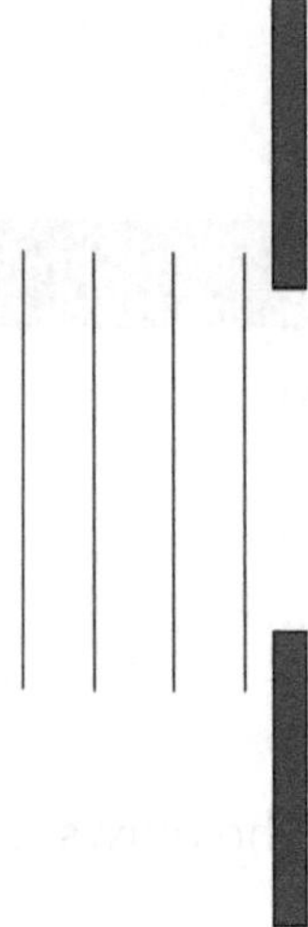

6 Copy and complete the following diagram showing how the rays of light will continue when they emerge from the lens below.

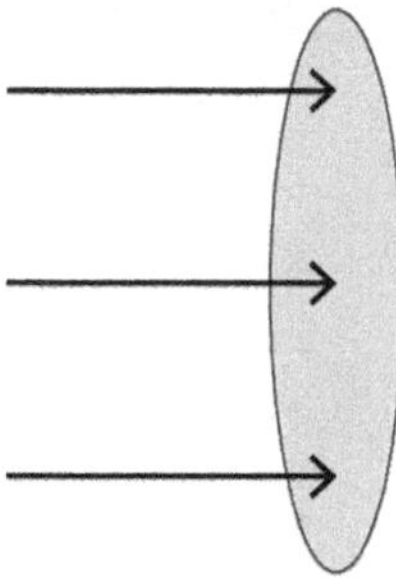

7 a) State what is meant by the term frequency.

b) The swimming pool for the 2014 Commonwealth games in Glasgow is 50 m long. You are sat at the side of the pool and observe 10 waves passing in 6 seconds. Ten waves take up the entire length of the pool. Using this information calculate:

(i) The frequency of the waves

(ii) The wavelength of the waves

(iii) The period of the waves

(iv) The wave speed

c) Are these waves transverse or longitudinal? Explain your answer.

8 A mountain rescue team use walkie-talkie radios to stay in contact with each other. Using your knowledge of waves, explain what type of waves would be used, discussing their wavelength.

9 a) Describe what all electromagnetic waves have in common.

b) Take a look at the table showing the family of electromagnetic waves. Copy and complete this table filling in the missing sections.

Radio	Microwaves		Visible	UV		Gamma

c) Some Gamma rays are found to have a frequency of 3.5×10^{19} Hz. Calculate the wavelength of these Gamma rays.

10 A 100 m sprint is started by an official with a starting pistol. The pistol produces a bang that indicates the start of the race. The sprint is won by 0.01 seconds. The winner was running in lane 1 next to the starting pistol and the runner who came second was in lane 8. For the purposes of this question consider the speed of sound to be 340 ms^{-1}.

a) Calculate the time difference between the athletes in lane 1 and lane 8 hearing the bang. The distance between the athletes is 7 m.

b) Using your knowledge of the speed of sound, discuss the issues that arise from using this method of starting a sprint. What could be done to eliminate this problem?

11 a) Which type of wave in the electromagnetic spectrum has the highest energy?

b) X-rays have a wavelength of 1×10^{-10} m. Calculate their energy.

c) Discuss an application of X-rays including the benefits and dangers.

12 a) Describe what is meant by the term refraction.

b) With reference to the normal, angle of incidence and angle of refraction, describe what happens when a ray of light travels from a medium of low density (for example, air) into a medium of high density (for example, glass).

c) With the aid of a diagram, describe how the common eye defect known as long sight can be corrected.

Area 6: Radiation

QUESTIONS

1 a) Describe a practical use of nuclear radiation.

b) What is meant by background radiation?

c) Give an example of a source of background radiation.

2 a) A radioactive source has a half-life of 12 hours. After 3 days, the source has an activity of 20 kBq. What was the initial radioactivity of the source?

b) A source has a total measured count rate of 360 Bq when background radiation is measured as 40 Bq. What will the total measured activity be in 9 days if the source has a half-life of 3 days?

3 Doctors use a source of radiation as a tracer to help them make a diagnosis.

a) Explain whether the source they use would be alpha, beta or gamma radiation.

b) Suggest a suitable half-life for the source, giving a reason for your answer.

4 Describe how you could use different materials of different thicknesses to determine if a source of nuclear radiation is emitting alpha, beta or gamma radiation.

5 Shortly after the nuclear power plant in Fukushima was damaged by a tsunami scientists had to measure the nuclear activity outside the plant.

a) Describe the measurements the scientists would have to take and discuss how they could calculate the activity.

b) What precautions must scientists take in order to minimise the danger of being exposed to nuclear radiation?

6 a) What is meant by an activity of 15 becquerels?

A technician working with alpha radiation becomes exposed to 1.5 mJ of the radiation. It is just her hand that is exposed and she measures her hand to have a mass of 300 g. The weighting factor of alpha radiation is 20.

b) Calculate the equivalent dose received by the technician.

7 a) Explain the term half-life.

b) Describe how you would measure the half-life of a source of radiation.

c) On a day where the background radiation is found to be 15 Bq, a source of radiation is found to have an activity of 1215 Bq. After a time of 48 hours the measured activity has dropped to 90 Bq. Calculate the half-life of this source.

8 Discuss the use of nuclear fusion as a method for generating electricity. You should mention the advantages and disadvantages of this method of generating electricity.